Human Factors Methods and Sports Science

Human Factors Methods and Sports Science
A Practical Guide

Paul M. Salmon
Neville A. Stanton
Adam C. Gibbon
Daniel P. Jenkins
Guy H. Walker

CRC Press
Taylor & Francis Group
Boca Raton London New York

CRC Press is an imprint of the
Taylor & Francis Group, an **informa** business

CRC Press
Taylor & Francis Group
6000 Broken Sound Parkway NW, Suite 300
Boca Raton, FL 33487-2742

© 2010 by Taylor and Francis Group, LLC
CRC Press is an imprint of Taylor & Francis Group, an Informa business

No claim to original U.S. Government works

Printed in the United States of America on acid-free paper
10 9 8 7 6 5 4 3 2 1

International Standard Book Number: 978-1-4200-7216-7 (Hardback)

This book contains information obtained from authentic and highly regarded sources. Reasonable efforts have been made to publish reliable data and information, but the author and publisher cannot assume responsibility for the validity of all materials or the consequences of their use. The authors and publishers have attempted to trace the copyright holders of all material reproduced in this publication and apologize to copyright holders if permission to publish in this form has not been obtained. If any copyright material has not been acknowledged please write and let us know so we may rectify in any future reprint.

Except as permitted under U.S. Copyright Law, no part of this book may be reprinted, reproduced, transmitted, or utilized in any form by any electronic, mechanical, or other means, now known or hereafter invented, including photocopying, microfilming, and recording, or in any information storage or retrieval system, without written permission from the publishers.

For permission to photocopy or use material electronically from this work, please access www.copyright.com (http://www.copyright.com/) or contact the Copyright Clearance Center, Inc. (CCC), 222 Rosewood Drive, Danvers, MA 01923, 978-750-8400. CCC is a not-for-profit organization that provides licenses and registration for a variety of users. For organizations that have been granted a photocopy license by the CCC, a separate system of payment has been arranged.

Trademark Notice: Product or corporate names may be trademarks or registered trademarks, and are used only for identification and explanation without intent to infringe.

Library of Congress Cataloging-in-Publication Data

Salmon, Paul, 1948-
　　Human factors methods and sports science : a practical guide / Paul Salmon, Adam Gibbon, and Neville Anthony Stanton.
　　　　p. cm.
　　Includes bibliographical references and index.
　　ISBN 978-1-4200-7216-7
　　1. Sports sciences. 2. Athletes--Training of. 3. Sports--Psychological aspects. 4. Sports--Physiological aspects. I. Gibbon, Adam. II. Stanton, Neville Anthony. III. Title.

GV558.S28 2010
613.7'1--dc22   2009023394

Visit the Taylor & Francis Web site at
http://www.taylorandfrancis.com

and the CRC Press Web site at
http://www.crcpress.com

Contents

List of Figures xxiii
List of Tables xxvii
Preface xxix
Acknowledgments xxxi
About the Authors xxxiii
Acronyms xxxv

Chapter 1 Introduction 1

    Introduction 1
    Human Factors Methods 3
    Application in Sport 5
    Structure of the Book 7

Chapter 2 Data Collection Methods 9

    Introduction 9
    Interviews 10
        Background and Applications 10
        Domain of Application 13
        Application in Sport 13
        Procedure and Advice (Semi-Structured Interview) 13
            Step 1: Clearly Define the Aim(s) of the Interview 13
            Step 2: Develop Interview Questions 13
            Step 3: Piloting the Interview 14
            Step 4: Modify Interview Procedure and/or Content 14
            Step 5: Select Participants 14
            Step 6: Conduct Interviews 14
            Step 7: Transcribe Data 14
            Step 8: Data Gathering 14
            Step 9: Data Analysis 15
        Advantages 15
        Disadvantages 15
        Approximate Training and Application Times 16
        Reliability and Validity 16
        Tools Needed 16
        Example 16
        Flowchart 16
        Recommended Texts 16
    Questionnaires 16
        Background and Applications 16
        Domain of Application 18
        Application in Sport 19
        Procedure and Advice 20
            Step 1: Clearly Define Study Aims and Objectives 20
            Step 2: Define Target Population 22

v

        Step 3: Questionnaire Construction 23
        Step 4: Piloting the Questionnaire 23
        Step 5: Questionnaire Administration 24
        Step 6: Data Analysis..24
        Step 7: Follow-Up Phase 25
    Advantages 25
    Disadvantages 25
    Related Methods 25
    Approximate Training and Application Times 25
    Reliability and Validity 25
    Tools Needed 26
    Example 26
    Flowchart 26
    Recommended Texts 26
Observational Study 26
    Domain of Application 28
    Application in Sport 29
    Procedure and Advice 29
        Step 1: Define Aims and Objectives 29
        Step 2: Define Scenario(s)..29
        Step 3: Develop Observation Plan 29
        Step 4: Pilot Observation 29
        Step 5: Conduct Observation 30
        Step 6: Data Analysis..30
        Step 7: Further Analysis 30
        Step 8: Participant Feedback 30
    Advantages 30
    Disadvantages 31
    Related Methods 31
    Approximate Training and Application Times 31
    Reliability and Validity 31
    Tools Needed 31
    Example 32
    Flowchart 32
    Recommended Texts 32

Chapter 3   Task Analysis Methods..35

Introduction 35
Hierarchical Task Analysis 37
    Background and Applications 37
    Domain of Application 37
    Application in Sport 40
    Procedure and Advice 40
        Step 1: Define Aims of the Analysis...40
        Step 2: Define Analysis Boundaries 40
        Step 3: Collect Data Regarding Task/System/Procedure/Device under
        Analysis 40
        Step 4: Determine and Describe Overall System/Task Goal 40
        Step 5: Determine and Describe Sub-Goals 41
        Step 6: Decompose Sub-Goals 41

Contents

      Step 7: Create Plans 41
      Step 8: Revise Analysis 41
      Step 9: Verify and Refine Analysis 42
   Advantages 42
   Disadvantages 42
   Related Methods 42
   Approximate Training and Application Times 42
   Reliability and Validity 43
   Tools Needed 43
   Example 43
   Flowchart 43
   Recommended Texts 43
Task Decomposition 47
   Background and Applications 47
   Domain of Application 48
   Application in Sport 48
   Procedure and Advice 48
      Step 1: Clearly Define Aims of the Analysis 48
      Step 2: Select Task Decomposition Categories 48
      Step 3: Construct HTA for the Task, Scenario, System, or Device under
      Analysis 48
      Step 4: Create Task Decomposition Table 49
      Step 5: Collect Data for Task Decomposition 49
      Step 6: Complete Task Decomposition Table 49
      Step 7: Propose Redesigns/Remedial Measures/Training Interventions, Etc. . 49
   Advantages 49
   Disadvantages 50
   Related Methods 50
   Approximate Training and Application Times 50
   Reliability and Validity 50
   Tools Needed 50
   Example 51
   Flowchart 51
   Recommended Text 51
Verbal Protocol Analysis 51
   Background and Applications 51
   Domain of Application 51
   Application in Sport 55
   Procedure and Advice 56
      Step 1: Clearly Define Aims of the Analysis 56
      Step 2: Define Task/Scenario under Analysis 56
      Step 3: Brief Participant(s) 56
      Step 4: Conduct Pilot Run 56
      Step 5: Undertake Scenario and Record Data 56
      Step 6: Transcribe Data 56
      Step 7: Encode Verbalisations 56
      Step 8: Devise Other Data Columns 57
      Step 9: Establish Inter- and Intra-Rater Reliability 57
      Step 10: Perform Pilot Study 57
      Step 11: Analyse Structure of Encoding 57
   Advantages 57

Disadvantages 57
Related Methods 58
Approximate Training and Application Times 58
Reliability and Validity 58
Tools Needed 58
Example 58
Flowchart 59
Recommended Texts 59
Operation Sequence Diagrams 59
Domain of Application 59
Application in Sport 59
Procedure and Advice 59
    Step 1: Clearly Define Aims of the Analysis 59
    Step 2: Define Task/Scenario under Analysis 61
    Step 3: Collect Data via Observational Study 61
    Step 4: Describe the Task or Scenario Using HTA 61
    Step 5: Construct the OSD Diagram 62
    Step 6: Add Additional Analyses Results to OSD 62
    Step 7: Calculate Operational Loading Figures 62
Advantages 62
Disadvantages 62
Related Methods 64
Approximate Training and Application Times 64
Reliability and Validity 64
Tools needed 64
Example 64
Flowchart 67
Recommended Texts 67

Chapter 4   Cognitive Task Analysis 69

Introduction 69
Cognitive Work Analysis 71
Domain of Application 74
Application in Sport 74
Procedure and Advice 75
    Step 1: Clearly Define Aims of the Analysis 75
    Step 2: Select Appropriate CWA Phase(s) 75
    Step 3: Construct Work Domain Analysis 75
    Step 4: Conduct Control Task Analysis 77
    Step 5: Conduct Strategies Analysis 77
    Step 6: Conduct Social Organisation and Cooperation Analysis 79
    Step 7: Conduct Worker Competencies Analysis 80
    Step 8: Review and Refine Outputs with SMEs 81
Advantages 81
Disadvantages 82
Related Methods 82
Approximate Training and Application Times 82
Reliability and Validity 83
Tools Needed 83
Example 83

Recommended Texts 85
Critical Decision Method 85
    Domain of Application 86
    Application in Sport 86
    Procedure and Advice 86
        Step 1: Clearly Define Aims of the Analysis...................................86
        Step 2: Identify Scenarios to Be Analysed 87
        Step 3: Select/Develop Appropriate CDM Interview Probes 87
        Step 4: Select Appropriate Participant(s)..88
        Step 5: Observe Scenario under Analysis or Gather Description of Incident ..88
        Step 6: Define Timeline and Critical Incidents 88
        Step 7: Conduct CDM Interviews..88
        Step 8: Transcribe Interview Data 88
        Step 9: Analyse Data as Required 88
    Advantages 88
    Disadvantages 89
    Related Methods 89
    Approximate Training and Application Times 89
    Reliability and Validity 90
    Tools Needed 90
    Example 90
    Flowchart..90
    Recommended Texts 90
Concept Maps...90
    Background and Applications 90
    Domain of Application 91
    Application in Sport 91
    Procedure and Advice 91
        Step 1: Clearly Define Aims of the Analysis...................................91
        Step 2: Identify Scenarios to Be Analysed 91
        Step 3: Select Appropriate Participant(s)..93
        Step 4: Observe the Task or Scenario under Analysis 93
        Step 5: Introduce Participants to Concept Map Method 93
        Step 6: Identify Focus Question...93
        Step 7: Identify Overarching Concepts..95
        Step 8: Link Concepts..95
        Step 9: Review and Refine Concept Map 95
    Advantages 95
    Disadvantages...97
    Related Methods...98
    Approximate Training and Application Times 98
    Reliability and Validity 98
    Tools Needed 98
    Example 98
    Flowchart..98
    Recommended Text 98
Applied Cognitive Task Analysis... 101
    Background and Applications 101
    Domain of Application 101
    Application in Sport 101
    Procedure and Advice 101

        Step 1: Define the Task under Analysis 101
        Step 2: Select Appropriate Participant(s) 102
        Step 3: Observe Task or Scenario under Analysis 102
        Step 4: Conduct Task Diagram Interview 102
        Step 5: Conduct Knowledge Audit Interview 102
        Step 6: Conduct Simulation Interview 103
        Step 7: Construct Cognitive Demands Table 103
    Advantages 103
    Disadvantages 104
    Related Methods 105
    Approximate Training and Application Times 105
    Reliability and Validity 106
    Tools Needed 106
    Flowchart 106
    Recommended Text 106

Chapter 5 Human Error Identification and Analysis Methods 109

Introduction 109
Defining Human Error 110
Error Classifications 110
    Slips and Lapses 111
    Mistakes 111
    Violations 111
Theoretical Perspectives on Human Error 112
    The Person Approach 113
    The Systems Perspective Approach 113
Human Error Methods 115
    Human Error Analysis Methods 115
    Human Error Identification Methods 115
        Taxonomy-Based Methods 116
        Error Identifier Methods 116
Accimaps 117
    Background and Applications 117
    Domain of Application 117
    Application in Sport 117
    Procedure and Advice 120
        Step 1: Data Collection 120
        Step 2: Identify Physical Process/Actor Activities Failures 120
        Step 3: Identify Causal Factors 120
        Step 4: Identify Failures at Other Levels 120
        Step 5: Finalise and Review Accimap Diagram 120
    Advantages 120
    Disadvantages 121
    Example 121
    Related Methods 121
    Approximate Training and Application Times 123
    Reliability and Validity 123
    Tools Needed 123
    Flowchart 123
    Recommended Texts 123

Contents

Fault Tree Analysis 123
Background and Application 123
Domain of Application 123
Application in Sport 124
Procedure and Advice 124
Step 1: Define Failure Event 124
Step 2: Collect Data Regarding Failure Event 125
Step 3: Determine Causes of Failure Event 125
Step 4: AND/OR Classification 125
Step 5: Construct Fault Tree Diagram 125
Step 6: Review and Refine Fault Tree Diagram 125
Advantages 125
Disadvantages 126
Related Methods 126
Approximate Training and Application Times 126
Reliability and Validity 126
Tools Needed 126
Example 126
Flowchart 128
Recommended Text 128
Systematic Human Error Reduction and Prediction Approach 128
Background and Applications 128
Domain of Application 128
Application in Sport 130
Procedure and Advice 130
Step 1: Conduct HTA for the Task or Scenario under Analysis 130
Step 2: Task Classification 130
Step 3: Identify Likely Errors 130
Step 4: Determine and Record Error Consequences 130
Step 5: Recovery Analysis 131
Step 6: Ordinal Probability Analysis 131
Step 7: Criticality Analysis 132
Step 8: Propose Remedial Measures 132
Step 9: Review and Refine Analysis 132
Advantages 132
Disadvantages 132
Related Methods 133
Approximate Training and Application Times 133
Reliability and Validity 133
Tools Needed 134
Example 134
Flowchart 134
Recommended Texts 134
Human Error Template 134
Background and Applications 134
Domain of Application 141
Application in Sport 142
Procedure and Advice 142
Step 1: Hierarchical Task Analysis (HTA) 142
Step 2: Human Error Identification 142
Step 3: Consequence Analysis 146

Step 4: Ordinal Probability Analysis	142
Step 5: Criticality Analysis	142
Step 6: Interface Analysis	142
Step 7: Review and Refine Analysis	143
Advantages	143
Disadvantages	143
Related Methods	143
Approximate Training and Application Times	143
Reliability and Validity	144
Tools Needed	144
Example	144
Flowchart	144
Recommended Texts	145
Task Analysis for Error Identification	**146**
Background and Applications	146
Domain of Application	146
Application in Sport	146
Procedure and Advice	147
Step 1: Construct HTA for Device under Analysis	147
Step 2: Construct State Space Diagrams	147
Step 3: Create Transition Matrix	147
Advantages	148
Disadvantages	149
Related Methods	149
Approximate Training and Application Times	149
Reliability and Validity	150
Tools Needed	150
Example	150
Flowchart	150
Recommended Texts	150
Technique for Human Error Assessment	**153**
Background and Applications	153
Domain of Application	153
Application in Sport	153
Procedure and Advice	154
Step 1: System Description	154
Step 2: Scenario Description	155
Step 3: Task Description and Goal Decomposition	155
Step 4: Error Analysis	156
Step 5: Design Implications/Recommendations	156
Advantages	156
Disadvantages	156
Related Methods	158
Approximate Training and Application Times	158
Reliability and Validity	158
Tools Needed	158
Flowchart	158
Recommended Texts	158

Chapter 6 Situation Awareness Assessment Methods 161
Introduction 161
Situation Awareness Theory 161
Individual Models of Situation Awareness 162
Team Models of Situation Awareness 164
Situation Awareness and Sport 166
Measuring Situation Awareness 167
Situation Awareness Requirements Analysis 168
Background and Applications 168
Domain of Application 169
Application in Sport 169
Procedure and Advice 169
Step 1: Define the Task(s) under Analysis 169
Step 2: Select Appropriate SMEs 169
Step 3: Conduct SME Interviews 169
Step 4: Conduct Goal-Directed Task Analysis 172
Step 5: Compile List of SA Requirements Identified 172
Step 6: Rate SA Requirements 172
Step 7: Determine SA Requirements 172
Step 8: Create SA Requirements Specification 172
Advantages 173
Disadvantages 173
Related Methods 173
Approximate Training and Application Times 173
Reliability and Validity 174
Tools Needed 174
Example 174
Flowchart 176
Recommended Texts 176
Situation Awareness Global Assessment Technique 177
Background and Applications 177
Domain of Application 177
Application in Sport 179
Procedure and Advice 179
Step 1: Define Analysis Aims 179
Step 2: Define Task(s) or Scenario under Analysis 179
Step 3: Conduct SA Requirements Analysis and Generate SAGAT Queries 179
Step 4: Brief Participants 179
Step 5: Conduct Pilot Run(s) 179
Step 6: Begin SAGAT Data Collection 179
Step 7: Freeze the Simulation 180
Step 8: Administer SAGAT Queries 180
Step 9: Query Response Evaluation and SAGAT Score Calculation 180
Step 10: Analyse SAGAT Data 180
Advantages 180
Disadvantages 181
Related Methods 181
Approximate Training and Application Times 181
Reliability and Validity 181
Tools Needed 182

    Example 182
    Flowchart 182
    Recommended Texts 182
Situation Present Assessment Method 182
    Background and Applications 182
    Domain of Application 182
    Application in Sport 183
    Procedure and Advice 183
        Step 1: Define Analysis Aims 183
        Step 2: Define Task(s) or Scenario under Analysis 185
        Step 3: Conduct SA Requirements Analysis and Generate Queries 185
        Step 4: Brief Participants 185
        Step 5: Conduct Pilot Runs 185
        Step 6: Undertake Task Performance 185
        Step 7: Administer SPAM Queries 185
        Step 8: Calculate Participant SA/Workload Scores 185
    Advantages 185
    Disadvantages 186
    Related Methods 186
    Training and Application Times 186
    Reliability and Validity 186
    Tools Needed 186
    Flowchart 187
    Recommended Text 187
Situation Awareness Rating Technique 187
    Background and Applications 187
    Domain of Application 187
    Application in Sport 187
    Procedure and Advice 187
        Step 1: Define Task(s) under Analysis 187
        Step 2: Selection of Participants 189
        Step 3: Brief Participants 189
        Step 4: Conduct Pilot Run 189
        Step 5: Performance of Task 189
        Step 6: Complete SART Questionnaires 189
        Step 7: Calculate Participant SART Scores 189
    Advantages 189
    Disadvantages 190
    Related Methods 190
    Approximate Training and Application Times 190
    Reliability and Validity 190
    Tools Needed 190
    Example 191
    Flowchart 191
    Recommended Text 191
Situation Awareness Subjective Workload Dominance 191
    Background and Applications 191
    Domain of Application 191
    Application in Sport 191
    Procedure and Advice 191
        Step 1: Define the Aims of the Analysis 191

Contents

Step 2: Define Task(s) under Analysis	191
Step 3: Create SWORD Rating Sheet	191
Step 4: SA and SA-SWORD Briefing	193
Step 5: Conduct Pilot Run	194
Step 6: Undertake Trial	194
Step 7: Administer SA-SWORD Rating Sheet	194
Step 8: Constructing the Judgement Matrix	194
Step 9: Matrix Consistency Evaluation	194
Advantages	194
Disadvantages	195
Related Methods	195
Approximate Training and Application Times	195
Reliability and Validity	195
Tools Needed	195
Flowchart	196
Recommended Text	196
Propositional Networks	196
Background and Applications	196
Domain of Application	196
Application in Sport	196
Procedure and Advice	198
Step 1: Define Analysis Aims	198
Step 2: Define Task(s) or Scenario under Analysis	198
Step 3: Collect Data Regarding the Task or Scenario under Analysis	198
Step 4: Define Concepts and Relationships between Them	198
Step 5: Define Information Element Usage	199
Step 6: Review and Refine Network	199
Step 7: Analyse Networks Mathematically	200
Advantages	200
Disadvantages	200
Related Methods	201
Approximate Training and Application Times	201
Reliability and Validity	201
Tools Needed	201
Example	201
Flowchart	201
Recommended Text	201

Chapter 7 Mental Workload Assessment Methods 207

Introduction	207
Mental Workload	207
Workload and Sport	209
Mental Workload Assessment	210
Primary and Secondary Task Performance Measures	212
Background and Applications	212
Domain of Application	212
Application in Sport	212
Procedure and Advice	212
Step 1: Define Primary Task under Analysis	212
Step 2: Define Primary Task Performance Measures	213

  Step 3: Design Secondary Task and Associated Performance Measures 213
  Step 4: Test Primary and Secondary Tasks 213
  Step 5: Brief Participants 213
  Step 6: Conduct Pilot Run 213
  Step 7: Undertake Primary Task Performance 213
  Step 8: Administer Subjective Workload Assessment Method 213
  Step 9: Analyse Data 216
 Advantages 216
 Disadvantages 216
 Related Methods 216
 Training and Application Times 217
 Reliability and Validity 217
 Tools Needed 217
 Example 217
 Flowchart 217
 Recommended Text 217
Physiological Measures 219
 Background and Applications 219
 Domain of Application 219
 Application in Sport 219
 Procedure and Advice 219
  Step 1: Define Task under Analysis 219
  Step 2: Select the Appropriate Measuring Equipment 220
  Step 3: Conduct Initial Testing of the Data Collection Procedure 220
  Step 4: Brief Participants 220
  Step 5: Fit Equipment 220
  Step 6: Conduct Pilot Run 220
  Step 7: Begin Primary Task Performance 220
  Step 8: Administer Subjective Workload Assessment Method 220
  Step 9: Download Data 221
  Step 10: Analyse Data 221
 Advantages 221
 Disadvantages 221
 Related Methods 221
 Training and Application Times 222
 Reliability and Validity 222
 Tools Needed 222
 Flowchart 222
 Recommended Text 222
NASA Task Load Index 222
 Background and Applications 222
 Domain of Application 224
 Application in Sport 224
 Procedure and Advice 224
  Step 1: Define Analysis Aims 224
  Step 2: Define Task(s) or Scenario under Analysis 224
  Step 3: Select Participants 224
  Step 4: Brief Participants 224
  Step 5: Conduct Pilot Run 224
  Step 6: Performance of Task under Analysis 225
  Step 7: Conduct Weighting Procedure 225

Contents

    Step 8: Conduct NASA-TLX Rating Procedure 225
    Step 9: Calculate NASA-TLX Scores 225
  Advantages 225
  Disadvantages 226
  Related Methods 226
  Approximate Training and Application Times 226
  Reliability and Validity 226
  Tools Needed 226
  Example 227
  Flowchart 227
  Recommended Texts 229
Subjective Workload Assessment Technique 229
  Background and Applications 229
  Domain of Application 229
  Application in Sport 229
  Procedure and Advice 229
    Step 1: Define Analysis Aims 229
    Step 2: Define Task(s) or Scenario under Analysis 230
    Step 3: Select Participants 230
    Step 4: Brief Participants 230
    Step 5: Undertake Scale Development 230
    Step 6: Conduct Pilot Run 230
    Step 7: Performance of Task under Analysis 231
    Step 8: Conduct SWAT Rating Procedure 231
    Step 9: Calculate SWAT Scores 231
  Advantages 231
  Disadvantages 231
  Related Methods 232
  Approximate Training and Application Times 232
  Reliability and Validity 232
  Tools Needed 232
  Flowchart 232
  Recommended Texts 232
The Subjective Workload Dominance Method 232
  Background and Applications 232
  Domain of Application 234
  Application in Sport 234
  Procedure and Advice 234
    Step 1: Define Analysis Aims 234
    Step 2: Define Task(s) or Scenario under Analysis 234
    Step 3: Create SWORD Rating Sheet 234
    Step 4: Select Participants 234
    Step 5: Brief Participants 235
    Step 6: Conduct Pilot Run 235
    Step 7: Undertake Task(s) under Analysis 235
    Step 8: Administration of SWORD Questionnaire 235
    Step 9: Construct Judgement Matrix 235
    Step 10: Evaluate Matrix Consistency 235
  Advantages 236
  Disadvantages 236
  Related Methods 236

        Approximate Training and Application Times 237
        Reliability and Validity 237
        Tools Needed 237
        Flowchart 237
        Recommended Texts 237
    Instantaneous Self-Assessment Method 237
        Background and Applications 237
        Domain of Application 238
        Application in Sport 239
        Procedure and Advice 239
            Step 1: Define Analysis Aims 239
            Step 2: Define Task(s) or Scenario under Analysis 239
            Step 3: Select Participants 239
            Step 4: Brief Participants 239
            Step 5: Conduct Pilot Run 240
            Step 6: Begin Task(s) under Analysis and Collect ISA Ratings 240
            Step 7: Construct Task Workload Profile 240
        Advantages 240
        Disadvantages 240
        Related Methods 240
        Training and Application Times 241
        Reliability and Validity 241
        Tools Needed 241
        Example 241
        Flowchart 241
        Recommended Text 241

Chapter 8   Teamwork Assessment Methods 243

    Introduction 243
    Teamwork 243
    Teamwork Assessment Methods 245
    Social Network Analysis 246
        Background and Applications 246
        Application in Sport 246
        Procedure and Advice 247
            Step 1: Define Analysis Aims 247
            Step 2: Define Task(s) or Scenario under Analysis 247
            Step 3: Collect Data 247
            Step 4: Validate Data Collected 247
            Step 5: Construct Agent Association Matrix 247
            Step 6: Construct Social Network Diagram 247
            Step 7: Analyse Network Mathematically 250
        Advantages 251
        Disadvantages 251
        Related Methods 251
        Approximate Training and Application Times 251
        Reliability and Validity 252
        Tools Needed 252
        Example 252
        Flowchart 255

Contents

- Recommended Texts ... 255
- Team Task Analysis ... 255
  - Background and Applications ... 255
  - Domain of Application ... 255
  - Application in Sport ... 255
  - Procedure and Advice (Adapted from Burke, 2004) ... 256
    - Step 1: Conduct Requirements Analysis ... 256
    - Step 2: Define Task(s) or Scenario under Analysis ... 257
    - Step 3: Identify Teamwork Taxonomy ... 257
    - Step 4: Conduct Coordination Demands Analysis ... 257
    - Step 5: Determine Relevant Taskwork and Teamwork Tasks ... 258
    - Step 6: Translation of Tasks into Knowledge, Skills, and Attitudes ... 258
    - Step 7: Link KSAs to Team Tasks ... 258
  - Advantages ... 258
  - Disadvantages ... 258
  - Related Methods ... 259
  - Approximate Training and Application Times ... 259
  - Tools Needed ... 259
  - Example ... 259
  - Recommended Text ... 259
- Coordination Demands Analysis ... 260
  - Background and Applications ... 260
  - Domain of Application ... 262
  - Application in Sport ... 262
  - Procedure and Advice ... 265
    - Step 1: Define Analysis Aims ... 265
    - Step 2: Define Task(s) under Analysis ... 265
    - Step 3: Select Appropriate Teamwork Taxonomy ... 266
    - Step 4: Data Collection ... 266
    - Step 5: Conduct an HTA for the Task under Analysis ... 266
    - Step 6: Taskwork/Teamwork Classification ... 266
    - Step 7: Construct CDA Rating Sheet ... 266
    - Step 8: Rate Coordination Levels ... 266
    - Step 9: Validate Ratings ... 266
    - Step 10: Calculate Summary Statistics ... 266
  - Advantages ... 267
  - Disadvantages ... 267
  - Related Methods ... 267
  - Approximate Training and Application Times ... 267
  - Reliability and Validity ... 268
  - Tools Needed ... 268
  - Example ... 268
  - Flowchart ... 268
  - Recommended Text ... 268
- Event Analysis of Systemic Teamwork ... 268
  - Background and Applications ... 268
  - Domain of Application ... 270
  - Application in Sport ... 270
  - Procedure and Advice ... 270
    - Step 1: Define Analysis Aims ... 270
    - Step 2: Define Task(s) under Analysis ... 270

        Step 3: Conduct Observational Study of the Task or Scenario under Analysis 271
        Step 4: Conduct CDM Interviews 271
        Step 5: Transcribe Data 271
        Step 6: Reiterate HTA 271
        Step 7: Conduct Coordination Demands Analysis 271
        Step 8: Construct Communications Usage Diagram 271
        Step 9: Conduct Social Network Analysis 271
        Step 10: Construct Operation Sequence Diagram 272
        Step 11: Construct Propositional Networks 272
        Step 12: Validate Analysis Outputs 272
    Advantages 272
    Disadvantages 272
    Related Methods 272
    Approximate Training and Application Times 273
    Reliability and Validity 273
    Tools Needed 273
    Example 273
    Flowchart 273
    Recommended Texts 273

Chapter 9   Interface Evaluation 275

Introduction 275
Interface Evaluation Methods 275
Checklists 278
    Background and Applications 278
    Domain of Application 279
    Application in Sport 279
    Procedure and Advice 279
        Step 1: Define Analysis Aims 279
        Step 2: Select Appropriate Checklist 280
        Step 3: Identify Set of Representative Tasks for the Device in Question 280
        Step 4: Brief Participants 280
        Step 5: Product/Device Familiarisation Phase 280
        Step 6: Check Item on Checklist against Product 280
        Step 7: Analyse Data 280
    Advantages 280
    Disadvantages 281
    Related Methods 281
    Approximate Training and Application Times 281
    Reliability and Validity 281
    Tools Needed 281
    Example 281
    Flowchart 284
    Recommended Texts 284
Heuristic Analysis 284
    Background and Applications 284
    Domain of Application 286
    Application in Sport 286
    Procedure and Advice 286

Contents

- Step 1: Define Aims of the Analysis 286
- Step 2: Identify Set of Representative Tasks for the Device in Question 286
- Step 3: Product/Device Familiarisation Phase 286
- Step 4: Perform Task(s) 286
- Step 5: Propose Design Remedies 286
- Advantages 287
- Disadvantages 287
- Related Methods 287
- Approximate Training and Application Times 287
- Reliability and Validity 287
- Tools Needed 287
- Example 287
- Flowchart 288
- Recommended Text 288
- Link Analysis 289
  - Background and Applications 289
  - Domain of Application 290
  - Application in Sport 290
  - Procedure and Advice 290
    - Step 1: Define Analysis Aims 290
    - Step 2: Define Task(s) or Scenario under Analysis 290
    - Step 3: Collect Data 290
    - Step 4: Validate Data Collected 290
    - Step 5: Construct Link Analysis Table 291
    - Step 6: Construct Link Analysis Diagram 291
    - Step 7: Offer Redesign Proposals/Analyse Link Analysis Data 292
  - Advantages 292
  - Disadvantages 292
  - Related Methods 292
  - Approximate Training and Application Times 292
  - Reliability and Validity 292
  - Tools Needed 293
  - Example 293
  - Flowchart 294
  - Recommended Texts 296
- Layout Analysis 297
  - Background and Applications 297
  - Domain of Application 297
  - Application in Sport 297
  - Procedure and Advice 297
    - Step 1: Define Aims of the Analysis 297
    - Step 2: Identify Set of Representative Tasks for the Device in Question 297
    - Step 3: Create Schematic Diagram 297
    - Step 4: Familiarisation with Device 297
    - Step 5: Arrange Interface Components into Functional Groupings 298
    - Step 6: Arrange Functional Groupings Based on Importance of Use 298
    - Step 7: Arrange Functional Groupings Based on Sequence of Use 298
    - Step 8: Arrange Functional Groupings Based on Frequency of Use 298
  - Advantages 299
  - Disadvantages 299
  - Related Methods 299

	Approximate Training and Application Times	299
	Reliability and Validity	299
	Tools Needed	299
	Example	300
	Flowchart	300
	Recommended Text	301

Interface Surveys 301
    Background and Application 301
        Control and Display Survey 302
        Labelling Surveys 303
        Coding Consistency Survey 303
        Operator Modification Survey 303
    Domain of Application 303
    Application in Sport 303
    Procedure and Advice 304
        Step 1: Define Analysis Aims..304
        Step 2: Define Task(s) or Scenario under Analysis.......................................304
        Step 3: Data Collection 304
        Step 4: Complete Appropriate Surveys..304
        Step 5: Propose Remedial Measures 304
    Advantages 304
    Disadvantages..304
    Related Methods 305
    Approximate Training and Application Times 305
    Reliability and Validity 305
    Tools Needed 305
    Example 305
    Flowchart 305
    Recommended Text 305

Chapter 10   Human Factors Methods Integration: Case Study 309

    Introduction 309
    Integrating Human Factors Methods..309
        Summary of Component Methods 310
    Methodology 312
        Participants 312
        Materials 313
        Procedure 313
        Results 314
    Discussion..316
    Conclusions 323

References 325

Index 339

List of Figures

FLOWCHART 2.1	Interviews flowchart.	22
FIGURE 2.1	System Usability Scale. *Source*: Adapted from Brooke, J. (1996).	27
FLOWCHART 2.2	Questionnaire flowchart.	28
FLOWCHART 2.3	Observational flowchart.	34
FIGURE 3.1	"Defeat opposition" soccer HTA extract with example decomposition.	44
FIGURE 3.2	"Tackle opposition player" decomposition.	45
FIGURE 3.3	"Score goal from penalty kick" decomposition.	46
FLOWCHART 3.1	HTA flowchart.	47
FLOWCHART 3.2	Task decomposition flowchart.	55
FIGURE 3.4	Runner VPA extract.	60
FLOWCHART 3.3	Verbal Protocol Analysis flowchart.	61
FIGURE 3.5	Standard OSD template.	63
FIGURE 3.6	Scrum HTA description.	65
FIGURE 3.7	Scrum OSD diagram.	66
FLOWCHART 3.4	Operation sequence diagram flowchart.	68
FIGURE 4.1	Decision ladder (showing leaps and shunts).	79
FIGURE 4.2	Strategies analysis simplified flow map.	80
FIGURE 4.3	Mapping actors onto strategies analysis output for SOCA.	80
FIGURE 4.4	SRK behavioural classification scheme.	81
FIGURE 4.5	Golf abstraction hierarchy.	84
FIGURE 4.6	Golf shot decision ladder.	85
FIGURE 4.7	Example approach shot strategies analysis flow map.	86
FLOWCHART 4.1	Critical decision method flowchart.	96
FIGURE 4.8	Concept map about concept maps. Adapted from Crandall et al. 2006.	97
FIGURE 4.9	Example concept map for identifying the factors considered by golfers when determining approach shot strategy.	99
FLOWCHART 4.2	Concept map flowchart.	100
FLOWCHART 4.3	ACTA flowchart.	107
FIGURE 5.1	Unsafe acts taxonomy. *Source*: Reason, J. (1990). *Human error*. New York: Cambridge University Press.	112

FIGURE 5.2 Reason's Swiss cheese systems perspective on error and accident causation. *Source*: Adapted from Reason, J. (2000) Human error: Models and management. *British Medical Journal* 320:768–70. 114

FIGURE 5.3 Hillsborough disaster Accimap. 122

FLOWCHART 5.1 Accimap flowchart. 124

FIGURE 5.4 Fault tree diagram of Jean Van de Velde's final-hole triple bogey in the 1999 British Open. 127

FLOWCHART 5.2 Fault tree flowchart. 129

FIGURE 5.5 SHERPA external error mode taxonomy. 131

FIGURE 5.6 Golf HTA extract. 135

FLOWCHART 5.3 SHERPA flowchart. 141

FIGURE 5.7 HET analysis extract. 144

FLOWCHART 5.4 HET flowchart. 145

FIGURE 5.8 HTA for boil kettle task. 147

FIGURE 5.9 State space TAFEI diagram. 148

FIGURE 5.10 HTA for "Use Forerunner device to complete sub 5.30-minute mile 10-mile run" scenario. 151

FIGURE 5.11 Garmin 305 Forerunner SSD. 152

FLOWCHART 5.5 TAFEI flowchart. 154

FLOWCHART 5.6 THEA flowchart. 159

FIGURE 6.1 Endsley's three-level model of situation awareness. *From* Endsley, Human Factors, 1995. With permission. 162

FIGURE 6.2 Smith and Hancock's perceptual cycle model of situation awareness. *Source*: Adapted from Smith, K. and Hancock, P. A. (1995). With permission. 164

FIGURE 6.3 HTA passing sub-goal decomposition. 174

FIGURE 6.4 SA requirements extract for "make pass" sub-goals. 175

FLOWCHART 6.1 SA requirements analysis flowchart. 178

FLOWCHART 6.2 SAGAT flowchart. 184

FLOWCHART 6.3 SPAM flowchart. 188

FIGURE 6.5 SART rating scale. 192

FLOWCHART 6.4 SART flowchart. 193

FLOWCHART 6.5 SA-SWORD flowchart. 197

FIGURE 6.6 Propositional network diagram about propositional networks. 198

FIGURE 6.7 Example relationships between concepts. 199

FIGURE 6.8 Soccer propositional network example. 202

List of Figures

FIGURE 6.9 Attacking midfielder in possession of ball propositional network; the shaded information elements represent the attacking midfielder's awareness. 203

FIGURE 6.10 Defender without possession of the ball propositional network; the shaded information elements represent the defender's awareness. 204

FLOWCHART 6.6 Propositional network flowchart. 205

FIGURE 7.1 Framework of interacting stressors impacting workload. *Source*: Adapted from Megaw, T. (2005). The definition and measurement of mental workload. In *Evaluation of human work*, eds. J. R. Wilson and N. Corlett, 525–53. Boca Raton, FL: CRC Press. 208

FLOWCHART 7.1 Primary and secondary task performance measures flowchart. 218

FLOWCHART 7.2 Physiological measures flowchart. 223

FIGURE 7.2 NASA-TLX pro-forma (Figue courtesy of NASA Ames Research Center). 227

FLOWCHART 7.3 NASA-TLX flowchart. 228

FLOWCHART 7.4 SWAT flowchart. 233

FLOWCHART 7.5 SWORD flowchart. 238

FLOWCHART 7.6 ISA flowchart. 242

FIGURE 8.1 Example social network diagram for five-a-side soccer team. 250

FIGURE 8.2 Social network diagram for 2006 FIFA World Cup Final game. 253

FIGURE 8.3 Social network diagrams for sub-teams and passing links between them. 253

FIGURE 8.4 Social network diagrams for different pitch areas. 254

FLOWCHART 8.1 Social network analysis flowchart. 256

FIGURE 8.5 Standard scrum formation. 260

FIGURE 8.6 Retrieve ball from scrum HTA. 261

FLOWCHART 8.2 Coordination demands analysis flowchart. 269

FIGURE 8.7 Network of networks approach to analysing distributed teamwork; figure shows example representations of each network, including hierarchical task analysis (task network), social network analysis (social network), and propositional network (knowledge network) representations. *Source*: Adapted from Houghton, R. J., Baber, C., Cowton, M., Stanton, M. A., and Walker, G. H. (2008). WESTT (Workload, Error, Situational Awareness, Time and Teamwork): An analytical prototyping system for command and control. *Cognition Technology and Work* 10(3):199–207. 270

FLOWCHART 8.3 Event analysis of systemic teamwork flowchart. 274

FIGURE 9.1 Garmin 305 Forerunner wrist unit. *Source*: www8.garmin.com, with permission. 278

FLOWCHART 9.1 Checklist flowchart. 285

FLOWCHART 9.2 Heuristics flowchart. 289

FIGURE 9.2 Link analysis diagram example for goal scored during Euro 2008 group game. ...291

FIGURE 9.3 Link diagram for goal scored in Euro 2008 group game. 294

FIGURE 9.4 Goals scored and mean number of passes leading to goals per team for the soccer European Championships 2008. 295

FIGURE 9.5 Link analysis diagram and tables showing comparison of player passing; the darker arrows depict the origin, destination, and receiving player for each successful pass made, and the lighter arrows depict the origin, destination, and intended receiving player for each unsuccessful pass. 295

FLOWCHART 9.3 Link analysis flowchart. 296

FIGURE 9.6 Garmin Forerunner 305 schematic diagram. 298

FIGURE 9.7 Interface reorganised based on importance of use during running events. 300

FIGURE 9.8 Interface reorganised based on sequence of use. 300

FIGURE 9.9 Interface reorganised based on frequency of use. 301

FIGURE 9.10 Interface redesign based on layout analysis. 301

FLOWCHART 9.4 Layout analysis flowchart. 302

FLOWCHART 9.5 Interface surveys flowchart. 308

FIGURE 10.1 Network of networks approach. *Source*: Adapted from Houghton, R. J., Baber, C., Cowton, M., Stanton, N. A., and Walker, G. H. (2008). WESTT (Workload, Error, Situational Awareness, Time and Teamwork): An analytical prototyping system for command and control. *Cognition Technology and Work* 10(3):199–207. 310

FIGURE 10.2 Network of network approach overlaid with methods applied during case study. 311

FIGURE 10.3 Network of network approach overlaid with questions potentially addressed. 312

FIGURE 10.4 Internal structure of integrated Human Factors methods framework. 313

FIGURE 10.5 Fell race HTA. 315

FIGURE 10.6 Task model of fell race. 316

FIGURE 10.7 Ascent and descent propositional networks for amateur runner. 318

FIGURE 10.8 Example social network diagram showing associations between runner, GPS device, other runners, and stewards. 320

FIGURE 10.9 Mean workload ratings for typical training run and race ascent and descent. 320

FIGURE 10.10 Ascent mean workload ratings per runner group. 320

FIGURE 10.11 Descent mean workload ratings per runner group. 321

FIGURE 10.12 Multi-perspective output example. 322

List of Tables

TABLE 1.1 Human Factors Methods and Their Potential Application in Major Sports 6
TABLE 2.1 Data Collection Methods Summary Table 11
TABLE 2.2 Critical Decision Method Interview Transcript for "Select Race Strategy" Decision Point 17
TABLE 2.3 Critical Decision Method Interview Transcript for "6 Miles—Check Condition and Revise Race Strategy as Appropriate" Decision Point 18
TABLE 2.4 Critical Decision Method Interview Transcript for "13 Miles—Check Condition and Revise Race Strategy as Appropriate" Decision Point 19
TABLE 2.5 Critical Decision Method Interview Transcript for "18 Miles—Decision Regarding Strategy for Remainder of the Race" 20
TABLE 2.6 Critical Decision Method Interview Transcript for "22 Miles—Decision Regarding Strategy for Remainder of the Race" 21
TABLE 2.7 Questionnaire Questions 24
TABLE 2.8 World Cup 2006 Game Passing Observational Transcript 33
TABLE 3.1 Task Analysis Methods Summary Table 38
TABLE 3.2 Task Decomposition Categories 49
TABLE 3.3 Score Goal from Penalty Kick Task Decomposition Output 52
TABLE 4.1 Cognitive Task Analysis Methods Summary Table 72
TABLE 4.2 Example WDA Prompts 78
TABLE 4.3 Critical Decision Method Probes 87
TABLE 4.4 Example "Ascent" CDM Output 92
TABLE 4.5 Example "Descent" CDM Output 94
TABLE 4.6 Example Simulation Interview Table 104
TABLE 4.7 Example Cognitive Demands Table 105
TABLE 5.1 Human Error Assessment Methods Summary Table 118
TABLE 5.2 Golf SHERPA Extract 136
TABLE 5.3 Transition Matrix 148
TABLE 5.4 Error Descriptions and Design Solutions 149
TABLE 5.5 Transition Matrix 153
TABLE 5.6 Scenario Description Template 155
TABLE 5.7 Example THEA Error Analysis Questions 157
TABLE 6.1 Situation Awareness Assessment Methods Summary Table 170

TABLE 6.2	Example SAGAT Queries for "Make Pass" Task	183
TABLE 7.1	Mental Workload Assessment Methods Summary Table	214
TABLE 7.2	SWAT Dimensions	230
TABLE 7.3	ISA Scale	239
TABLE 8.1	Teamwork Models	245
TABLE 8.2	Teamwork Assessment Methods Summary Table	248
TABLE 8.3	Social Network Agent Association Matrix Example	250
TABLE 8.4	Sociometric Status Statistics	254
TABLE 8.5	Teamwork Taxonomy	257
TABLE 8.6	CDA Results for Scrum Task	262
TABLE 8.7	Scrum Task KSA Analysis	263
TABLE 9.1	Interface Evaluation Methods Summary Table	276
TABLE 9.2	Visual Clarity Checklist Evaluation Extract	282
TABLE 9.3	System Usability Problems Checklist Evaluation Extract	283
TABLE 9.4	Garmin Forerunner Device Heuristic Evaluation Example	288
TABLE 9.5	Control Survey Analysis for Garmin Forerunner Training Device	306
TABLE 9.6	Display Survey Analysis for Garmin Forerunner Training Device	307
TABLE 9.7	Labelling Survey Analysis for Garmin Forerunner Training Device	307
TABLE 10.1	CDM Probes	314
TABLE 10.2	Example Elite Runner "Ascent" CDM Output	317
TABLE 10.3	Information Element Usage by Runner Group	319

Preface

Two of my many loves in life are Sport and Human Factors methods. Having begun academic life way back as an undergraduate studying for a degree in Sports Science, and having since progressed to a research career in the discipline of Human Factors and Ergonomics, I am continually struck by the similarities in the concepts being investigated in both disciplines. Accordingly, this book began life as a result of a discussion between Professor Neville Stanton and I on the utility of using social network analysis, a currently popular Human Factors method for analysing the associations between entities in networks, to analyse passing performance in team sports such as soccer. As the discussion continued, more and more Human Factors methods capable of contributing to sports performance description and evaluation were thrown into the conversation, and more and more intriguing possibilities for cross-disciplinary interaction were identified. At the end of the conversation, I had managed to persuade Neville (or had he managed to persuade me?) that provided the work could be undertaken as a personal project, as a labour of love so to speak, in our own spare time, we could write a book on how Human Factors methods could be applied in a sporting context.

I therefore cajoled a collection of close friends and colleagues who know a lot more about Sport (Adam Gibbon) and Human Factors methods (Prof. Stanton, Dr. Walker and Dr. Jenkins) than I do, into working on the project. The project initially began with a review of the Human Factors literature in order to identify the Human Factors concepts and methods that could be studied and applied, with worthwhile outputs, in a sporting context. Following this, the chosen methods were applied, in our spare time and more often than not involving ourselves and/or sporting friends and associates (often those who we run with, and play soccer and golf with) as willing participants or subject matter experts, in a variety of sporting scenarios. Refreshingly, and perhaps a sign of the utility of such research in a sporting context, all the people we engaged with gave their consent enthusiastically, and were extremely interested in the outputs produced and the potential uses to which they could be put—particularly where it helped their own game. We owe a great debt of gratitude to all those sports men and women who freely and voluntarily helped make this book possible. This book therefore provides an overview of those Human Factors concepts that are applicable in a sporting context, and detailed guidance on how to apply them. We hope that the material in this book will provide those wishing to apply such methods, and study such concepts in a sporting context, with sufficient guidance and background information to do so.

This book has been constructed so that students, practitioners, and researchers with interest in one particular area of Human Factors can read the chapters non-linearly independently from one another. Each category of Human Factors methods is treated as a separate chapter containing an introduction to the area of interest and an overview of the range of applicable methods, including practical guidance on how to apply them, within that category area. Each of the chapters is therefore self-contained, so those wanting to explore a particular topic area can choose to simply read the chapter relevant to their needs. For example, if a reader wishes to investigate situation awareness, the concept that deals with how actors develop and maintain sufficient levels of awareness to perform optimally in complex sociotechnical systems, then they would proceed to the situation awareness chapter, read the introduction for background and theoretical perspectives, and then select the method that is most suited to their analysis needs. By following the guidance presented, they can then apply their chosen method in the field. We feel that this book will be of most use to those who are involved in the evaluation of sports performance and products, and those who are seeking information to inform the design of sports products, the development of coaching and tactical interventions, and the development of Sports Science and Human Factors theory and methodologies. This book will also be useful for students wishing to learn about Human Factors in general and about the application of structured Human Factors methods. They can apply the methods to their own area of study and use the material within the book to prepare case studies, coursework, and report assignments.

The prospect of Human Factors methods being applied in new domains by researchers from other disciplines is a fascinating one, and testing theory and methods across domains can only advance knowledge in both fields. Toward this end, it is our hope that this book provides Sports Scientists with the methods necessary to investigate core Human Factors concepts in a sporting context.

Dr. Paul M. Salmon

Acknowledgments

We owe a great debt of gratitude to all those sportsmen and -women who freely and voluntarily helped make this book possible. Special thanks go to all of the runners from the running club who were involved in the case study described in the final chapter. Many thanks go also to a number of subject matter experts who have gone over and above the call of duty to review and refine analyses, discuss ideas, and provide invaluable domain knowledge, opinions, and viewpoints. We would also like to thank Kirsten Revell for designing the book cover, and Linda Paul for all of her help and assistance throughout the work undertaken and the production of this book. Thanks go also to Steven Thomas for assistance in collecting data and refining some of the ideas presented.

About the Authors

Paul M. Salmon, PhD
Human Factors Group
Monash University Accident Research Centre
Monash University
Victoria, Australia

Paul Salmon is a senior research fellow within the Human Factors Group at the Monash University Accident Research Centre and holds a BSc in Sports Science, an MSc in Applied Ergonomics, and a PhD in Human Factors. Paul has eight years experience of applied Human Factors research in a number of domains, including the military, aviation, and rail and road transport, and has worked on a variety of research projects in these areas. This has led to him gaining expertise in a broad range of areas, including situation awareness, human error, and the application of Human Factors methods, including human error identification, situation awareness measurement, teamwork assessment, task analysis, and cognitive task analysis methods. Paul has authored and co-authored four books and numerous peer-reviewed journal articles, conference articles, and book chapters, and was recently awarded the 2007 Royal Aeronautical Society Hodgson Prize for a co-authored paper in the society's *Aeronautical Journal* and, along with his colleagues from the Human Factors Integration Defence Technology Centre (HFI DTC) consortium, the Ergonomics Society's President's Medal in 2008.

Neville A. Stanton
Transportation Research Group, University of Southampton
School of Civil Engineering and the Environment
Highfield, Southampton, United Kingdom

Professor Stanton holds a chair in Human Factors in the School of Civil Engineering and the Environment at the University of Southampton. He has published over 140 peer-reviewed journal papers and 14 books on Human Factors and Ergonomics. In 1998, he was awarded the Institution of Electrical Engineers Divisional Premium Award for a co-authored paper on Engineering Psychology and System Safety. The Ergonomics Society awarded him the Otto Edholm medal in 2001 and the President's Medal in 2008 for his contribution to basic and applied ergonomics research. In 2007, The Royal Aeronautical Society awarded Professor Stanton and colleagues the Hodgson Medal and Bronze Award for their work on flight deck safety. Professor Stanton is an editor of the journal *Ergonomics* and is on the editorial boards of *Theoretical Issues in Ergonomics Science* and the *International Journal of Human Computer Interaction*. Professor Stanton is a fellow and chartered occupational psychologist registered with the British Psychological Society, and a fellow of the Ergonomics Society. He has a BSc (Hons) in Occupational Psychology from the University of Hull, an MPhil in Applied Psychology, and a PhD in Human Factors from Aston University in Birmingham.

Adam C. Gibbon
Head of Physical Education
Staindrop Enterprise College
Staindrop, Darlington
County Durham, United Kingdom

Adam Gibbon is head of Physical Education at Staindrop Enterprise College and holds a BSc in Sports Science. Adam has over seven years' experience in teaching and coaching a range of sports, and was recently graded as an outstanding teacher by the Office for Standards in Education, Children's Services and Skills (OFSTED), the UK teaching inspectorate. Adam has an interest in the analysis of sports performance in both team and individual sports, and has practical experience in the use of human factors methods, including task analysis, cognitive task analysis, teamwork assessment, and mental workload assessment methods for this purpose. Adam is also a keen sportsman, having previously being involved in professional football, and now a keen marathon and fell race runner, golfer, and cyclist.

Daniel P. Jenkins, PhD
Sociotechnic Solutions Ltd.
2 Mitchell Close
St. Albans, UK
AL1 2LW

Dan Jenkins graduated in 2004, from Brunel University with an M.Eng (Hons) in mechanical engineering and design, receiving the University Prize for the highest academic achievement in the school. As a sponsored student, Dan finished his work at Brunel with more than two years of experience as a design engineer in the automotive industry. Upon graduation, Dan went to work in Japan for a major car manufacturer, facilitating the necessary design changes to launch a new model in Europe. In 2005, Dan returned to Brunel University taking up the full-time role of research fellow in the Ergonomics Research Group, working primarily on the Human Factors Integration Defence Technology Centre (HFI-DTC) project. Dan studied part-time on his PhD in human factors and interaction design – graduating in 2008, receiving the 'Hamilton Prize' for the Best Viva in the School of Engineering and Design. In March 2009, Dan founded Sociotechnic Solutions Limited, a company specialising in the design and optimisation of complex sociotechnical systems. Dan has authored and co-authored numerous journal paper, conference articles, book chapters and books. Dan and his colleagues on the HFI DTC project were awarded the Ergonomics Society's President's Medal in 2008.

Guy H. Walker, PhD
School of the Built Environment
Heriot-Watt University
Edinburgh, Scotland
EH14 4AS

Guy Walker has a BSc Honours degree in psychology from the University of Southampton and a PhD in human factors from Brunel University. His research interests are wide ranging, spanning driver behaviour and the role of feedback in vehicles, railway safety and the issue of signals passed at danger, and the application of sociotechnical systems theory to the design and evaluation of military command and control systems. Guy is the author or co-author of more than forty peer-reviewed journal articles and several books. This volume was produced during his time as senior research fellow within the HFI DTC. Along with his colleagues in the research consortium, Guy was awarded the Ergonomics Society's President's Medal for the practical application of ergonomics theory. Guy currently resides in the School of the Built Environment at Heriot-Watt University in Edinburgh, Scotland, working at the cross-disciplinary interface between engineering and people.

Acronyms

The following is a reference list of acronyms used within this book.

ACTA	Applied Cognitive Task Analysis
ADS	Abstraction Decomposition Space
AH	Abstraction Hierarchy
CARS	Crew Awareness Rating Scale
CDA	Coordination Demands Analysis
CDM	Critical Decision Method
CIT	Critical Incident Technique
Comms	Communications
ConTA	Control Task Analysis
CREAM	Cognitive Reliability and Error Analysis Method
CRM	Crew Resource Management
CTA	Cognitive Task Analysis
CUD	Communications Usage Diagram
CWA	Cognitive Work Analysis
DM	Decision Making
DRAWS	Defence Research Agency Workload Scales
DSA	Distributed Situation Awareness
EAST	Event Analysis of Systemic Teamwork
EEG	Electroencephalogram
EEM	External Error Mode
FA	Football Association
FAA	Federal Aviation Administration
FCR	Fire Control Radar
FIFA	Fédération Internationale de Football Association
FWDs	Forwards
H	Hooker
HAZOP	Hazard and Operability
HCI	Human Computer Interaction
HEI	Human Error Identification
HEIST	Human Error Identification in Systems Tool
HET	Human Error Template
HFACS	Human Factors Analysis and Classification System
HR	Heart Rate
HRV	Heart Rate Variability
HSE	Health and Safety Executive
HSF	Horizontal Situational Format
HTA	Hierarchical Task Analysis
HUD	Head-Up Display
ICAM	Incident Cause Analysis Method
ISA	Instantaneous Self-Assessment
KSA	Knowledge, Skills, and Attitudes
MA	Mission Analysis
MACE	Malvern Capacity Estimate
MART	Malleable Attentional Resources Theory
NASA TLX	National Aeronautics and Space Administration Task Load Index

No. 8	Number 8
OSD	Operation Sequence Diagram
PC	Personal Computer
PFA	Professional Footballers Association
QUIS	Questionnaire for User Interface Satisfaction
RNASA TLX	Road National Aeronautics and Space Administration Task Load Index
SA	Situation Awareness
SAGAT	Situation Awareness Global Assessment Technique
SARS	Situation Awareness Rating Scales
SART	Situation Awareness Rating Technique
SA-SWORD	Situation Awareness-Subjective Workload Dominance
SH	Scrum Half
SHERPA	Systematic Human Error Reduction and Prediction Approach
SME	Subject Matter Expert
SNA	Social Network Analysis
SPAM	Situation Present Assessment Method
SOCA	Social Organisation and Co-operation Analysis
SRK	Skill, Rule, Knowledge
SSD	State Space Diagram
SUMI	Software Usability Measurement Inventory
SUS	System Usability Scale
SWAT	Subjective Workload Assessment Technique
SWORD	Subjective Workload Dominance
TAFEI	Task Analysis For Error Identification
THEA	Technique for Human Error Assessment
TOTE	Test-Operate-Test-Exit
TRACEr	Technique for Retrospective Analysis of Cognitive Error
TTA	Team Task Analysis
UEFA	Union of European Football Associations
WCA	Worker Competencies Analysis
VPA	Verbal Protocol Analysis
WDA	Work Domain Analysis
WESTT	Workload Error Situation Awareness Time and Teamwork

1 Introduction

INTRODUCTION

It is the 1996 soccer European Championship semifinals and, with the score tied at 5-5, Gareth Southgate, England's centre-half, steps up to take England's sixth penalty kick in a sudden death shoot-out against Germany. With a place in the European Championship final at stake, Southgate has to score at all costs; missing the penalty gives Germany the chance of scoring with its next shot and securing a place in the final, knocking England out of the competition. The German goalkeeper Andreas Köpke stands between Southgate, the ball, and the goal. Awaiting the referee's instruction, Southgate makes his decision regarding power and placement of the penalty; in doing so he uses his training, experience, observation of the goalkeeper's behaviour on the previous five penalties, advice given to him by teammates and his manager, and attempts to read Köpke's likely strategy. The 75,000-plus crowd waits in anticipation, with the German contingent attempting to distract Southgate from his task.

It is the final round of golf's 1999 British Open at Carnoustie, Forfarshire, and Frenchman Jean Van de Velde stands on the final tee. Leading the 72-hole tournament by three shots, Van de Velde knows that he needs only a double bogey 6 or better on the par 4 hole (which he has birdied twice in the previous three rounds) to become the first Frenchman to win the tournament in over 90 years. Having never before won a major, nor having led one on the final hole, Van de Velde has to decide what strategy he should use to play out the final hole in six shots or fewer. Placing his ball on the tee, he muses over his tee shot; should he play easy and take a mid-iron down the left or right of the fairway, or should he take a driver and try to play as far down the hole as he possibly can? The first strategy will allow him to lay up with his approach shot and pitch onto the green in three, leaving him with three putts for victory, whereas the latter will leave the green reachable in two shots, giving him four putts for victory.

Both scenarios bear all of the hallmarks of the complex domains in which Human Factors researchers work. Human Factors, the study of human performance in sociotechnical systems, has previously been defined as "the scientific study of the relationship between man and his working environment" (Murell, 1965); "the study of how humans accomplish work-related tasks in the context of human-machine systems" (Meister, 1989); and "applied information about human behaviour, abilities, limitations and other characteristics to the design of tools, machines, tasks, jobs and environments" (Sanders and McCormick, 1993). Human Factors is therefore concerned with human capabilities and limitations, human–machine interaction, teamwork, tools, machines and material design, environments, work, organisational design, system performance, efficiency, effectiveness, and safety.

In the scenarios described, critical cognitive and physical tasks are being performed in a dynamic, complex, collaborative system consisting of multiple humans and artefacts, under pressurised, complex, and rapidly changing conditions. Highly skilled, experienced, and well-trained individuals walk a fine line between task success and failure, with only slightly inadequate task execution leading to the latter. Accordingly, all manner of Human Factors concepts are at play, including naturalistic decision making, situation awareness, expertise, human error, teamwork, physical and mental workload, stress, trust, communication, and distraction.

Southgate steps up and places the ball low and to the keeper's right hand side; his execution is not perfect, and Köpke, guessing correctly, dives to his right and saves the ball. Next, Andreas Möller steps up for Germany, and following his own decision-making and course of action selection process, drives the ball high and powerfully down the middle of the goal, sending the English goalkeeper, David Seaman, the wrong way. Germany is through to the final, and England is out, defeated once again on penalties at the semifinal stage of a major professional soccer tournament.

Van De Velde chooses to take a driver off the tee in an attempt to get as far down the 18th hole as possible. Struck badly, the shot goes wide to the right and ends up in the rough of the 17th hole fairway. Following this, a series of bad decisions and poorly executed shots results in Van de Velde making a triple bogey 7 and ending the tournament in a three-way tie for first place, resulting in a four-hole playoff, which he eventually loses to Paul Lawrie.

In the aftermath of both incidents, there is much debate and discussion over why Southgate and Van de Velde had failed. Why did they make the decisions that they did? Did their lack of experience of similar scenarios affect their performance? Did factors such as crowd, pressure, expectation, and other players detract them from their task? Exactly what was going through each player's mind as they prepared for, and took, their shots? Had they had sufficient appropriate training for such situations? All were pertinent questions; however, the majority of which were answered, as is often the case in spectator sports, with only conjecture, hearsay, and personal opinion.

As Human Factors researchers, we have a range of valid, reliable, and scientifically supported methods for describing and evaluating human performance and the concepts underpinning it in complex settings such as those described above. The impetus for this book comes from our contention that most Human Factors methods are highly applicable in a sporting context, and the notion that they can be used to answer some of the questions described above. For example, although sports scientists have hypothesised about why penalties are missed (e.g., Jordet, 2009) and how success can be achieved in penalty shoot-outs (e.g., McGarry and Franks, 2000), it is notable that structured Human Factors methods, despite being highly applicable, have not yet been used in this area.

Traditionally, Human Factors methods have been applied in the safety critical domains, such as the military, nuclear power, aviation, road transport, energy distribution, and rail domains; however, there is great scope for applying these methods in a sporting context. Further, advances in scientific theory and methodology are often brought about by cross-disciplinary interaction (Fiore and Salas, 2006, 2008), whereby methods and theories developed and applied in one domain are applied in another. In this book, we argue that Human Factors methods can be applied, with significant gains to be made, within the sporting domains. Accordingly, we present an overview of the core Human Factors concepts and guidance on the methods available to study them, with a view to these methods being applied in a sporting context.

The purpose of this book is therefore to present a range of Human Factors methods for describing, representing, and evaluating human, team, and system performance in sports. Traditionally, the application of Human Factors (and Ergonomics) methods in the sporting domains has focused on the biomechanical (e.g., Dixon, 2008; Lees, Rojas, Cepero, Soto, and Gutierrez, 2000; Rojas, Cepero, Ona, and Gutierrez, 2000), physiological (e.g., Rahnama, Lees, and Bambaecichi, 2005; Tessitore, Meeusen, Tiberi, Cortis, Pagano, and Capranica, 2005), environmental (e.g., Noakes, 2000; Sparks, Cable, Doran, and MacLaren, 2005), and equipment (e.g., Lake, 2000; Purvis and

Tunstall, 2004; Webster, Holland, Sleivert, Laing, and Niven, 2005) related aspects of sports performance. Accordingly, methods from the realm of Physical Ergonomics have most commonly been applied and there has been only limited uptake of Cognitive Ergonomics or Human Factors methods (as we will call them from now on) in sports science circles. More recently, however, significant interest has begun to be shown in the psychological or cognitive aspects of sporting performance, and sports scientists are beginning to look at various aspects of cognition and sports performance, some of which Human Factors researchers deal with on a daily basis. For example, naturalistic decision making (e.g., Macquet and Fluerance, 2007), situation awareness (e.g., James and Patrick, 2004), and human error (e.g., Helsen, Gilis, and Weston, 2006; Oudejans, Bakker, and Beek, 2007) are examples of popular Human Factors concepts that have recently been investigated in a sporting context.

It is hoped that this book promotes cross-disciplinary interaction between the Human Factors and Sports Science disciplines. Despite obvious similarities in the concepts requiring investigation within sport and the more typical Human Factors domains, traditionally there has been a lack of cross-disciplinary interaction between Human Factors and Sports Science researchers. Fiore and Salas, for example, point out the surprising paucity of cross-disciplinary interaction between sports scientists and researchers working in the area of military psychology (Fiore and Salas, 2008), and between researchers in team cognition and sports psychologists (Fiore and Salas, 2006). This is not to say the potential for both groups learning from the other does not exist. A recent special issue of the *Military Psychology* journal (Fiore and Salas, 2008), for example, was devoted to communicating contemporary sports psychology research and methods to researchers working in the area of military psychology, with the intention of promoting cross-disciplinary interaction between the two domains. Similarly, a recent special issue of the *International Journal of Sports Psychology* was used to provide a medium for team cognition researchers to discuss their research in the context of sports psychology, with a view to enhancing cross-disciplinary interaction between the two fields.

Among other things, one of the reasons for the lack of cross-disciplinary interaction between sports scientists and human factors researchers appears to be a lack of appreciation (in both directions) of the methods used in both disciplines. Increasing other researchers' awareness of our methods, in terms of what they do, how they are applied, and what outputs they produce, is therefore one way of enhancing cross-disciplinary research. Accordingly, the aim of this book is, first, to introduce some of the fundamental Human Factors concepts which are likely to be of interest to sports scientists; second, to identify those Human Factors methods which can be applied for studying these concepts in the sports domain; third, to provide guidance on how to apply the methods identified; and fourth and finally, to present example applications in a sporting context of the methods described.

HUMAN FACTORS METHODS

Structured methods form a major part of the Human Factors discipline. A recent review identified well over 100 Human Factors methods (Stanton, Salmon, Walker, Baber, and Jenkins, 2005), and the International Encyclopaedia of Human Factors and Ergonomics (Karwowski, 2001) has an entire section devoted to various methods and techniques. Human Factors methods are used to describe, represent, and evaluate human activity within complex sociotechnical systems. These methods focus on human interaction with other humans, products, devices, or systems, and cover a variety of issues ranging from the physical and cognitive aspects of task performance, errors, decision making, situation awareness, device usability, time, and physical and mental workload. They are applied by researchers for various reasons, including to inform system and product design and redesign; to evaluate existing systems, devices, procedures, and training programmes; for performance evaluation; for theoretical development purposes; and for training and procedure design.

For the purposes of this book, the many Human Factors methods available can be categorised as:

1. *Data collection methods.* The starting point in any Human Factors analysis, be it for system design or evaluation or for theoretical development, involves describing existing or analogous systems via the application of data collection methods (Diaper and Stanton, 2004). These methods are used by researchers to gather specific data regarding a task, device, system, or scenario, and the data obtained are used as the input for the Human Factors analyses methods described below.
2. *Task analysis methods.* Task analysis methods (Annett and Stanton, 2000) are used to describe tasks and systems and typically involve describing activity in terms of the goals and physical and cognitive task steps required. Task analysis methods focus on "what an operator … is required to do, in terms of actions and/or cognitive processes to achieve a system goal" (Kirwan and Ainsworth, 1992, p. 1). Also included in this category for the purposes of this book are process charting methods, which are a representational form of task analysis method that use standard symbols to depict a task or sequence of events.
3. *Cognitive task analysis methods.* Cognitive Task Analysis (CTA) methods (Schraagen, Chipman, and Shalin, 2000) focus on the cognitive aspects of task performance and are used for "identifying the cognitive skills, or mental demands, needed to perform a task proficiently" (Militello and Hutton, 2000, p. 90) and describing the knowledge, thought processes, and goal structures underlying task performance (Schraagen, Chipman, and Shalin, 2000). CTA method outputs are used for a variety of different purposes, including, amongst other things, to inform the design of new technology, systems, procedures, and processes; for the development of training procedures and interventions; for allocation of functions analysis; and for the evaluation of individual and team performance within complex sociotechnical systems.
4. *Human error identification/analysis methods.* In the safety critical domains, a high proportion (often over 70%) of accidents and incidents are attributed to human error. Human error identification methods (Kirwan, 1992a, 1992b, 1998a, 1998b) use taxonomies of human error modes and performance shaping factors to predict any errors that might occur during a particular task. They are based on the premise that, provided one has an understanding of the task being performed and the technology being used, one can identify the errors that are likely to arise during the man–machine interaction. Human error analysis approaches are used to retrospectively classify and describe the errors, and their causal factors, that occurred during a particular accident or incident.
5. *Situation awareness measures.* Situation awareness refers to an individual's, team's, or system's awareness of "what is going on" during task performance. Situation awareness measures (Salmon, Stanton, Walker, and Jenkins, 2009) are used to measure and/or model individual, team, or system situation awareness during task performance.
6. *Mental workload measures.* Mental workload represents the proportion of operator resources that are demanded by a task or series of tasks. Mental workload measures are used to determine the level of operator mental workload incurred during task performance.
7. *Team performance measures.* Teamwork is formally defined by Wilson, Salas, Priest and Andrews (2007, p. 5) as "a multidimensional, dynamic construct that refers to a set of interrelated cognitions, behaviours, and attitudes that occur as team members perform a task that results in a coordinated and synchronised collective action." Team performance measures are used to describe, analyse, and represent various facets of team performance, including the knowledge, skills, and attitudes underpinning team performance, team cognition, workload, situation awareness, communications, decision making, collaboration and coordination.
8. *Interface evaluation methods.* A poorly designed interface can lead to unusable products, user frustration, user errors, inadequate performance, and increased performance times.

Introduction

Interface evaluation approaches (Stanton and Young, 1999) are used to assess the interface of a product or device; they aim to improve interface design by understanding or predicting user interaction with the device in question. Various aspects of an interface can be assessed, including layout, usability, colour coding, user satisfaction, and error potential.

Other forms of Human Factors methods are also available, such as performance time prediction and analysis methods (e.g., critical path analysis and the keystroke level model) and system and product design approaches (e.g., allocation of functions analysis, storyboarding, scenario-based design); however, since they do not fit in with the scope of this book we have not described them here.

APPLICATION IN SPORT

All of the different categories of methods described above can potentially make a significant contribution within the sporting domains (indeed some of them already do). Data collection methods are already prominent, since they form the basis for any empirical scientific study. Task analysis methods are also applicable, since they are used to describe and represent systems and the activity that goes on within them. This description is required for understanding performance in any domain, and acts as the description on which most Human Factors methods are applied. The outputs of cognitive task analysis methods are likely to be of particular interest in sports domains, since they can be used to describe the cognitive processes (e.g., decision making) underlying expert sports performance. Human error identification and analysis methods can be used to predict and/or analyse the different errors that sports performers, coaches, and officiators make, which in turn can be used to inform the development of countermeasures and remedial measures. Situation awareness, the concept that describes how individuals, teams, and systems develop and maintain sufficient awareness of what is going on during task performance, is a pertinent line of inquiry in the sporting domains since it represents what it is exactly that performers need to know in a given sport in order to perform optimally. Also of interest is how the introduction of new methods, coaching interventions, technologies, and strategies affects performer situation awareness, and what the differences between the situation awareness levels achieved by sports performers of differing abilities are. Similarly, the concept of mental workload can be studied in order to determine how different events, technologies, devices, strategies, and procedures affect the levels of mental workload experienced by sports performers. Workload optimisation is critical for enhancing human performance (e.g., Sebok, 2000; Young and Stanton, 2002), and the application of workload assessment methods through the product design process is critical for this purpose. Measures of different teamwork constructs are of obvious interest, particularly in relation to the differences between successful and unsuccessful teams in the sporting domains. Finally, interface evaluation methods are likely to be of significant use to sports product designers. Such approaches can be used to evaluate and redesign existing or design concept sports products, such as training devices (e.g., Global Positioning System [GPS]-based running watches) and performance aids (e.g., digital caddie devices).

The classes of methods described above can therefore all potentially make a significant contribution within the sporting domains, which in turn can lead to theoretical and methodological advances in both the Human Factors and sport science arena. In our view, Human Factors analyses methods outputs could potentially be used for the following reasons in a sporting context:

1. *Theoretical development.* Applications across domains are required for theoretical development. Applying Human Factors methods within the sporting domains will lead to theoretical developments both in the fields of Human Factors and Sports Science. For example, concepts such as situation awareness, human error, workload, and decision making all represent key concepts that remain largely unexplored in a sporting context.
2. *Methodological development.* Methodological development occurs based on continued application, testing, and validation in different domains. Applying Human Factors methods,

or variations of them, in a sporting context will lead to the development and validation of existing and new forms of analysis methodologies for Sports Science and Human Factors researchers.

3. *Product and interface design.* Human Factors methods have a critical role to play in the design of products and interfaces; such methods are designed to improve product design by understanding or predicting user interaction with those devices (Stanton and Young, 1999). Applying Human Factors methods in a sporting context will lead to significant contributions to the sports product design process, be it for identifying user situation awareness requirements, specifying optimal interface layouts, or minimising device interaction times and user errors.

4. *Product and interface evaluation.* As disappointing as the standard post product/system design and development Human Factors evaluation is, it means that Human Factors methods are good at evaluating operational products and producing insightful suggestions for how they can be improved. Existing sports products, devices, and interfaces will thus benefit via evaluation with structured Human Factors methods.

5. *Coaching and training intervention development.* Human Factors methods have also played a significant role in the development of training programmes for the complex safety critical domains. Applying such methods in a sporting context will lead to recommendations for training and coaching interventions designed to improve performance.

6. *Performance evaluation and tactical development.* The Human Factors methods described are highly suited to performance evaluation and the subsequent reorganisation of systems so that performance levels are improved. Using Human Factors methods for evaluating sports performance will lead to suggestions for ways in which performance can be improved, such as the specification of new tactics, equipment, or organisation, and also for ways in which opposition team performance can be combated.

Of course, not all of the Human Factors methods described are applicable to all sports. Table 1.1 presents a selection of major sports and a judgement as to whether the different methods from within each category identified can be applied, with a useful output, in each sporting domain. Table 1.1

TABLE 1.1
Human Factors Methods and Their Potential Application in Major Sports

	Football	Rugby	Golf	Tennis	Running	Motorsport	Basketball	Boxing	Cycling	Hockey
Data collection	✓	✓	✓	✓	✓	✓	✓	✓	✓	✓
Task analysis	✓	✓	✓	✓	✓	✓	✓	✓	✓	✓
Cognitive task analysis	✓	✓	✓	✓	✓	✓	✓	✓	✓	✓
Human error identification and analysis	✓	✓	✓	✓	✓	✓	✓	✓	✓	✓
Situation awareness assessment	✓	✓	✓	✓	✓	✓	✓	✓	✓	✓
Mental workload assessment	✓	✓	✓	✓	✓	✓	✓	✓	✓	✓
Teamwork assessment	✓	✓	✓	✓	✓	✓	✓	✓	✓	✓
Interface evaluation	✗	✓	✓	✓	✓	✓	✓	✗	✓	✗

makes it clear that there is likely to be significant utility associated with the application of most forms of Human Factors methods in a sporting context. Of the different methods categories identified, it is only interface evaluation approaches that may not be applicable in some of the sporting domains in which technological devices are not currently used by players and officials, such as soccer and basketball; however, with the introduction of referee assistance technologies proving to be successful in some sports, such as Hawkeye in tennis and the use of video replays in rugby, it is likely that a requirement for interface evaluation within other sports involving referees will emerge (i.e., to evaluate new technological systems such as goal line technology and referee replays in soccer).

STRUCTURE OF THE BOOK

This book has been constructed so that students, practitioners, and researchers with interest in one particular area of Human Factors can read the chapters non-linearly and independently from one another. Each category of Human Factors methods described above is treated as a separate chapter containing an introduction to the area of interest (e.g., for human error identification methods, an introduction to the concept of human error is provided; for situation awareness measurement methods an introduction to the concept of situation awareness is provided; and so on), and an overview of the range of applicable methods within that category area.

Each method presented is described using an adapted form of standard Human Factors methods description criteria that we have found useful in the past (e.g., Stanton, Hedge, Brookhuis, Salas, and Hendrick, 2004; Stanton et al., 2005). An overview of the methods description criteria is presented below.

1. *Name and acronym*—the name of the method and its associated acronym
2. *Background and applications*—provides a short introduction to the method, including a brief overview of the method and its origins and development
3. *Domain of application*—describes the domain that the method was originally developed for and applied in and any domains in which the method has since been applied
4. *Application in sport*—denotes whether the method has been applied to the assessment of sports performance and gives an overview of recommended application areas in a sporting context
5. *Procedure and advice*—describes the step-by-step procedure for applying the method as well as general points of advice
6. *Flowchart*—presents a flowchart depicting the procedure that analysts should follow when applying the method
7. *Advantages*—lists the main advantages associated with using the method
8. *Disadvantages*—lists the main disadvantages associated with using the method
9. *Example output*—presents an example, or examples, of the outputs derived from applications of the method in question
10. *Related methods*—lists any closely related methods, including contributory methods that should be applied in conjunction with the method, other methods to which the method acts as an input, and similar methods
11. *Approximate training and application times*—estimates of the training and application times are provided to give the reader an idea of the commitment required when using the method
12. *Reliability and validity*—any evidence, published in the academic literature, on the reliability or validity of the method is cited
13. *Tools needed*—describes any additional tools (e.g., software packages, video and audio recording devices, flipcharts) required when using the method
14. *Recommended texts*—a bibliography lists recommended further reading on the method and the surrounding topic area

The criteria work on three levels. First, they provide a detailed overview of the method in question. Second, they provide researchers with some of the information that they may require when selecting an appropriate method to use for a particular analysis effort (e.g., associated methods, example outputs, flowchart, training and application times, tools needed, and recommended further reading). And third, they provide detailed guidance, in a step-by-step format, on how to apply the chosen method.

In the final chapter an example case study, utilising a method from each of the different methods categories described, is presented. Although example applications are presented for each of the methods described throughout the book, the purpose of the case study presented in the final chapter is to demonstrate how Human Factors methods can be applied together in an integrated manner. This has a number of compelling advantages, because not only does the integration of existing methods bring reassurance in terms of a validation history, but it also enables the same data to be analysed from multiple perspectives. These multiple perspectives, as well as being inherent in the object that is being described and measured (i.e., sporting scenarios), also provides a form of internal validity. Assuming that the separate methods integrate on a theoretical level, then their application to the same data set offers a form of "analysis triangulation" (Walker, Gibson, Stanton, Baber, Salmon, and Green, 2006).

The potential for cross-disciplinary interaction between Human Factors researchers and sports scientists has already been recognised (e.g., Fiore and Salas, 2006, 2008), and testing theory and methods across domains can only advance knowledge in both fields. Toward this end, it is our hope that this book provides sports scientists with the methods necessary to investigate core Human Factors concepts in a sporting context.

2 Data Collection Methods

INTRODUCTION

In any form of research, when we want to understand, analyse, and/or describe something in detail, we use structured and valid approaches to collect data. Human Factors is no different; once a Human Factors analysis has been scoped in terms of aims and expected outcomes (i.e., defining hypotheses and identifying research questions that the analysis is intended to answer) researchers use structured data collection methods to collect valid data regarding the system, activity, personnel, or device that the analysis is focused upon. This data is then used to inform the impending Human Factors analyses. The importance of reliable and valid data collection methods is therefore manifest; they are the cornerstone of any Human Factors analysis effort and, in addition to providing significant insights into the area of study themselves, provide the input data for the range of different methods that Human Factors researchers use.

Data collection methods, from a Human Factors point of view, are typically used to collect data regarding the nature of human activity within sociotechnical systems, and therefore focus on the interaction between humans, as well as between humans and artefacts. This includes the activities being performed (e.g., what the activity actually is, what goals and task steps are involved, how it is performed, what is used to perform the activity, how well the activity is performed), the individuals or teams performing the activity (e.g., their organisation and interactions with artefacts and one another; their physical and mental condition; their decision-making strategies, situation awareness, workload, and the errors that they make), the devices used (e.g., usability and errors made), and the system in which the activity is performed (e.g., environmental data, such as temperature, noise, and lighting conditions).

The importance of having an accurate representation of the system, personnel, or activity under analysis should not be underestimated. Such representations are a necessary prerequisite for further Human Factors analysis efforts. There are various data collection methods available—of course, it could be argued that every Human Factors method has some form of data collection component. For the purposes of this book, we focus on the three most commonly used data collection methods within Human Factors: interviews, questionnaires, and observational study. Typically, these three approaches are used to collect the data that is used as the basis for additional Human Factors analyses such as task analysis, cognitive task analysis, human error identification, usability and interface evaluation, and teamwork assessment.

Interviews are commonly used when researchers have good levels of access to Subject Matter Experts (SMEs) for the area of study, and involve the use of focused questions or probes to elicit information regarding a particular topic or area of interest. Interviews can be structured (e.g., involving a rigid, predefined series of questions), semi-structured (e.g., using a portion of predefined questions, but with the added flexibility of pursuing other areas of interest), or unstructured (e.g., no predefined structure or questions). Various interview-based Human Factors methods exist, each providing their own predefined questions or interview "probes." For example, the Critical Decision Method (CDM; Klein, Calderwood, and McGregor, 1989), Applied Cognitive Task Analysis (ACTA; Militello and Hutton, 2000), and the Critical Incident Technique (CIT; Flanagan, 1954) all use interviews to evaluate the cognitive elements of task performance.

Questionnaires are another popular data collection approach and involve participants responding, either verbally or on paper, to a series of predefined, targeted questions regarding a particular subject area. A range of established Human Factors questionnaires already exists. For example, the Questionnaire for User Interface Satisfaction (QUIS: Chin, Diehl, and Norman, 1988), the Software Usability Measurement Inventory (SUMI; Kirakowski, 1996), and the System Usability Scale (SUS; Brooke, 1996) are all examples of questionnaires designed specifically to assess the usability of a particular device or interface. Further, questionnaire-based approaches are also used in the assessment of situation awareness (e.g., Taylor, 1990) and mental workload (e.g., Hart and Staveland, 1988).

Observational study is another popular data collection approach that is used to gather data regarding the observable aspects (e.g., physical and verbal) of task performance. Observational study typically forms the starting point of any investigation, and the data derived often form the input for various other Human Factors methods, such as task analysis, human error identification and analysis, process charting, and teamwork assessment methods. Various forms of observational study exist, including direct observation, whereby analysts directly observe the task or scenario in question being performed in its natural habitat; participant observation, whereby analysts observe activities while engaging in them themselves; and remote observation, whereby analysts observe activities from a remote site (i.e., via video camera link-up). Due to the difficulties in gaining access to elite sports performers, observational study offers perhaps the most suitable means of collecting data for the analysis of elite sports performance.

The selection of data collection method to be used is dependent upon the aims of the analysis and the opportunities for access to SMEs from within the domain in question. For example, often interviews with SMEs may be not be possible, and so observational study and/or questionnaires will have to suffice. Similarly, there may be no opportunity for questionnaires to be distributed and completed, so observational study may be the only means of data collection available. Annett (2003) argues that data collection should comprise observation and interviews at the very least, and both Annett (2003) and Kieras (2003) argue for the least intrusive method of observation that circumstances permit. A summary of the data collection methods described in this chapter is presented in Table 12.1.

INTERVIEWS

BACKGROUND AND APPLICATIONS

Interviews provide a flexible approach for gathering specific data regarding a particular topic. Interviews involve the use of questions or probes to elicit information regarding a particular topic or area of interest. Within Human Factors they have been used extensively to gather data regarding all manner of subjects, including decision making (e.g., Klein, Calderwood, and McGregor, 1989), usability (Baber, 1996), situation awareness (Matthews, Strater, and Endsley, 2004), teamwork (Klinger and Hahn, 2004), and command and control (Riley, Endsley, Bolstad, and Cuevas, 2006). There are three main forms of interview available: structured interviews, semi-structured interviews, and unstructured interviews, although group interview methods, such as focus groups, are also often used by Human Factors researchers. Here is a brief description of each interview method:

1. *Structured interviews.* Structured interviews involve the use of predefined questions or "probes" designed to elicit specific information regarding the subject under analysis. In a structured interview, the interview content in terms of questions and their order is predetermined and is adhered to quite rigidly; no scope for discussion outside of the predefined areas of interest is usually permitted. The interviewer uses the predefined probes to elicit the data required, and does not take the interview in any other direction. Due to their rigid nature, structured interviews are probably the least popular form of interview used during Human Factors data collection efforts; however, they are particularly useful if only limited time is available for data collection purposes, since they serve to focus data collection considerably.

Data Collection Methods

TABLE 2.1
Data Collection Methods Summary Table

Name	Domain	Application in Sport	Training Time	Application Time	Tools Needed	Main Advantages	Main Disadvantages	Outputs
Interviews	Generic	Data collection	High	High	Audio and video recording equipment Pen and paper Word processing software	1. Flexible approach that can be used to assess range of issues 2. Interviewer has high degree of control over data collection process 3. Offers a good return in terms of data collected in relation to time invested	1. Highly time consuming to design, apply, and analyse 2. Subject to a range of biases 3. Quality of data collected is highly dependent upon skill of interviewer and quality of interviewee	Interview transcript
Questionnaires	Generic	Data collection	Low	Medium	Audio and video recording equipment Pen and paper Word processing software	1. Flexible approach that can be used to assess range of issues 2. Requires little training 3. Offers high degree of control over the data collection process since targeted questions can be designed a priori	1. Can be highly time consuming to design and analyse 2. Questionnaire design is more of an art than a science (Wilson and Corlett, 1995) and may require a high level of training and experience 3. Responses can be rushed, non-committal, and subject to various biases, such as prestige bias	Participant responses to targeted questions
Observational study	Generic	Data collection	Low	High	Audio and video recording equipment Pen and paper Word processing software Observer Pro	1. Provides real-life insight into task performance 2. Can be used to investigate a range of issues 3. Observational study data are used as the input to various other methods, including task analysis, social network analysis, and error analysis methods	1. Observational data are prone to various biases 2. The data analysis component can be hugely time consuming for large, complex tasks involving numerous actors 3. Offers very little experimental control to the analyst	Observational transcript of activities observed

2. *Semi-structured interviews.* In a semi-structured interview, a portion of the questions and their order is predetermined; however, a degree of flexibility is added which allows the interviewer to use additional questions, explore topics of interest further as they arise, and even shift the focus of the interview as they see fit during the interview itself. As a corollary of this, semi-structured interviews often uncover data regarding new or unexpected issues. The use of predefined questions allows semi-structured interviews to retain the focused nature of structured interviews, but the added flexibility allows them to be a much more usable and useful data collection tool. This focused yet flexible nature has led to semi-structured interviews becoming the most popular form of interview for data collection purposes.
3. *Unstructured interviews.* When using unstructured interviews, there is no predefined structure or questions. The interviewer is given full freedom to explore, on an ad-hoc basis, different topics of interest as he or she sees fit. The high level of flexibility is attractive, however, in order to ensure useful data are gathered the use of skilled interviewers is required, and the data transcription process is often lengthy due to the large amount of data gathered. Of the three different forms of interview, unstructured interviews are the least attractive, since their unstructured nature often leads to key information not being collected.
4. *Focus groups.* Group interviews are also heavily used by Human Factors researchers. One popular form of group interview is focus groups, which involve the use of group discussions to canvass consensus opinions from groups of people. A focus group typically involves a researcher introducing topics and facilitating discussion surrounding the topic of interest rather than asking specific questions.

The questions used during interviews are also of interest. Typically, three different types of question are used: closed questions, open-ended questions, and probing questions. A brief description of each is presented below.

1. *Closed questions.* Closed questions are designed to gather specific information and typically invoke Yes/No answers only. An example of a closed question would be, "Did you agree with your coach's tactics during the second half of the game?" The question is designed to elicit a Yes/No response and the interviewee does not elaborate on his or her chosen answer unless the interviewer probes further.
2. *Open-ended questions.* An open-ended question is used to gather more detail than the simple Yes/No response of a closed question. Open-ended questions are designed so that the interviewee gives more detail in their response than merely yes or no; they allow interviewees to answer in whatever way they wish and to elaborate on their answer. For example, an open-ended question approach to the coach's tactics topic used for the closed question example would be something like, "What did you think about the coach's tactics during the second half of the game?" By allowing the interviewee to elaborate upon the answers given, open-ended questions elicit more insightful data than closed questions. Whereas a closed question might reveal that a player did not agree with his coach's tactics, an open-ended question would reveal that the player did not agree with the tactics and also provide an insight into why the player did not agree with them. On the downside, open-ended questions are likely to lead to more time-consuming interviews, produce significantly more data than closed questions, and the data gathered take significantly longer to transcribe and analyse.
3. *Probing questions.* A so-called probing question is normally introduced directly after an open-ended or closed question to gather targeted data regarding the topic of discussion. Examples of probing questions following on from a question regarding a player's agreement with his or her coach's tactics would be, "Why did you not agree with the coach's

Data Collection Methods

tactics in the second half?", "What tactics do you think would have been more appropriate?", and "What effect did the tactics used have on your own performance and the overall performance of the whole team?"

It is recommended that interviewers should begin with a specific topic, probe the topic until it has been exhausted, and then move on to another topic. Stanton and Young (1999) advocate the use of an open-ended question, followed by a probing question and then a closed question. According to Stanton and Young, this cycle of open-ended, probe, and then closed questions should be maintained throughout the interview.

DOMAIN OF APPLICATION

Interviews are a generic procedure that can be used in any domain to collect data regarding any subject.

APPLICATION IN SPORT

Interviews can be used to collect data on a wide range of sporting issues, including the cognitive aspects of performance (e.g., decision making, problem solving, situation awareness) and sports performer's subjective opinions on performance, equipment, and training. Accordingly, they have been heavily used in the past for data collection purposes in various sporting domains. For example, Macquet and Fleurance (2007) used interviews to assess expert badminton player decision-making strategies, Smith and Cushion (2006) used semi-structured interviews to investigate the in-game behaviour of top-level professional soccer coaches, and Hanton, Fletcher, and Coughlan (2005) used interviews to investigate the nature of stress in elite sports performers.

PROCEDURE AND ADVICE (SEMI-STRUCTURED INTERVIEW)

Step 1: Clearly Define the Aim(s) of the Interview

Initially, before the design of the interview begins, the analyst should clearly define what the aims and objectives of the interview are. Without clearly defined aims and objectives, interviews can lack focus and inappropriate or inadequate data may be obtained. Clear specification of the aims and objectives ensures that interviews are designed appropriately and that the interview questions used are wholly relevant.

Step 2: Develop Interview Questions

Once the aims and objectives of the interview are clearly defined, development of appropriate interview questions can begin. As pointed out previously, the cyclical use of open-ended, probing, and closed questions is advocated as the best approach. Therefore, cycles of questions for each topic of interest should be developed. For example, for an interview focusing on golfer shot selection, one cycle of questions regarding the use of information for shot selection purposes could be:

- *Open-ended question*: "What information did you use to inform shot selection?"
- *Probing questions*: "Where did the information come from?", "Which piece of information was the most important in determining your final shot selection?", "Was any of the information incorrect or erroneous?", "Is there any other information that you would have liked when making your shot selection?"
- *Closed question*: "Were you happy with your shot selection?"

Once the initial set of questions is developed, they should be reviewed and refined as much as is possible. It is particularly useful at this stage to pass the questions to other researchers for their comments.

Step 3: Piloting the Interview

Once the questions have been developed, reviewed, refined, and placed in an appropriate order, the next step involves conducting pilot runs of the interview. This is a particularly important aspect of interview design; however, it is often ignored. Piloting interviews allows problems with the question content and ordering to be ironed out. Also, questions not initially thought of are often revealed during pilot testing. Piloting interviews can be as simple as performing a trial interview with a colleague, or more sophisticated, involving the conduct of a series of pilot runs with a test participant set. The data collected during interview piloting is also useful, as it gives an indication of the kind of data that will be gathered, which allows analysts to modify or change the interview questions if the appropriate data was not forthcoming during piloting.

Step 4: Modify Interview Procedure and/or Content

Following the piloting phase, any changes to the interview procedure or questions should be made. This might include the removal of redundant or inappropriate questions, rewording of existing questions, or addition of new questions. If time permits, the modified interview should also be subject to a pilot run.

Step 5: Select Participants

If the participant set has not already been determined, the next step requires that appropriate participants be selected for the study in question. This may or may not be constrained by the nature of the study and the participant sample available (i.e., an investigation into the injury profiles of male amateur soccer players under the age of 30). Normally, a representative sample from the target population is used.

Step 6: Conduct Interviews

The interviewer plays a key role in the overall quality of data collected. The interviewer(s) used should be familiar with the aims of the analysis and the subject area, should be confident, and should communicate clearly and concisely. Establishing a good rapport with the interviewee is also critical. Things for the interviewer to avoid include being overbearing during the interview, using technical jargon or acronyms, and misleading, belittling, embarrassing, or insulting the interviewee. Conducting the interview itself involves using a cycle of open-ended, probe, and closed questions to exhaust a topic of interest (Stanton and Young, 1999). The interviewer should persist with one particular topic until all avenues of interest are explored, and then move on to the next topic of interest. All interviews should be recorded using either audio or visual recording equipment. It is also recommended that a second analyst take notes of interest during the interview.

Step 7: Transcribe Data

Once all interviews are completed, the data should be transcribed. This should take place immediately after the interviews are completed, as it can be difficult to return to interview data after a significant period. Transcribing interview data involves using interview notes and recordings to create a full transcript of each interview; everything said by both the interviewer and interviewee should be transcribed, and notes of interest can be added. For interviews, the data transcription process is typically a time-consuming and laborious process. Due to the time saved, it is often useful to pay somebody (e.g., undergraduate/postgraduate students, temporary agency workers) to transcribe the data.

Step 8: Data Gathering

Data gathering involves reviewing the interview transcripts in order to pick out data of interest (i.e., the sort of data that the study is looking for); this is known as the "expected data." Once all of the expected data are gathered, the interview transcripts should be analysed again, but this time to identify any "unexpected data," that is, any data that are of interest to the study but that were not expected to be gained during the interviews.

Step 9: Data Analysis

The final step involves analysing the data using appropriate statistical tests. The form of analysis used is dependent upon the aims of the study but typically involves converting the interview transcripts into numerical form in readiness for statistical analysis. A good interview will always involve planning so that the data are collected with a clear understanding of how subsequent analyses will be performed. A good starting point for data analysis is to perform some form of content analysis on the interview transcripts, i.e., divide the transcription into specific concepts. One can then determine whether the data collected can be reduced to some numerical form, e.g., counting the frequency with which certain concepts are mentioned by different interviewees, or the frequency with which concepts occur together. Alternatively, if the interview transcripts are not amenable to reduction in numerical form and it is not possible to consider statistical analysis, it is common practice to look for common themes and issues within the data.

ADVANTAGES

1. Interviews are a flexible approach that can be used to gather data on any topic of interest.
2. Interviews have been used extensively within the sporting domain for all manner of purposes.
3. Interviews offer a good return in terms of data collected in relation to time invested.
4. Interviews are particularly good for revealing participant opinions and subjective judgements (Wilson and Corlett, 2004).
5. Interviews offer a high degree of control over the data collection process. Targeted interview questions can be designed a priori and interviewers can direct interviews as they see fit.
6. Interview data can be treated in a variety of ways, including statistically.
7. Structured interviews offer consistency and thoroughness (Stanton and Young, 1999).
8. A range of Human Factors interview methods and probes focusing on a variety of concepts already exist. For example, the Critical Decision Method (Klein, Calderwood, and McGregor, 1989) is an interview methodology that focuses on the cognitive processes underlying decision making.

DISADVANTAGES

1. Designing, conducting, transcribing, and analysing interviews is a highly time-consuming process, and the large amount of time required often limits the number of participants that can be used.
2. A high level of training may be required for analysts with no experience in conducting interviews.
3. The reliability and validity of interviews is difficult to address.
4. The quality of the data collected is heavily dependent upon the skill of the interviewer and the quality of the interviewees used.
5. Interviews are susceptible to a range of interviewer and interviewee biases.
6. Participants often do not have sufficient free time to fully engage in interviews. Often interviewers are given only a short period of time to collect their data.
7. Depending on the subject area, interviewees may be guarded with their responses for fear of reprisals.

Approximate Training and Application Times

In conclusion to a study comparing 12 Human Factors methods, Stanton and Young (1999) report that interviews had the greatest training time of all of the methods tested. The application time for interviews is also typically high. Typically, interviews take anywhere between 10 and 60 minutes each; however, the data transcription, gathering, and analysis components require considerable time to complete.

Reliability and Validity

The reliability and validity of interviews is difficult to address; however, when comparing 12 Human Factors methods, Stanton and Young (1999) found the reliability and validity of interviews to be poor.

Tools Needed

At its most basic, an interview can be conducted with pen and paper only; however, it is recommended that audio and/or visual recording devices are also used to record the interview. For transcription and data gathering purposes, a word processing package such as Microsoft Word is required. For the data analysis component, a statistical analysis package such as SPSS is normally used.

Example

Interview transcripts taken from a CDM interview focusing on marathon runner decision making during the course of a marathon event are presented. The purpose of conducting the CDM interviews was to examine the nature of athlete decision making during marathon running. A series of predefined interview probes (adapted from O'Hare, Wiggins, Williams, and Wong, 2000) were used. Initially, the interviewer and interviewee decomposed the marathon event into the following five key decision points: select race strategy prior to race beginning, reconsider/revise race strategy at 6 miles, reconsider/revise race strategy at 13 miles, reconsider/revise race strategy at 18 miles, and decide on finishing strategy at 22 miles. CDM interview transcripts for each decision point are presented in Tables 2.2 to 2.6.

Flowchart

(See Flowchart 2.1.)

Recommended Texts

Lehto, J. R., and Buck, M. (2007). *Introduction to Human Factors and ergonomics for engineers*. Boca Raton, FL: CRC Press.

McClelland, I., and Fulton Suri, J. F. (2005). Involving people in design. In *Evaluation of human work: A practical ergonomics methodology*, eds. J. R. Wilson and E. Corlett, 281–334. Boca Raton, FL: Taylor & Francis.

QUESTIONNAIRES

Background and Applications

Questionnaires are another popular data collection approach. Questionnaires have been used within Human Factors in many forms to collect data regarding a plethora of issues, including usability, user opinions, error, decision making, and situation awareness. The use of questionnaires involves participants responding, either verbally or on paper, to a series of predefined, targeted questions regarding a

TABLE 2.2
Critical Decision Method Interview Transcript for "Select Race Strategy" Decision Point

Goal specification	*What were your specific goals during this part of the race?*
	• My goal was to select an appropriate race strategy that would allow me to run a PB on the day (i.e., in the conditions that were present).
Decisions	*What decisions did you make during this part of the race?*
	• I needed to make a decision regarding the strategy to use during the marathon. Should I go out quick and attempt to maintain a fast pace, or should I take it steady and attempt a quicker finish?
Cue identification	*What features were you looking for when you formulated your decisions?*
	• My training performance, my performance in half marathons and other training events, my previous times and pace (from my training system), and the weather conditions on the day.
	How did you know that you needed to make the decisions? How did you know when to make the decisions?
	• I knew that I needed to have a race strategy (pace in minutes per mile, drinks, and gels) in order to achieve the best time possible
Expectancy	*Were you expecting to make these sorts of decisions during the course of the event? Describe how this affected your decision-making process.*
	• Yes.
Conceptual	*Are there any situations in which your decisions would have turned out differently?*
	• Yes, if my body was feeling different; also the weather has a significant impact, for example, if it was very hot weather I know that my pace would have to be slower.
Influence of uncertainty	*At any stage, were you uncertain about either the reliability or the relevance of the information that you had available?*
	• Yes, I was uncertain about the weather and also about my knowledge of the course and terrain, such as where there were big ascents and descents.
Information integration	*What was the most important piece of information that you used to make your decisions?*
	• Training performance (from my training system).
Situation awareness	*What information did you have available to you at the time of the decisions?*
	• My physical condition, my training performance, the weather, course information, advice from other athletes (who had run this course before).
Situation assessment	*Did you use all of the information available to you when making decisions?*
	• Yes.
	Was there any additional information that you might have used to assist you in making decisions?
	• Yes, race times from previous years, more accurate information regarding my own physical condition (on the race day it felt like more of a gut feeling than anything else), more detailed information regarding the weather, more detailed information regarding the course (i.e., where it is more difficult, where the fast parts are, etc.).
Options	*Were there any other alternatives available to you other than the decisions you made?*
	• Yes, I could have chosen a different race strategy (quicker or slower).
Decision blocking—stress	*Was their any stage during the decision-making process in which you found it difficult to process and integrate the information available?*
	• No.
Basis of choice	*Do you think that you could develop a rule, based on your experience, that could assist another person to make the same decisions successfully?*
	• No.
Analogy/generalisation	*Were you at any time reminded of previous experiences in which similar/different decisions were made?*
	• No.

TABLE 2.3
Critical Decision Method Interview Transcript for "6 Miles—Check Condition and Revise Race Strategy as Appropriate" Decision Point

Goal specification	*What were your specific goals during this part of the race?*
	• To check my race strategy and check that I was on course for under 3 hours, 30 minutes and 1 hour, 40 minutes for half marathon.
Decisions	*What decisions did you make during this part of the race?*
	• I needed to decide whether to speed up/slow down or stay at the same pace.
Cue identification	*What features were you looking for when you formulated your decisions?*
	• Pace (minutes per mile from my training system), previous experiences at this point in half marathons, training performance, how I felt physically.
	How did you know that you needed to make the decisions? How did you know when to make the decisions?
	• I knew I had to keep on top of my pace and race strategy in order to maximise my performance.
Expectancy	*Were you expecting to make these sorts of decisions during the course of the event? Describe how this affected your decision-making process.*
	• Yes, I knew I had to check pace and how I was feeling physically throughout the race.
Conceptual	*Are there any situations in which your decisions would have turned out differently?*
	• Yes, if I had found the first 6 miles very difficult or very easy.
Influence of uncertainty	*At any stage, were you uncertain about either the reliability or the relevance of the information that you had available?*
	• No.
Information integration	*What was the most important piece of information that you used to make your decisions?*
	• Minutes per mile pace, distance (ran and remaining).
Situation awareness	*What information did you have available to you at the time of the decisions?*
	• Minutes per mile pace, distance (ran and remaining), physical condition.
Situation assessment	*Did you use all of the information available to you when making decisions?*
	• Yes.
	Was there any additional information that you might have used to assist you in making decisions?
	• No.
Options	*Were there any other alternatives available to you other than the decisions you made?*
	• Yes, a different race strategy.
Decision blocking—stress	*Was their any stage during the decision-making process in which you found it difficult to process and integrate the information available?*
	• No.
Basis of choice	*Do you think that you could develop a rule, based on your experience, that could assist another person to make the same decisions successfully?*
	• No.
Analogy/generalisation	*Were you at any time reminded of previous experiences in which similar/different decisions were made?*
	• Yes, previous half marathons and training runs.

particular subject area. The questions used can be either closed or open-ended questions (see Interviews section). A range of established Human Factors questionnaires already exist; however, depending on the aims and objectives of the study, specific questionnaires can also be developed from scratch.

DOMAIN OF APPLICATION

Questionnaires are a generic procedure that can be used in any domain to collect data regarding any subject.

TABLE 2.4
Critical Decision Method Interview Transcript for "13 Miles—Check Condition and Revise Race Strategy as Appropriate" Decision Point

Goal specification	*What were your specific goals during this part of the race?*
	• To keep my current pace going so that I was on target for sub 3 hours, 30 minutes finishing time.
Decisions	*What decisions did you make during this part of the race?*
	• I needed to decide whether to speed up/slow down or stay at the same pace in order to maximise my time.
Cue identification	*What features were you looking for when you formulated your decisions?*
	• Pace in minutes per mile, physical condition, knowledge of previous half marathons.
	How did you know that you needed to make the decisions? How did you know when to make the decisions?
	• I knew what time I wanted to go through 13 miles.
Expectancy	*Were you expecting to make these sorts of decisions during the course of the event? Describe how this affected your decision-making process.*
	• Yes.
Conceptual	*Are there any situations in which your decisions would have turned out differently?*
	• If I had gone out too quickly or too slowly or if I was finding it too difficult or too easy.
Influence of uncertainty	*At any stage, were you uncertain about either the reliability or the relevance of the information that you had available?*
	• No.
Information integration	*What was the most important piece of information that you used to make your decisions?*
	• Distance (ran and remaining) and minutes per mile pace.
Situation awareness	*What information did you have available to you at the time of the decisions?*
	• Time, minutes per mile pace, distance (ran and remaining), how I felt physically.
Situation assessment	*Did you use all of the information available to you when making decisions?*
	• Yes.
	Was there any additional information that you might have used to assist you in making decisions?
	• No.
Options	*Were there any other alternatives available to you other than the decisions you made?*
	• Yes, I could have opted to finish at the half marathon point or to speed up or slow down pace-wise.
Decision blocking—stress	*Was their any stage during the decision-making process in which you found it difficult to process and integrate the information available?*
	• No.
Basis of choice	*Do you think that you could develop a rule, based on your experience, that could assist another person to make the same decisions successfully?*
	• No.
Analogy/generalisation	*Were you at any time reminded of previous experiences in which similar/different decisions were made?*
	• Yes, previous half marathon events and training runs.

APPLICATION IN SPORT

Due perhaps to their ease of application and flexibility, questionnaires have long been used for data collection in the sporting domains. For example, Tsigilis and Hatzimanouil (2005) used a self-report questionnaire to examine the influence of risk factors on injury in male handball players, and Lemyre, Roberts, and Stray-Gundersen (2007) used a questionnaire to examine the relationship between motivation, overtraining, and burnout in elite athletes. There is scope to use questionnaires in a sporting context to collect data regarding a range of issues of interest, including device usability,

TABLE 2.5
Critical Decision Method Interview Transcript for "18 Miles—Decision Regarding Strategy for Remainder of the Race

Goal specification	*What were your specific goals during this part of the race?*
	• To keep my pace up for a sub 3-hour, 30-minute time.
Decisions	*What decisions did you make during this part of the race?*
	• Had to decide on strategy required in order to achieve a sub 3-hour, 30-minute time.
Cue identification	*What features were you looking for when you formulated your decisions?*
	• Distance remaining, physical condition, current mile per minute pace.
	How did you know that you needed to make the decisions? How did you know when to make the decisions?
Expectancy	*Were you expecting to make these sorts of decisions during the course of the event? Describe how this affected your decision-making process.*
	• Yes.
Conceptual	*Are there any situations in which your decisions would have turned out differently?*
	• If I was going too quickly or too slowly pace-wise, or if I was finding it too difficult.
Influence of uncertainty	*At any stage, were you uncertain about either the reliability or the relevance of the information that you had available?*
	• No.
Information integration	*What was the most important piece of information that you used to make your decisions?*
	• Pace, distance remaining, time.
Situation awareness	*What information did you have available to you at the time of the decisions?*
	• Pace, distance ran and remaining, and time.
Situation assessment	*Did you use all of the information available to you when making decisions?*
	• Yes.
	Was there any additional information that you might have used to assist you in making decisions?
	• No.
Options	*Were there any other alternatives available to you other than the decisions you made?*
	• Yes, to speed up, slow down, or stop!
Decision blocking—stress	*Was their any stage during the decision-making process in which you found it difficult to process and integrate the information available?*
	• No.
Basis of choice	*Do you think that you could develop a rule, based on your experience, that could assist another person to make the same decisions successfully?*
	• No.
Analogy/generalisation	*Were you at any time reminded of previous experiences in which similar/different decisions were made?*
	• Yes, how I felt at this point in training runs.

situation awareness, decision making, workload, and human error. Due to difficulties associated with gaining the level of access required for interviews, questionnaires potentially offer a useful approach for gathering data from elite sports performers.

PROCEDURE AND ADVICE

Step 1: Clearly Define Study Aims and Objectives

Although somewhat obvious, the clear definition of study aims and objectives is often ignored or not adequately achieved. Wilson and Corlett (1995), for example, suggest that this component of questionnaire construction is often neglected and the data collected typically reflect this. It is vital that,

TABLE 2.6
Critical Decision Method Interview Transcript for "22 Miles—Decision Regarding Strategy for Remainder of the Race

Goal specification	*What were your specific goals during this part of the race?*
	• To keep going to the finish, my goal of achieving sub 3 hours, 30 minutes was now forgot about.
Decisions	*What decisions did you make during this part of the race?*
	• I needed to decide on the best strategy that would enable me to finish.
Cue identification	*What features were you looking for when you formulated your decisions?*
	• Physical condition, pain levels (in legs), mental condition, distance remaining.
	How did you know that you needed to make the decisions? How did you know when to make the decisions?
	• I knew that I needed to keep on running to finish.
Expectancy	*Were you expecting to make these sorts of decisions during the course of the event? Describe how this affected your decision-making process.*
	• No. As I had never run the full distance before, I was unsure.
Conceptual	*Are there any situations in which your decisions would have turned out differently?*
	• Yes. If I were running quicker or feeling physically and mentally better I would have kept going for a sub 3-hour, 30-minute finish.
Influence of uncertainty	*At any stage, were you uncertain about either the reliability or the relevance of the information that you had available?*
	• No.
Information integration	*What was the most important piece of information that you used to make your decisions?*
	• Physical and mental condition, decision on whether I could continue or not.
Situation awareness	*What information did you have available to you at the time of the decisions?*
	• Pace, distance remaining, time, pain levels, physical condition, mental condition.
Situation assessment	*Did you use all of the information available to you when making decisions?*
	• Yes.
	Was there any additional information that you might have used to assist you in making decisions?
	• No.
Options	*Were there any other alternatives available to you other than the decisions you made?*
	• Yes, I could have decided to stop or to walk.
Decision blocking—stress	*Was their any stage during the decision-making process in which you found it difficult to process and integrate the information available?*
	• No.
Basis of choice	*Do you think that you could develop a rule, based on your experience, that could assist another person to make the same decisions successfully?*
	• No.
Analogy/generalisation	*Were you at any time reminded of previous experiences in which similar/different decisions were made?*
	• No, as I had never run this distance before.

before any questionnaire or study design begins, the aims and objectives of the overall study are clearly defined. Researchers should go further than merely stating the goal of research. Questions to pose during this step include what is the research question(s), why are we undertaking this research, what data is required, what types of question will be used to collect the data, what data are expected, what analysis procedures will we use, and what do we want to achieve through this study? Without clearly defined aims and objectives, questionnaires can lack focus and inappropriate or inadequate data may be gathered. Clear specification of the aims and objectives ensures that questionnaires are designed appropriately and that the questions used are wholly relevant.

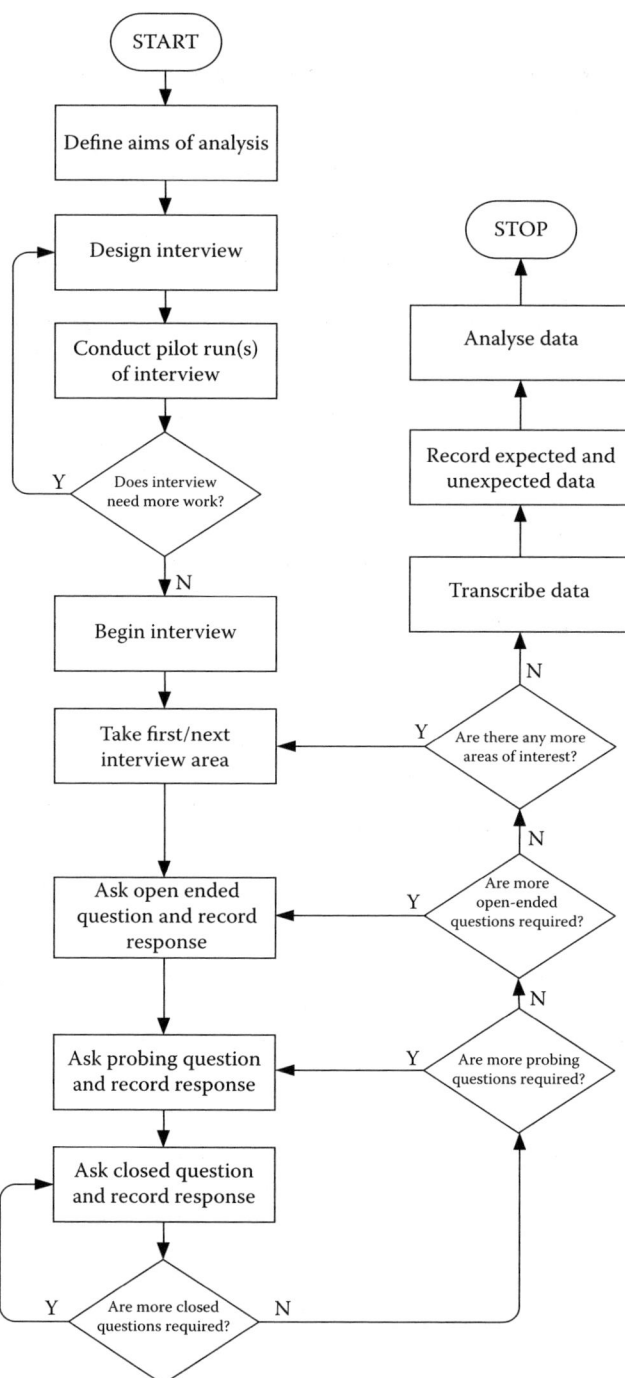

FLOWCHART 2.1 Interviews flowchart.

Step 2: Define Target Population

Once the objectives of the study are clearly defined, the analyst should define the sample population to which the questionnaire will be administered. Again, the definition of the participant population should go beyond simply describing a group of personnel, such as "soccer players" and should be as exhaustive as possible, including defining age groups, different positions, and different organisations.

The sample size should also be determined at this stage. Sample size is dependent upon the aims of the study and the amount of time and resources that are available for data analysis.

Step 3: Questionnaire Construction

Questionnaire construction is complex and involves deciding on the appropriate type of questions to use, how the questionnaire is to be administered, and the format and layout of the questionnaire. A questionnaire typically comprises four parts: an introduction, a participant information section, an information section, and an epilogue (Wilson and Corlett, 1995). The introduction should contain information that informs the participant who you are, what the purpose of the questionnaire is, and what the results are going to be used for. One must be careful to avoid putting information in the introduction that may bias the participant in any way. The participant information part of the questionnaire normally contains multiple-choice questions requesting demographic information about the participant, such as age, sex, occupation, and experience. The information part of the questionnaire is the most important part, as it contains the questions designed to gather the data related to the initial objectives of the study. There are various types of question that can be used in the information part of the questionnaire, and the type used is dependent upon the analysis aims and the type of data required. Where possible, the type of question used should be consistent (i.e., if the first few questions are multiple choice, then all of the questions should be kept as multiple choice). The different types of questions available are displayed in Table 2.7. Each question used should be short in length and worded clearly and concisely, using relevant language. It is also important to consider the data analysis component when constructing this part of the questionnaire. For instance, if there is little time available for data analysis, then the use of open-ended questions should be avoided, as they are time consuming to collate and analyse. If time is limited, then closed questions should be used, as they offer specific data that is quick to collate and analyse. Questionnaire size is also an important issue; too large and participants will not complete the questionnaire, and yet a very short questionnaire may seem worthless and could suffer the same fate. Optimum questionnaire length is dependent upon the participant population, but it is generally recommended that questionnaires should be no longer than two pages (Wilson and Corlett, 1995).

Step 4: Piloting the Questionnaire

Once the questionnaire construction stage is complete, a pilot run of the questionnaire should be undertaken (Wilson and Corlett, 1995). This is a critical part of the questionnaire design process, yet it is often neglected due to various reasons, such as time and financial constraints. During this step, the questionnaire is evaluated by its potential user population, SMEs, and/or by other Human Factors researchers. This allows any problems with the questionnaire to be removed before the critical administration phase. Typically, numerous problems are encountered during the piloting stage, such as errors within the questionnaire, redundant questions, and questions that the participants simply do not understand or find confusing. Wilson and Corlett (1995) recommend that questionnaire piloting should comprise three stages:

1. *Individual criticism.* The questionnaire should be administered for a critique to several colleagues who are experienced in questionnaire construction, administration, and analysis.
2. *Depth interviewing.* Following individual criticism, the questionnaire should be administered to a small sample of the intended population. Once they have completed the questionnaire, the participants should be subjected to an interview regarding the questionnaire and the answers that they provided. This allows the analyst to ensure that the questions were fully understood and that appropriate data is likely to be obtained.
3. *Large sample administration.* The redesigned questionnaire should then be administered to a large sample of the intended population. This allows the analyst to ensure that appropriate data is being collected and that the time required for data analysis is within that available for the study. Worthless questions are also highlighted during this stage. The

TABLE 2.7
Questionnaire Questions

Type of Question	Example Question	When to Use
Multiple choice	On approximately how many occasions have you made the wrong decision in a one-on-one situation? (0–5, 6–10, 11–15, 16–20, More than 20)	When the participant is required to choose a specific response
Rating scales	I found the tactics unnecessarily complex. (Strongly Agree [5], Agree [4], Not Sure [3], Disagree [2], Strongly Disagree [1])	When subjective data regarding participant opinions are required
Paired associates (bipolar alternatives)	Which of the two tasks A + B subjected you to the most mental workload? (A or B)	When two alternatives are available to choose from
Ranking	Rank, on a scale of 1 (Very Poor) to 10 (Excellent) your performance during the game.	When a numerical rating is required
Open-ended questions	What did you think of the coach's tactics?	When data regarding participants' own opinions about a certain subject are required, i.e., subjects compose their own answers
Closed questions	Which of the following errors have you made in a one-on-one situation? (Action Mistimed, Action Omitted, Wrong Selection)	When the participant is required to choose a specific response
Filter questions	Have you ever committed an error while taking a conversion kick? (Yes or No. If Yes, go to question 10. If No, go to question 15.)	To determine whether participant has specific knowledge or experience. To guide participant past redundant questions

likely response rate can also be predicted based on the number of questionnaires returned during this stage.

Step 5: Questionnaire Administration

Once the questionnaire has been successfully piloted and refined accordingly, it is ready to be administered. Exactly how the questionnaire is administered is dependent upon the aims and objectives of the analysis and the target population. For example, if the target population can be gathered together at a certain time and place, then the questionnaire could be administered at this time, with the analyst(s) present. This ensures that the questionnaires are completed. However, gathering the target population in one place at the same time is often difficult to achieve and so questionnaires are often administered by post. Although this is quick and incurs little effort and cost, the response rate is often very low, with figures typically around the 10% mark. Procedures to address poor response rates are available, such as offering payment on completion, the use of encouraging letters, offering a donation to charity upon return, contacting non-respondents by telephone, and sending shortened versions of the initial questionnaire to non-respondents, all of which have been shown in the past to improve response rates, but almost all involve substantial extra cost. More recently, the Internet is becoming popular as a way of administrating questionnaires.

Step 6: Data Analysis

Once all (or a sufficient amount) of the questionnaires have been returned or collected, the data analysis process can begin. This is a lengthy process and is dependent upon the analysis needs. Questionnaire data are typically computerised and analysed statistically. The type of statistical analysis performed is dependent upon the aims and objectives of the study.

Step 7: Follow-Up Phase

Once the data are analysed and conclusions are drawn, the participants who completed the questionnaire should be informed regarding the outcome of the study. This might include a thank-you letter and an associated information pack containing a summary of the research findings.

ADVANTAGES

1. Questionnaires can be used in any domain to collect large amounts of data on any topic of interest.
2. The use of postal or on-line questionnaires significantly increases the participant sample size.
3. Administering and analysing questionnaires requires very little training.
4. When questionnaires are appropriately designed and tested, the data collection and analysis phase is relatively quick and straightforward.
5. Questionnaires offer a good return in terms of data collected in relation to time invested.
6. Questionnaires offer a high degree of control over the data collection process since targeted questions can be designed a priori.
7. Once the questionnaire is designed, very few resources are required to administer it.

DISADVANTAGES

1. The process of designing, piloting, and administering a questionnaire, and then analysing the data obtained is time-consuming.
2. Questionnaire design is more of an art than a science (Wilson and Corlett, 1995) and requires a high level of training, experience, and skill on behalf of the analysts involved.
3. The reliability and validity of questionnaires is questionable.
4. For postal questionnaires, response rates are typically very low (e.g., around 10%).
5. Questionnaire responses can be rushed, non-committal, and subject to various biases, such as prestige bias.
6. Questionnaires can be limited in terms of what they can collect.

RELATED METHODS

There are various Human Factors questionnaires available, covering a range of concepts including usability, situation awareness, decision making, workload, and cognition. Different types of questionnaires include rating scales, paired comparison, and ranking questionnaires.

APPROXIMATE TRAINING AND APPLICATION TIMES

Questionnaire design is more of an art than a science (Wilson and Corlett, 1995). Practice makes perfect, and researchers normally need numerous attempts before they become proficient at questionnaire design (Openheim, 2000). For this reason, the training time associated with questionnaires is high. Although the actual time required to complete questionnaires is typically minimal, the lengthy process of questionnaire design and data analysis renders the total application time as high.

RELIABILITY AND VALIDITY

The reliability and validity of questionnaires is questionable. Questionnaires are prone to a series of different biases and often suffer from social desirability bias, whereby participants respond in the way in which they feel analysts or researchers want them to. Questionnaire answers can also be rushed and non-committal, especially when the analysts or researchers are not present. In a study comparing 12 Human Factors methods, Stanton and Young (1999) reported an acceptable level of inter-rater reliability, but unacceptable levels of intra-rater reliability and validity.

Tools Needed

Although traditionally a paper-based method requiring no more than pen and paper, questionnaires are increasingly being administered electronically on the Internet or via telephone (McClelland and Fulton Suri, 2005). For electronic administration, an Internet domain and specialist on-line questionnaire design software is required. For analysing questionnaire data, statistical analysis software programmes such as SPSS are typically used.

Example

Various questionnaires have been developed specifically for the evaluation of Human Factors concepts, including usability, situation awareness, teamwork, and human error. For example, the Questionnaire for User Interface Satisfaction (QUIS: Chin, Diehl, and Norman, 1988), the Software Usability Measurement Inventory (SUMI; Kirakowski, 1996), and the System Usability Scale (SUS; Brooke, 1996) are all examples of questionnaires designed specifically to assess the usability of a particular device or interface. The SUS questionnaire offers a quick, simple, and low cost approach to device usability assessment, and consists of 10 usability statements that are rated by participants on a Likert scale of 1 (strongly disagree with statement) to 5 (strongly agree with statement). Answers are coded and a total usability score is derived for the product or device under analysis. Each item on the SUS scale is given a score between 0 and 4. The items are scored as follows (Stanton and Young, 1999):

- The score for odd numbered items is the scale position (e.g., 1, 2, 3, 4, or 5) minus 1.
- The score for even numbered items is 5 minus the scale position.
- The sum of the scores is then multiplied by 2.5.

The final figure represents a usability score for the device under analysis and should range between 0 and 100. The SUS questionnaire proforma is presented in Figure 2.1.

Flowchart

(See Flowchart 2.2.)

Recommended Texts

Lehto, J. R., and Buck, M. (2007). *Introduction to Human Factors and ergonomics for engineers*. Boca Raton, FL: CRC Press.
McClelland, I., and Fulton Suri, J. (2005). Involving people in design. In *Evaluation of human work*, 3rd ed., eds. J. R. Wilson and E. N. Corlett, 281–333. Boca Raton, FL: CRC Press.

OBSERVATIONAL STUDY

Observational study is used to gather data regarding the observable aspects (e.g., physical and verbal) of task performance. Within Human Factors, observational study typically forms the starting point of any investigation; Annett (2003), for example, argues that, at the very least, data collection should comprise observation and SME interviews. Further, most of the methods in this book

Data Collection Methods

	Strongly disagree 1	2	3	4	Strongly agree 5
1. I think that I would like to use this system frequently					
2. I found the system unnecessarily complex					
3. I thought the system was easy to use					
4. I think that I would need the support of a technical person to be able to use this system					
5. I found the various functions in this system were well integrated					
6. I thought there was too much inconsistency in this system					
7. I would imagine that most people would learn to use this system very quickly					
8. I found the system very cumbersome to use					
9. I felt very confident using the system					
10. I needed to learn a lot of things before I could get going with this system					

FIGURE 2.1 System Usability Scale. *Source*: Adapted from Brooke, J. (1996).

require an observation of the task or system under analysis as an initial step in the process. Various forms of observational study exist including direct observation, whereby analysts directly observe the task or scenario in question being performed in its natural habitat; participant observation, whereby analysts observe activities while engaging in them themselves; and remote observation, whereby analysts observe activities from a remote site (i.e., via video camera link-up). The utility of the observational study method lies in the range of data that can be collected. Drury (1990), for example, highlights the following five key types of information that are derived from observational study: sequence of activities, duration of activities, frequency of activities, fraction of time spent in states, and spatial movement.

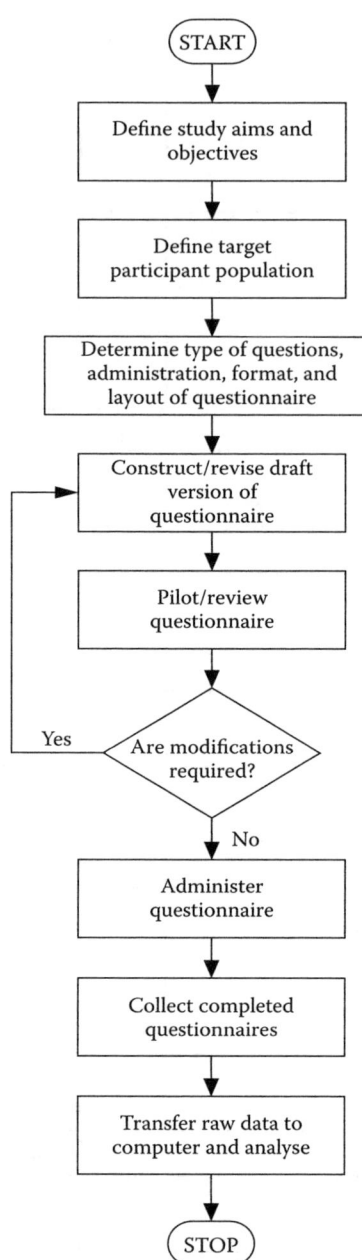

FLOWCHART 2.2 Questionnaire flowchart.

DOMAIN OF APPLICATION

A generic approach, observational study is used extensively in a wide range of domains. The authors, for example, have undertaken observational studies within the domains of military (Stanton et al., 2006), energy distribution (Salmon, Stanton, Walker, Jenkins, Baber, and McMaster, 2008), air traffic control (Walker, Stanton, Baber, Wells, Gibson, Young, Salmon, and Jenkins, in press), rail (Walker et al., 2006), and aviation (Stewart, Stanton, Harris, Baber, Salmon, Mock, Tatlock, Wells, and Kay, 2008).

Application in Sport

The use of observational study is extremely common within Sports Science. Unlike the traditional Human Factors domains, professional sports are, by nature, spectator sports and so there are no restrictions on access to live performance. Indeed most professional sports are transmitted regularly on TV and so observational data can be easily obtained. As a corollary, observational study has been used for all manner of purposes within a Sports Science context. Andersson, Ekblom, and Kustrup (2008), for example, observed professional Swedish soccer games in order to examine the movement patterns (e.g., distance covered, sprints, etc.) and ball skills (tackles, headers, passes, etc.) of elite football players on different playing surfaces (turf versus natural grass). Notational analysis (Hughes and Franks, 1997; James, 2006), a structured observational study method used in the sports domain, has also been used extensively to analyse various aspects of sports performance, for example, during soccer (e.g., James, 2006), rugby (Hughes and Franks, 2004), and tennis (O'Donoghue and Ingram, 2001) matches.

Procedure and Advice

Step 1: Define Aims and Objectives

The first step involves clearly defining the aims and objectives of the observation. This should include identifying what task or scenario is to be observed, in which environment the observation will take place, which participants will be observed, what data are required, and what data are expected.

Step 2: Define Scenario(s)

Once the aims and objectives of the observation are clearly defined, the scenario(s) to be observed should be defined and described further. Normally, the analyst(s) have a particular task or scenario in mind. It is often useful at this stage to use task analysis methods, such as Hierarchical Task Analysis (HTA), to describe the scenario under analysis. Often organisations have existing descriptions or standard operating procedures that can be used for this purpose.

Step 3: Develop Observation Plan

Next, the analysis team should proceed to plan the observation. They should consider what they are hoping to observe, what they are observing, and how they are going to observe it. Depending upon the nature of the observation, access to the system in question should be gained first. This might involve holding meetings with the organisation in question, and is typically a lengthy process. Any recording tools should be defined and the length of observations should be determined. An observational transcript detailing the different types of data required should also be developed. The physical aspects of the observation also require attention, including the placement of observers and video and audio recording equipment. To make things easier, a walkthrough or test observation of the system/environment/scenario under analysis is recommended. This allows the analyst(s) to become familiar with the task in terms of activity conducted, the time taken, and location, as well as the system under analysis.

Step 4: Pilot Observation

In any observational study, a pilot observation is critical. This allows the analysis team to assess any problems with the observation, such as noise interference, aspects of the task that cannot be recorded, or problems with the recording equipment. The quality of data collected can also be tested, as well as any effects upon task performance that may result from the presence of observers. Often it is found during this step that certain data cannot be collected or that the quality of some data (i.e., video or audio recordings) is poor; or it emerges that other data not previously thought of are likely to be of interest. If major problems are encountered, the observation may have to be redesigned. Steps 2 to 4 should be repeated until the analysis team is happy that the quality of the data collected will be sufficient for the study requirements.

Step 5: Conduct Observation

Once the observation has been designed and tested, the team should proceed with the observation(s). Typically, data are recorded visually using video and audio recording equipment and an observational transcript is created by the analysts during the observation. The transcript categories used are determined by the aims of the analysis. The observation should end when the required data are collected or when the task or scenario under analysis is completed.

Step 6: Data Analysis

Once the observation is complete, the data analysis procedure begins. Typically, the starting point of the data analysis phase involves typing up the observation notes or transcripts made during the observation. Depending upon the analysis requirements, the team should then proceed to analyse the data in the format that is required, such as frequency of tasks, verbal interactions, and sequence of tasks. When analysing visual data, typically user behaviours are coded into specific groups. The Observer Pro software package is typically used to aid the analyst in this process.

Step 7: Further Analysis

Once the initial process of transcribing and coding the observational data is complete, further analysis of the data begins. Depending upon the nature of the analysis, observation data are used to inform a number of different Human Factors analyses, such as task analysis, teamwork assessment, error analysis, and communications analysis. Typically, observational data are used to develop a task analysis (e.g., HTA) of the task or scenario under analysis, which is then used to inform other Human Factors analyses.

Step 8: Participant Feedback

Once the data have been analysed and conclusions have been drawn, the participants involved should be provided with feedback of some sort. This could be in the form of a feedback session or a letter to each participant. The type of feedback used is determined by the analysis team.

ADVANTAGES

1. Observational study provides real-life insight into task performance, since performance is observed in its natural habitat.
2. Most professional sports are televised, enabling easy access to observational data.
3. Various aspects of sporting performance can be analysed via observational study, such as movement patterns, behaviours exhibited, communications, passing, errors, goals scored, etc.
4. Observation has a long history of use within Human Factors for analysing various aspects of performance.
5. Observation has also been used extensively within Sports Science for analysis of sports performance.
6. The data gathered are objective.
7. Recorded data can be revisited for in-depth analysis.
8. Observation offers a good return in terms of data collected in relation to time invested.
9. The data derived from observational study inform various Human Factors analyses, such as task analysis, human error identification and analysis, social network analysis, teamwork assessment, etc.

Disadvantages

1. Observational study generates large amounts of data.
2. Observational study can be intrusive, although this is likely to be less of a problem in the sporting arena.
3. Observational data are prone to various biases.
4. The quality of the data collected is dependent on the skill of the observer. Inexperienced observers often attempt to record everything and miss key aspects of task performance.
5. The data analysis component can be hugely time consuming for large, complex tasks involving numerous actors. Kirwan and Ainsworth (1992), for example, point out that one hour of recorded data can take approximately eight hours to transcribe.
6. It is difficult to study the cognitive aspects of task performance (i.e., situation awareness, decision making, error causality, and cognitive workload) using observational study alone.
7. Observational studies can be difficult to arrange due to access problems; however, this is less likely to be a problem in the sporting domains.
8. Observational study offers very little experimental control to the analyst.
9. For complex collaborative tasks, it is likely that a large team of analysts will be required to conduct the observation.

Related Methods

Various forms of observational study exist, including direct, indirect, participant, and remote observation. Observational study data are typically used as the input for various Human Factors analysis methods, including task analysis, process charting, human error identification and analysis, social network analysis, and teamwork assessment methods. Within Sports Science, observational study is typically used as the basis to undertake task analyses such as notational analysis and activity analysis.

Approximate Training and Application Times

The training time for observational study methods is typically low (aside from participant observation, which requires some form of training in the job or task that is to be observed). Despite this, a significant amount of experience is required before analysts become proficient with the method. The overall application time of the method is typically high due to the time-consuming data transcription and analysis phases.

Reliability and Validity

Despite having high face (Drury, 1990) and ecological (Baber and Stanton, 1996b) validity, various issues can potentially limit the reliability and validity of observational study methods. These include problems with causality (i.e., observing errors but misunderstanding why they occurred), biases, construct validity, and internal and external validity, all of which can potentially manifest themselves unless the appropriate precautions are taken (Baber and Stanton, 1996b).

Tools Needed

On a simplistic level, observational study can be undertaken using pen and paper only. However, for complex tasks it is recommended that visual and audio recording equipment is used to record all activities being observed. For data analysis purposes, a PC with a word processing package such as

Microsoft Word is required. Specific observational study software packages, such as Observer Pro, are also available. For indirect observation, a television and a video/DVD player are required.

EXAMPLE

An observational study of the passing performance of the England International Soccer Team during the 2006 World Cup and qualification for the 2008 European Championships was undertaken. The observational study component involved observing video-recorded television coverage of the games under analysis. Two analysts observed the games and recorded data regarding the passes made between the England soccer players. This included which player was making the pass, which player the pass was intended to go to, whether or not the pass reached its intended recipient, the area of the pitch in which the pass was initiated (e.g., defensive, middle, or attacking third), and the area of the pitch in which the pass was received. The data collected were used to conduct social network analyses on the England Soccer Team's passing during the games analysed (see Social Network Analysis, Example section in Chapter 8). An extract of the observational transcript from the England versus Ecuador World Cup 2006 game is presented in Table 2.8.

FLOWCHART

(See Flowchart 2.3.)

RECOMMENDED TEXTS

Baber, C., and Stanton, N. A. (1996). Observation as a technique for usability evaluations. In *Usability evaluation in industry*, eds. P. W. Jordan, B. Thomas, B. A. Weerdmeester, and I. McClelland, 85–94. London: Taylor & Francis.

Drury, C. (1990). Methods for direct observation of performance. In *Evaluation of human work*, 2nd ed., eds. J. R. Wilson and E. N. Corlett, 45–68). London: Taylor & Francis.

TABLE 2.8
World Cup 2006 Game Passing Observational Transcript

England versus Ecuador
Word Cup 2006, Second Round, Sunday 25th June 2006
Score: England 1 – 0 Ecuador
Beckham, 60
Possession: England 51%, Ecuador 49%

Pass From	Pass To	Zone From	Zone To	Completed?	Score
HAR	XXX	1	—	No	0–0
TER	XXX	1	—	No	0–0
ROO	XXX	1	—	No	0–0
HAR	XXX	3	—	No	0–0
HAR	FER	1	1	Yes	0–0
FER	COL	1	1	Yes	0–0
COL	JCO	1	2	Yes	0–0
COL	XXX	2	—	No	0–0
BEC	XXX	2	—	No	0–0
BEC	XXX	3	—	No	0–0
LAM	XXX	3	—	No	0–0
GER	COL	2	2	Yes	0–0
COL	XXX	2	—	No	0–0
COL	LAM	2	2	Yes	0–0
LAM	TER	2	1	Yes	0–0
TER	FER	1	1	Yes	0–0
FER	HAR	1	2	Yes	0–0
HAR	FER	2	1	Yes	0–0
FER	TER	1	1	Yes	0–0
TER	COL	1	2	Yes	0–0
COL	LAM	2	2	Yes	0–0
LAM	XXX	2	—	No	0–0
CAR	XXX	2	—	No	0–0
TER	FER	1	—	No	0–0
FER	CAR	1	2	Yes	0–0
CAR	XXX	2	—	No	0–0
FER	GER	2	—	No	0–0
GER	BEC	2	—	No	0–0
BEC	XXX	2	—	No	0–0
ROB	XXX	1	3	Yes	0–0
HAR	GER	2	—	No	0–0
GER	BEC	2	—	No	0–0
BEC	HAR	2	3	Yes	0–0
HAR	BEC	3	2	Yes	0–0

34 Human Factors Methods and Sports Science: A Practical Guide

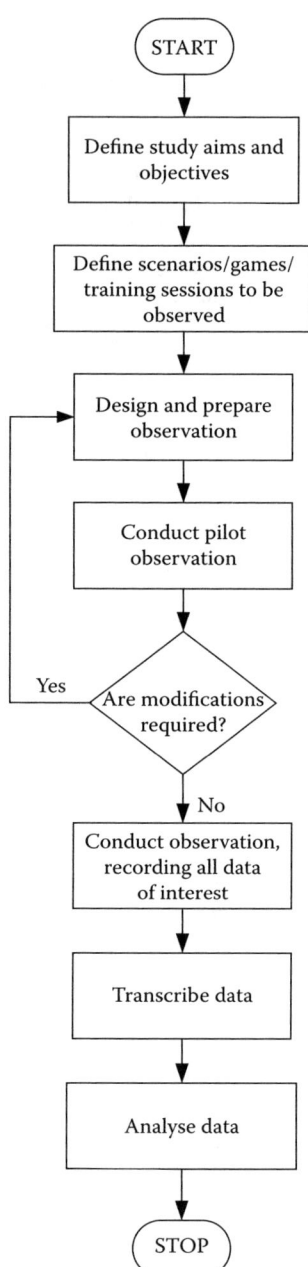

FLOWCHART 2.3 Observational flowchart.

3 Task Analysis Methods

INTRODUCTION

Methods of collecting, classifying, and interpreting data on human performance in work situations lie at the very heart of Ergonomics and Human Factors (Annett and Stanton, 2000). Accordingly, task analysis approaches are the most commonly used and well-known form of Human Factors methods. They are used to collect data about and describe tasks, systems, and devices in terms of the physical and cognitive activities involved during task performance; they focus on "what an operator ... is required to do, in terms of actions and/or cognitive processes to achieve a system goal" (Kirwan and Ainsworth, 1992, p. 1). The outputs from task analysis methods are used for a variety of purposes, including task and system design and evaluation, interface design, training programme design and evaluation, allocation of functions analysis, error identification and analysis, and procedure design. Their popularity is such that there are over 100 task analysis approaches described in the literature (Diaper and Stanton, 2004), and they have been applied in all manner of domains, including the military (e.g., Matthews, Strater, and Endsley, 2004), aviation (e.g., Stanton, Harris, Salmon, Demagalski, Marshall, Young, Dekker, and Waldmann, 2006), driving (e.g., Walker, Stanton, and Young, 2001), healthcare (e.g., Lane, Stanton, and Harrison, 2007), public technology (e.g., Adams and David, 2007; Stanton and Stevenage, 1998), music (e.g., Hodgkinson and Crawshaw, 1985), and sport (Doggart, Keane, Reilly, and Stanhope, 1992).

The origins of task analytic approaches go as far back as the early 1900s to the so-called scientific management movement of that time. Scientific management methods, advocated by the likes of Frederick Taylor and the Gilbreths, were used to analyse tasks in order to investigate more efficient ways in which to undertake them. They were primarily concerned with the analysis of physical work so that new, more economical work methods could be proposed (Annett and Stanton, 2000). Early methods focused on how the work was performed, what was needed to perform the work, why the work was performed in this way, and how the work could be improved (Stanton, 2006). Due to the changing nature of industrial work processes around the 1950s and 1960s, which meant that tasks were becoming more cognitive in nature, new task analysis approaches, such as Hierarchical Task Analysis (HTA; Annett, Duncan, Stammers, and Gray, 1971) emerged, which focused also on the cognitive elements of task performance, such as cues, decisions, feedback, and knowledge (Annett and Stanton, 2000).

The popularity of task analytic approaches is partly due to their flexibility and the resultant utility of their outputs. The main benefit is that they structure information about the task in question (Annett and Stanton, 2000). Further, not only is the process enlightening in terms of allowing the analyst to develop a deep understanding of the task in question, but also the outputs form the basis for various other Human Factors analyses approaches, such as human error identification, interface evaluation, and process charting methods. For example, the Systematic Human Error and Prediction Approach (SHERPA; Embrey, 1986) is a human error identification approach that uses task analysis data as its input; whereas operation sequence diagrams are built based on the task steps identified in a task analysis of the activity in question. Typically, a task analysis of some sort is required for any form of Human Factors analysis, be it usability evaluation, error identification, or performance evaluation.

Although their utility is assured, task analysis methods are not without their flaws. The resource usage associated with task analysis efforts can be considerable both in terms of data collection and in developing, reiterating, and refining the task analysis itself. For complex collaborative systems, in particular, the time invested can be considerable and the outputs can be large, complex, and unwieldy. The process can also be laborious, with many iterations often being required before an appropriate representation of the activity in question is developed. Further, for complex tasks and systems, video and audio recording equipment and access to SMEs is often a necessity.

Task analysis is one form of methodology that has received attention from sports scientists, albeit most often under slightly different guises. Notational analysis (e.g., James, 2006), for example, is a popular task analytic approach that is used by sports scientists for analysing different aspects of performance via recording different events that occur during sports activities. According to Hughes and Franks (2004; cited in James, 2006) notational analysis is primarily concerned with the analysis of movement, technical and tactical evaluation, and statistical compilation. Similarly, activity analysis is often used to identify the different physical activities required during a sporting endeavour. Martin, Smith, Tolfrey, and Jones (2001), for example, used activity analysis to analyse the movement activities of referees, and their frequency and duration, during English Premiership Rugby Union games. Martin et al. (2001) recommended that such analyses be used to design physical training programmes. Of course, these two approaches differ from task analysis approaches as we know them, since they are descriptive in nature and describe the activity that did occur, whereas Human Factors task analysis approaches can be either normative (describe activity as it should occur), descriptive, or even formative (describing activity as it could occur).

For the potential contribution that task analysis approaches can make within sport one only has to look at the range of different applications to which such approaches have been put by Human Factors researchers. For example, HTA (Annett et al., 1971), the most popular of all task analysis approaches, has been used by Human Factors researchers for a range of purposes, including system design and evaluation (e.g., Stanton, Jenkins, Salmon, Walker, Rafferty, and Revell, in press), interface design and evaluation (e.g., Hodgkinson and Crawshaw, 1985; Shepherd, 2001; Stammers and Astley, 1987), job design (e.g., Bruseberg and Shepherd, 1997), training programme design and evaluation (e.g., Piso, 1981), human error prediction (e.g., Stanton et al., 2006; event analysis Lane, Stanton, and Harrison, 2007), as well as analysis (e.g., Adams and David, 2007), team task analysis (e.g., Annett, 2004; Walker et al., 2006), allocation of functions analysis (e.g., Marsden and Kirby, 2004), workload assessment (Kirwan and Ainsworth, 1992), and procedure design (Stanton, 2006). Transferring these applications to a sporting context identifies a range of useful applications to which sports task analyses could potentially be put, including coaching and training design and evaluation, the design and evaluation of tactics, the design and evaluation of sports technology and performance aids, error prediction, role allocation, and performance evaluation. The approach seems to be particularly suited to coaching and training design. Kidman and Hanrahan (2004), for example, suggest that task analysis is important for coaching design as it identifies the purpose of a particular task and then breaks the task in question into sub-tasks. They suggest that conducting task analyses is a key component of sports coaching. Citing the example of a rugby coach wishing to teach a drop kick, Kidman and Hanrahan (2004) recommend that a drop kick task analysis should be completed and the component sub-tasks and coaching requirements should then be specified from the task analysis output. In addition, the requirement for initial task analysis descriptions to inform most forms of Human Factors analyses was articulated above. Therefore, if Human Factors methods are to be applied in sporting contexts, then it is imperative that task analysis approaches, such as HTA, are used initially.

Many task analysis approaches exist. This book focuses on four popular task analysis methods: HTA (Annett et al., 1971), task decomposition (Kirwan and Ainsworth, 1992), Verbal Protocol Analysis (VPA), and Operation Sequence Diagrams (OSDs; Kirwan and Ainsworth, 1992). HTA is by far the most commonly used task analysis approach and is essentially a descriptive task analysis method that is used to describe systems, goals, and tasks. It works by decomposing activities into a hierarchy of goals, subordinate goals, operations, and plans, which allows systems and goals, sub-goals, and activities to be described exhaustively. Using HTA outputs as its input, task decomposition (Kirwan and Ainsworth, 1992) is used to further analyse component task steps by looking specifically at the different aspects involved, including the equipment used, the actions required, feedback, and errors made. VPA is used to generate information regarding the processes, cognitive and physical, that individual operators use to perform tasks, and involves operators "thinking aloud" (i.e., describing how they are undertaking tasks) as they perform the task under analysis. Finally, OSDs are a representational task analysis approach that use standardised symbols to graphically represent collaborative activities and the interactions between team members (and technology, if required). The output of an OSD graphically depicts the task, including the different types of tasks performed (e.g., action, request, receipt of information, delay, transport), and the interaction between operators over time. A summary of the task analysis methods described is presented in Table 3.1.

HIERARCHICAL TASK ANALYSIS

Background and Applications

The popularity of HTA is unparalleled; it is the most commonly used method, not just out of task analysis methods, but also out of all Human Factors and ergonomics methods (Annett, 2004; Kirwan and Ainsworth, 1992; Stanton, 2006). The "task" in HTA, however, is something of a misnomer. HTA does not in fact analyse tasks; rather, it is concerned with goals (an objective or end state) and these are hierarchically decomposed (Annett and Stanton, 1998). Inspired by the work of Miller and colleagues on "plans and the structure of behaviour" (Miller, Galanter, and Pribram, 1960), HTA was developed in the 1960s in response to a need to better understand complex cognitive tasks (Annett, 2004). The changing nature of industrial work processes around that time meant that tasks were becoming more cognitive in nature, and approaches that could be used to describe and understand them were subsequently required. HTA was unique at the time in that, in addition to the physical tasks being performed, it also attempted to describe the cognitive processes that were required to achieve the goals specified (something that is often overlooked by antagonists of the method). The HTA method therefore represented a significant departure from existing approaches of the time, since it focused on goals, plans, and cognitive processes rather than merely the physical and observable aspects of task performance.

Stanton (2006) describes the heavy influence of control theory on the HTA methodology and demonstrates how the test-operate-test-exit (TOTE) unit (central to control theory) and the notion of hierarchical levels of analysis are similar to HTA representations (plans and sub-goal hierarchy). HTA itself works by decomposing systems into a hierarchy of goals, subordinate goals, operations, and plans, which allows systems and goals, sub-goals, and activities to be described exhaustively. It is important to note here that an "operator" may be a human or a technological operator (e.g., system artefacts such as equipment, devices, and interfaces). HTA outputs therefore specify the overall goal of a particular system, the sub-goals to be undertaken to achieve this goal, the operations required to achieve each of the sub-goals specified, and the plans, which are used to ensure that the goals are achieved. The plans component of HTA are especially important since they specify the sequence and under what conditions different sub-goals have to be achieved in order to satisfy the requirements of a superordinate goal.

The HTA process is simplistic, involving collecting data about the task or system under analysis (through methods such as observation, questionnaires, interviews with SMEs, walkthroughs, user trials, and documentation review, to name but a few), and then using these data to decompose and describe the goals, sub-goals, and tasks involved.

Domain of Application

HTA was originally conceived for the chemical processing and power generation industries (Annett, 2004); however, due to its generic nature and flexibility the method has since been applied in all

TABLE 3.1
Task Analysis Methods Summary Table

Name	Domain	Application in Sport	Training Time	App. Time	Input Methods	Tools Needed	Main Advantages	Main Disadvantages	Outputs
Hierarchical Task Analysis (HTA; Annett, Duncan, Stammers, and Gray, 1971)	Generic	Task description; Task analysis; Coaching design, and evaluation; Tactics design, evaluation, and selection; Sports product design and evaluation; Performance evaluation; Error prediction; Allocation of functions	Low	High	Observational study; Interviews; Questionnaires; Walkthrough	Audio and video recording equipment; Pen and paper; Word processing software; Task analysis software, e.g., the HTA tool	1. Extremely popular and flexible approach that can be used for all manner of purposes 2. Considers the physical and cognitive elements of task performance along with context 3. HTA outputs act as the input to most forms of Human Factors methods	1. Provides mainly descriptive information and has little direct input into design 2. For complex tasks and systems the analysis may be highly time consuming and may require many iterations 3. Can be a laborious method to apply	Goal-based description of activity
Task decomposition	Generic	Task description; Task analysis; Coaching design and evaluation; Tactics design, evaluation, and selection; Sports product design and evaluation; Performance evaluation; Error prediction	Low	High	Hierarchical Task Analysis; Observational study; Interviews; Questionnaires; Walkthrough	Audio and video recording equipment; Pen and paper; Word processing software; Task analysis software, e.g., the HTA tool	1. Highly comprehensive 2. Offers a structured approach for analysing, in detail, the task steps involved in a particular scenario 3. Extremely flexible, allowing the analyst(s) to choose the decomposition categories based upon analysis requirements	1. Can be highly time consuming to apply 2. The analysis often becomes repetitive and laborious 3. The initial data collection phase may be highly time consuming, especially for complex tasks and systems	Detailed analysis of the component sub-tasks involved in a particular process

Task Analysis Methods

Method	Type	Uses	Training time	Related methods	Tools needed	Advantages	Disadvantages	Output
Verbal Protocol Analysis (VPA; Walker, 2004)	Generic	Data collection Cognitive task analysis Situation awareness assessment Identification of strategies Coaching design and evaluation	Low	Observational study	Audio and video recording equipment Pen and paper Word processing software Observer Pro	1. VPA provides a rich data source 2. All manner of concepts can be analysed, including situation awareness, decision making, teamwork, human error, and workload 3. Can be used to compare the processes adopted by sports performers of differing expertise and skill levels	1. Highly time consuming to apply 2. The data analysis phase is often extremely laborious and time consuming 3. There are various issues associated with the use of verbal protocol data. For example, Noyes (2006) reports that some mental processes cannot be verbalised accurately	Verbal transcript of task performance List of key concepts/themes related to task performance
Operation Sequence Diagrams (OSD; Kirwan and Ainsworth, 1992)	Generic	Task description Task analysis Coaching design and evaluation Tactics design, evaluation, and selection Performance evaluation	Low	Hierarchical Task Analysis Observational study	Pen and paper Microsoft Visio	1. Provides an exhaustive analysis of the task in question 2. Particularly useful for representing collaborative activities 3. Also useful for demonstrating the relationship between tasks, technology, and team members	1. Can be hugely time consuming to apply, particularly for large, complex tasks involving multiple team members 2. Without any supporting software, OSD construction is overly time consuming and laborious 3. OSDs can become large, unwieldy, cluttered, and confusing	Graphical representation of activity

manner of domains, ranging from safety critical domains, such as the military (Stanton et al., in press), healthcare (Lane et al. 2007), road transport (Walker, Stanton, and Young, 2001), and aviation domains (Stanton et al., 2006), to others such as public technology (Adams and David, 2007; Stanton and Stevenage, 1998), and music (Hodgkinson and Crawshaw, 1985).

APPLICATION IN SPORT

While there are no published applications of the HTA method in a sporting context, it is clearly suited to the analyses of sports tasks, activities, and systems since it deals with the goal-related physical and cognitive elements of task performance. The output could potentially be used for all manner of purposes, including coaching intervention design and evaluation, sports product design, performance prediction and evaluation, and tactics design and selection. Further, the requirement of most other Human Factors methods for initial HTA descriptions to act as input data means that sporting applications of the method are imperative if the other methods described in this book are to be applied within the sporting arena.

PROCEDURE AND ADVICE

Step 1: Define Aims of the Analysis

The first step in conducting an HTA involves clearly defining what the overall purpose of the analysis is. HTA has been used in the past for a wide range of purposes, including system design, interface design, operating procedure design, workload analysis, human error identification and analysis, and training design, to name only a few. Defining the aims of the analysis also includes clearly defining the system, task, procedure, or device that is to be the focus of the HTA.

Step 2: Define Analysis Boundaries

Once the purpose and focus of the analysis is clearly defined, the analysis boundaries should be considered. Depending on the purpose of the analysis, the analysis boundaries may vary. For example, for analyses focusing on training design for individuals, or the prediction of errors that a particular sports performer is likely to make, the system boundary should be drawn around the tasks performed by the individual in question. On the contrary, if the analysis is focusing on the level of coordination between sports team members, then the overall team task should be analysed. Annett (2004), Shepherd (2001), and Stanton (2006) all emphasise the importance of constructing HTAs appropriate to the analyses' overall purpose.

Step 3: Collect Data Regarding Task/System/Procedure/Device under Analysis

Once the task under analysis is clearly defined, specific data regarding the task should be collected. The data collected during this process are used to inform the development of the HTA and to verify the output once completed. It is important here to utilise as many sources of data as possible (Kirwan and Ainsworth, 1992; Shepherd, 2001; Stanton, 2006). Data regarding the goals, physical and cognitive activities, technology used, interactions between man and between man and technology, decision making, temporal aspects, environmental conditions, and task constraints should be collected. A number of data sources can be used, including direct and indirect (i.e., video recordings) observation, interviews with SMEs, walkthrough/talk-through analysis, verbal protocol analysis, standard operating procedure documents, training manuals, and simulation. It is also important that the sources of data used to inform the analysis are clearly documented (Stanton, 2006).

Step 4: Determine and Describe Overall System/Task Goal

The process of constructing the HTA output begins with specification of the overall goal of the activity or system under analysis. For example purposes, we will use the goal of scoring after a

Task Analysis Methods

penalty during a game of soccer. The overall goal in this case would be "Score goal via penalty kick." The overall goal should be numbered as "0".

Step 5: Determine and Describe Sub-Goals

Once the overall goal has been specified, the next step is to decompose it into meaningful sub-goals (usually four or five are recommended but this is not rigid), which together form the goals required to achieve the overall goal. For example, the overall goal of "Score goal via penalty kick" can be meaningfully decomposed into the following sub-goals: 1) Place ball; 2) Check goal and goalkeeper; 3) Determine kick strategy; 4) Make run; 5) Strike ball as intended; and 6) Observe outcome.

Step 6: Decompose Sub-Goals

Next, the analyst should decompose each of the sub-goals identified during step five into further sub-goals. All of the lower level goals should be an appropriate expansion of the higher goals (Patrick, Spurgeon, and Shepherd, 1986). This process should go on until all sub-goals are exhausted. For example, the sub-goal "determine kick strategy" is broken down into the following sub-goals: "determine placement," "determine power," and "visualise shot." The level of decomposition used is dependent upon the analysis; however, Stanton (2006) suggests that analysts should try to keep the number of immediate sub-goals under any superordinate goal to a minimum (between 3 and 10). Stanton (2006) points out that, rather than turn the analysis into a procedural list of operations, the goal hierarchy should contain clusters of operations that belong together under the same goal. The decomposition of goals and sub-goals is a highly iterative process and often involves several reviews and revisions until the decomposition is appropriate. A commonly cited heuristic for ending decomposition is the PxC rule (e.g., Annett, 2004; Stanton, 2006); however, since analysts often have trouble quantifying these aspects, it is more often than not recommended that the decomposition should be stopped before it continues beyond a point where it will be useful (Stanton, 2006). For example, for team communications analysis purposes, sub-goal decomposition should stop at the point of interactions between team members (e.g., transmission of information between team members). Alternatively, for error prediction purposes, sub-goal decomposition should continue to the point where the interaction between humans and devices is described sufficiently for errors arising from the interaction to be identified (i.e., press this button, read this display).

Step 7: Create Plans

Plans are used to specify the sequence and conditions under which different sub-goals have to be achieved in order to satisfy the requirements of the superordinate goal; they represent the conditions under which sub-goals are triggered and exited. Various forms of plans exist. The most commonly used include linear (i.e., do 1 then 2 then 3), simultaneous (i.e., do 1 and 2 and 3 simultaneously), selection (i.e., choose one from 1 to 3), and non-linear plans (i.e., do 1 to 3 in any order). For more complex tasks, custom plans, often including a mixture of the four types described, are also used. Stanton (2006) points out that plans contain the context in which sub-goals are triggered, including time, environmental conditions, completion of other sub-goals, system state, and receipt of information. Once the plan is completed, the agent returns to the superordinate level.

Step 8: Revise Analysis

As pointed out earlier, HTA is a highly reiterative process that requires multiple reviews and revisions of its output. Stanton (2006) points out that the first pass is never going to be acceptable, regardless of its purpose. It is therefore recommended that, once a first pass has been drafted, it is subject to various reviews and revisions depending on the time and purpose of the analysis (i.e., level of detail required). As a general rule of thumb, Stanton (2006) suggests that simple analyses require at least three iterations, whereas analyses that are more complex require at least 10 iterations.

Step 9: Verify and Refine Analysis

Once an acceptable draft HTA is created, it should be reviewed by SMEs and subsequently refined based on their feedback. The involvement of SMEs at this stage is critical to ensure the validity of the output.

ADVANTAGES

1. HTA is an extremely flexible approach that can be used for all manner of purposes.
2. HTA outputs act as the input to various Human Factors methods, including human error identification, interface design and evaluation, training design, allocation of function, and teamwork assessment methods.
3. The output provides a comprehensive description of the task or system under analysis.
4. The process of constructing an HTA is enlightening itself, enabling the analyst to develop an in-depth understanding of the task or system in question.
5. HTA is a generic approach that can be applied in any sporting domain or context.
6. The analyst has control over the analysis boundaries and the level of decomposition used.
7. In most cases HTA is a quick and simplistic approach to use.
8. HTA is the most commonly used of all Human Factors methods and has been applied in all manner of domains for a wide range of purposes.
9. Various software tools exist that expedite the process (e.g., Salmon, Stanton, Walker, Jenkins, and Farmillo, 2009).

DISADVANTAGES

1. When used in isolation HTA provides mainly descriptive information only and outputs have little direct input into design.
2. For complex tasks and systems the analysis may be highly time consuming and may require many iterations.
3. HTA can be a laborious method to apply, especially for complex tasks and systems.
4. The initial data collection phase may also be time consuming, especially for complex tasks and systems.
5. The reliability of the approach is questionable; for example, different analysts may produce very different decompositions for the same task or system.
6. Applying HTA has been described as "more art than science" (Stanton, 2006) and a great deal of practice is required before proficiency in the method is achieved.

RELATED METHODS

Various data collection methods are used to gather the data required to construct an HTA, including interviews, questionnaires, observation, walkthrough, and verbal protocol analysis. Following completion, the HTA output can be used to inform all manner of Human Factors analysis methods, including process charting, human error identification, cognitive task analysis, situation awareness assessment, teamwork assessment, workload assessment, and timeline analysis.

APPROXIMATE TRAINING AND APPLICATION TIMES

The training time for HTA is likely to be high. Although the training itself can be undertaken in a short period of time, in order to ensure that analysts become proficient in the approach, a substantial amount of practical experience is required. Stanton and Young (1999) report that the training time for HTA is substantial, and Annett (2004) describes a study by Patrick, Gregov, and Halliday (2000) which found that students given only a few hours' training in the method produced unsatisfactory results for a simplistic task, but that performance improved significantly with further training. The application time incurred when applying HTA is dependent upon the task or system being analysed. For simplistic tasks it is likely to be low; however, for more complex tasks and systems it is likely to be considerable. For example, Salmon, Stanton, Jenkins, and Walker (in press) report that the development of an HTA for an Apache helicopter 3-phase mission-planning task took approximately 30 hours, with the associated data collection procedure involving around 3 days' worth of observational study, walkthrough analysis, and SME interviews.

Reliability and Validity

The reliability and validity of HTA is difficult to assess (Annett, 2004). Notwithstanding this, it is apparent that although the validity of the approach is high, the reliability is low. For example, in conclusion to a study comparing 12 Human Factors methods, Stanton and Young (1999) reported that HTA achieved an acceptable level of validity but a poor level of reliability. The level of reliability achieved by HTA seems to be problematic; in particular, intra-analyst reliability and inter-analyst reliability levels may often be poor.

Tools Needed

HTA can be carried out using pen and paper only; however, HTA outputs are typically developed using some form of software drawing tool application, such as Microsoft Visio or Microsoft Word. Further, a number of HTA software tools exist (e.g., Bass, Aspinall, Walters, and Stanton, 1995; Salmon, Stanton, Walker, Jenkins, and Farmillo, 2009).

Example

An example HTA for a soccer game was produced. The data collection phase for the soccer HTA involved direct observation of a series of recorded English Premier League games and international soccer matches that took place as part of the World Cup 2006 and the European Championship 2008. The HTA produced was subsequently refined as a result of discussions with one SME. Extracts from the HTA, which was originally developed in draft form using Microsoft Notepad and then drawn using Microsoft Visio, are presented in Figures 3.1 through 3.3. An extract of the overall HTA, including example decompositions, is presented in Figure 3.1. Depending on the analysis requirements, the HTA description can be decomposed to a low level of analysis. For example, the sub-goal "1.2 Tackle opposition player" decomposition is presented in Figure 3.2 and the sub-goal "Score goal from penalty kick" decomposition is presented in Figure 3.3.

Flowchart

(See Flowchart 3.1.)

Recommended Texts

Annett, J. (2004). Hierarchical task analysis. In *The handbook of task analysis for human-computer interaction*, eds. D. Diaper and N. A. Stanton, 67–82. Mahwah, NJ: Lawrence Erlbaum Associates.
Stanton, N. A. (2006). Hierarchical task analysis: Developments, applications, and extensions. *Applied Ergonomics* 37:55–79.

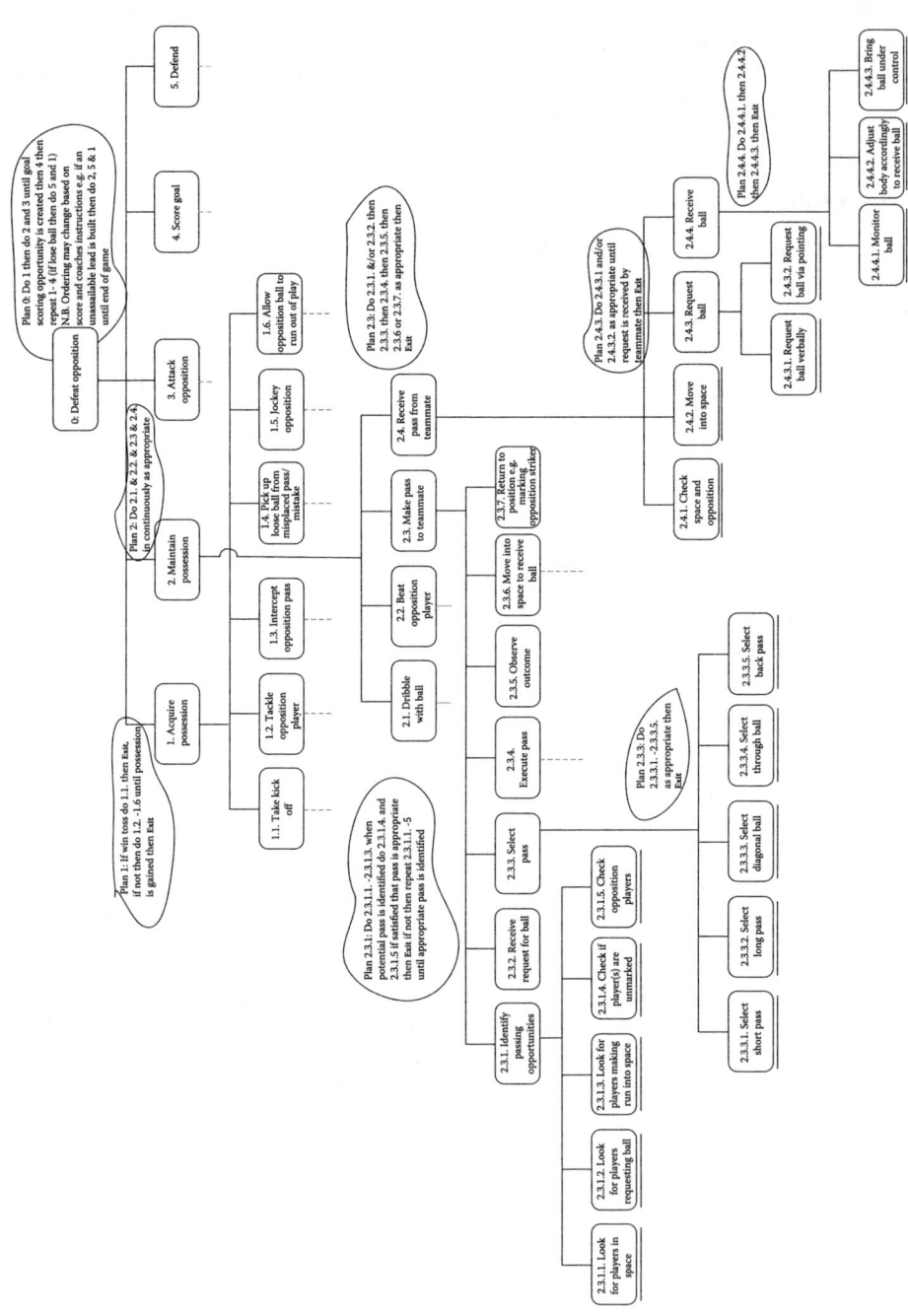

FIGURE 3.1 "Defeat opposition" soccer HTA extract with example decomposition.

Task Analysis Methods

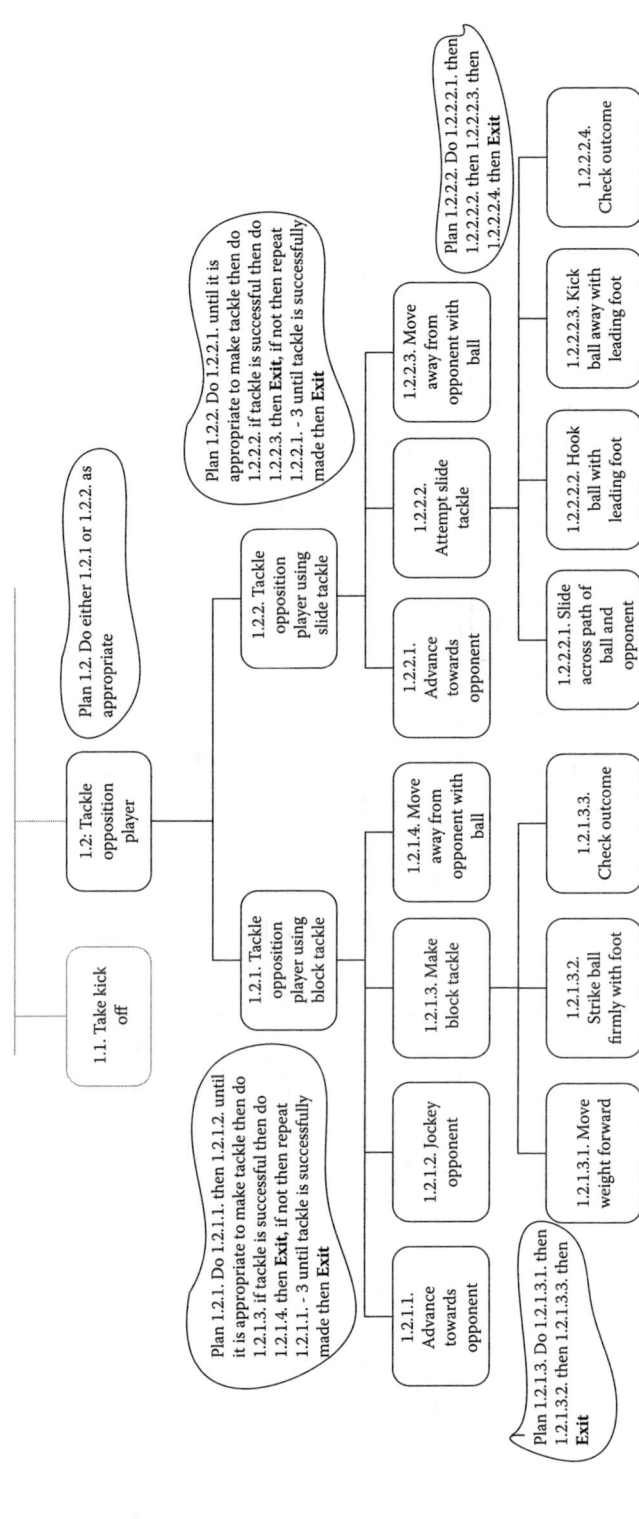

FIGURE 3.2 "Tackle opposition player" decomposition.

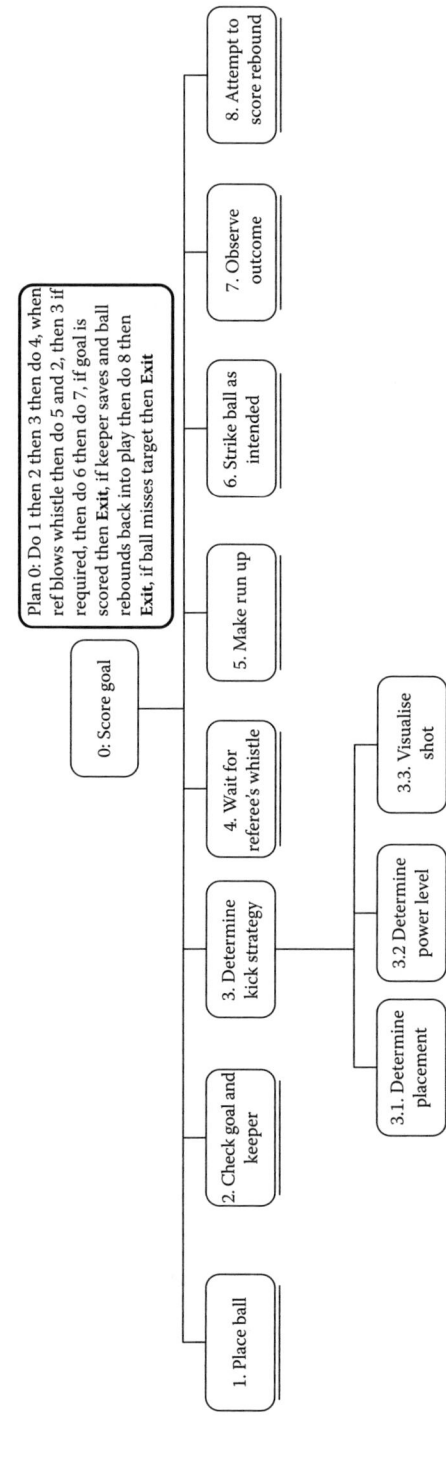

FIGURE 3.3 "Score goal from penalty kick" decomposition.

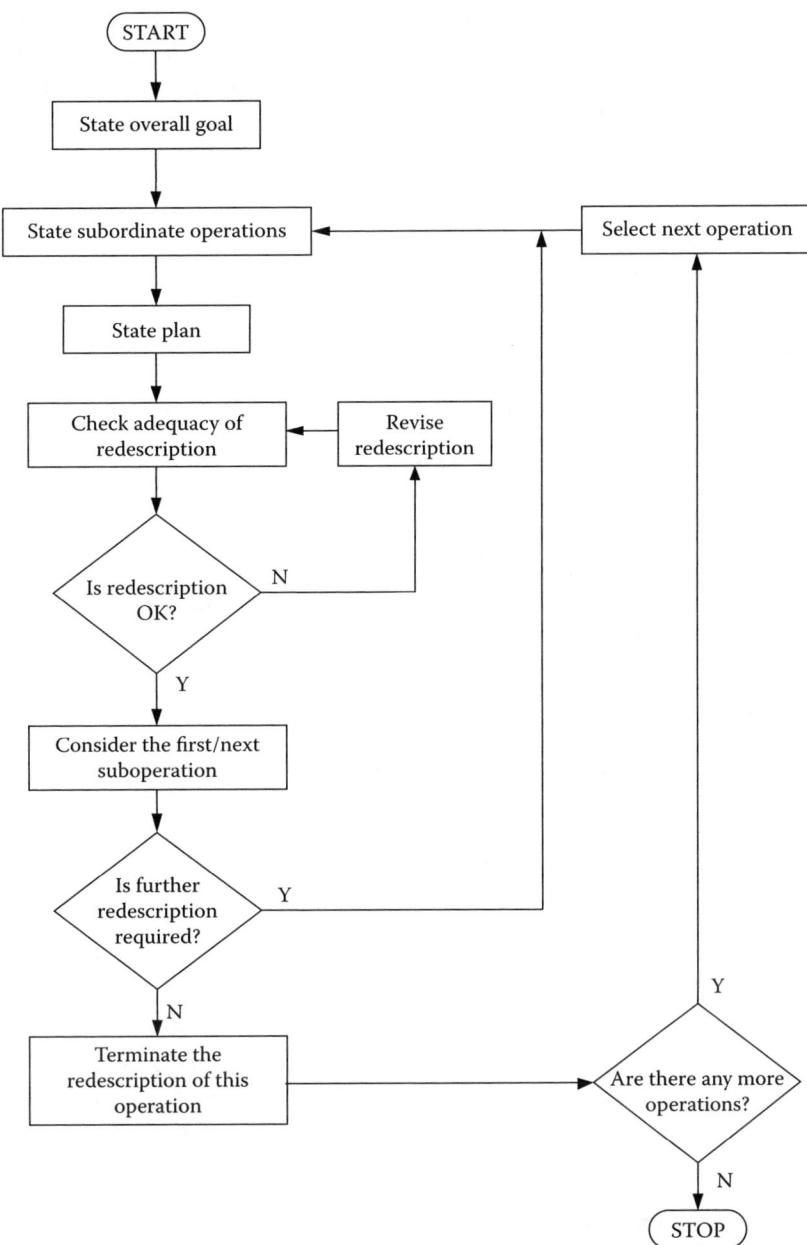

FLOWCHART 3.1 HTA flowchart.

TASK DECOMPOSITION

BACKGROUND AND APPLICATIONS

The task decomposition methodology (Kirwan and Ainsworth, 1992) is used to gather detailed information regarding a particular task, scenario, or system. When a lower level of analysis is required than that provided by HTA, the task decomposition method can be used to analyse, in detail, goals, tasks, and sub-tasks. It involves further decomposing each of the operations described in an HTA using a series of decomposition categories. Depending on the purpose of the analysis, tasks are decomposed to describe a variety of task-related features, including goals, physical and cognitive actions required, devices and interface components used, time taken, errors made, feedback, and decisions required. According to Kirwan and Ainsworth (1992), the method was first used by Miller (1953; cited in Kirwan and Ainsworth, 1992), who recommended that tasks be decomposed using the following decomposition categories:

- Task description
- Sub-task(s)
- Cues initiating action
- Controls used
- Decisions
- Typical errors
- Response
- Criterion of acceptable performance
- Feedback

Domain of Application

The task decomposition methodology is a generic approach that can be used in any domain. Applications presented in the literature include in the aviation domain (Stanton et al., 2005) and the military (Salmon, Jenkins, Stanton, and Walker, in press).

Application in Sport

While there are no published applications of the approach in a sporting context, it is clearly suited to the analyses of sports tasks, activities, and systems, particularly those that involve technology or devices of some sort. It is likely to be particularly useful for sports product and coaching intervention design and evaluation.

Procedure and Advice

Step 1: Clearly Define Aims of the Analysis

It is important to first clearly define what the overall aims of the task decomposition analysis are. Without clearly defined aims and objectives, the analysis can lack focus and inadequate or inappropriate data may be gathered. Clear specification of the aims and objectives ensures that the task decomposition analysis is well designed and that the appropriate task decomposition categories are used.

Step 2: Select Task Decomposition Categories

Once the aims of the analysis are clearly defined, the next step involves selecting a series of appropriate decomposition categories that will ensure that the analysis aims are met by the output produced. The selection of task decomposition categories is entirely dependent upon the analysis requirements. Kirwan and Ainsworth (1992) identify three main decomposition categories: descriptive, organisation specific, and modelling. A series of task decomposition categories are presented in Table 3.2.

Step 3: Construct HTA for the Task, Scenario, System, or Device under Analysis

Before decomposing tasks in detail, an initial task description is required. For this purpose, it is recommended that an HTA is constructed for the task, scenario, system, or device under analysis. This involves using the procedure described in the HTA section, including data collection, and construction and validation of the HTA.

TABLE 3.2
Task Decomposition Categories

Task Description	Critical Values	Sub-Tasks
Activity/behaviour type	Job aids required	Communications
Task/action verb	Actions required	Coordination requirements
Function/purpose	Decisions required	Concurrent tasks
Sequence of activity	Responses required	Task outputs
Requirements for undertaking task	Task complexity	Feedback
Initiating cue/event	Task criticality	Consequences
Information	Amount of attention required	Problems
Skills/training required	Performance on task	Likely/typical errors
Personnel requirements/manning	Time taken/time permitted	Errors made
Hardware features	Required speed	Error consequences
Location	Required accuracy	Adverse conditions
Controls used	Criterion of response adequacy	Hazards
Displays used	Other activities	

Source: Kirwan, B., and Ainsworth, L. K. (1992). *A guide to task analysis.* London, UK: Taylor & Francis.

Step 4: Create Task Decomposition Table

Once the HTA description has been completed and verified by appropriate SMEs, a task decomposition table should be constructed. This involves taking the task steps of interest from the HTA description (usually the bottom level operations) and entering them into a table containing a column for each task decomposition category. Each task step description should be included in the table, including the numbering from the HTA.

Step 5: Collect Data for Task Decomposition

Using the task decomposition table as a guide, the analyst should next work through each category, collecting appropriate data and performing the appropriate analyses for each one. This might involve observing the tasks in question, reviewing associated documentation such as standard operating procedures (SOPs), training and user manuals, or holding discussions with SMEs.

Step 6: Complete Task Decomposition Table

Once sufficient data are collected, the analyst should complete the task decomposition table. Each decomposition category should be completed for each task step under analysis. The level of detail involved is dependent upon the analysis requirements.

Step 7: Propose Redesigns/Remedial Measures/Training Interventions, Etc.

Normally, task decomposition analysis is used to propose redesigns, remedial measures, countermeasures, etc., for the system under analysis. The final step involves reviewing the decomposition analysis and providing redesign/remedial measure/countermeasure/training intervention suggestions for each of the problems identified.

ADVANTAGES

1. Task decomposition offers a structured approach for analysing, in detail, the task steps involved in a particular scenario or process or with a particular device or system.

2. The task decomposition approach is extremely flexible, allowing the analyst(s) to choose the decomposition categories used based upon the analysis requirements (analysts can even create their own decomposition categories).
3. Since the approach is generic it can be applied in any domain.
4. Task decomposition offers high levels of comprehensiveness.
5. The approach is easy to learn and apply.
6. It provides a much more detailed task description than most task analysis approaches.
7. Since the analyst(s) has control over the categories adopted, any aspect of the tasks in question can be analysed.
8. The approach normally produces redesign/remedial measure/countermeasure/training suggestions.

Disadvantages

1. Task decomposition can be highly time consuming in application.
2. The analysis often becomes repetitive and laborious, especially for large tasks.
3. The initial data collection phase may also be time consuming, especially for complex tasks and systems.
4. Without involving specialists (e.g., designers, sports scientists) in the analysis process the output may be more descriptive than anything else and may offer little in the way of guidance for improving task performance or device design.

Related Methods

The task decomposition approach relies heavily on various data collection methods, such as observation, interviews, and questionnaires. HTA is also typically used to provide the initial task/system description on which the decomposition analysis is based.

Approximate Training and Application Times

The time associated with training in the task decomposition approach alone is minimal; however, it is worth remembering that the analyst also requires experience in various other methods, such as interviews, observational study, and HTA. Due to its exhaustive nature, application times are typically high. Kirwan and Ainsworth (1992) point out that the method's main disadvantage is the large amount of time associated with collecting the data on which the task decomposition analysis is based.

Reliability and Validity

Although the analysis is based largely on analyst and SME subjective judgement, it is structured somewhat by the use of task decomposition categories. For this reason the reliability of the method, when the same decomposition categories are applied, is likely to be high. The validity of the approach, provided an accurate HTA underpins the analysis, is also likely to be high in most cases.

Tools Needed

Task decomposition analysis can be carried out using pen and paper only; however, the data collection phase often requires video and audio recording equipment. Since task decomposition outputs are often so large, a drawing package such as Microsoft Visio may also be used to construct them (since they may be too large for standard Microsoft Word table presentation).

Task Analysis Methods

EXAMPLE

A task decomposition analysis was undertaken for the task "Score goal from penalty kick," for which the HTA is presented in Figure 3.3. For the purposes of this example, the following decomposition categories were selected:

- Task description
- Activity/behaviour type
- Function
- Initiating event/cue
- Information
- Skills required
- Training required
- Decisions required
- Actions required
- Performance shaping factors
- Communications
- Typical errors

The task decomposition was undertaken by one Human Factors analyst and one soccer SME. The output is presented in Table 3.3.

FLOWCHART

(See Flowchart 3.2.)

RECOMMENDED TEXT

Kirwan, B., and Ainsworth, L. K. (1992). *A guide to task analysis*. London, UK: Taylor & Francis.

VERBAL PROTOCOL ANALYSIS

BACKGROUND AND APPLICATIONS

Verbal Protocol Analysis (VPA), also commonly referred to as "think aloud" protocol analysis, is used to describe tasks from the point of view of the individual performing them. It is used to elicit verbalisations of the processes, both cognitive and physical, that an individual undertakes in order to perform the task under analysis. Noyes (2006) points out that VPA is as close as we can get to observing people's thoughts. Conducting VPA involves creating a written transcript of operator behaviour as he or she thinks aloud while undertaking the task under analysis. The verbal transcript obtained is then coded and analysed in order to generate insight regarding the processes used during task performance. VPA is commonly used to investigate the cognitive processes associated with complex task performance and has been used to date in various domains, including the military (Salmon et al., "Distributed awareness," 2009), process control (Bainbridge, 1972; Vicente, 1999), the Internet (Hess, 1999), road transport (Walker, Stanton, and Young, 2001), and sport (McPherson and Kernodle, 2007).

DOMAIN OF APPLICATION

VPA is a generic approach that can be applied in any domain.

TABLE 3.3
Score Goal from Penalty Kick Task Decomposition Output

Task Step	Task Description	Activity/ Behaviour Type	Function	Initiating Event/Cue	Information	Skills Required	Decisions Required	Actions Required	Training Required	Performance Shaping Factors	Typical Errors
1. Place ball	Player places the ball on the penalty spot prior to taking penalty	Cognitive selection of optimum ball placement Physical placement of ball	To place the ball in an optimum position which maximises efficiency of penalty kick	Award of penalty by referee	Whistle from referee Penalty spot Personal preference on placement Ball position	Physical placement of ball	What is the optimum placement of the ball for this penalty kick? What is my preference regarding ball placement?	Retrieve ball Determine optimum ball placement Place ball Check ball placement	Practice penalty kicks with different ball placements, e.g., valve up	Ground (surface condition) Weather Pitch makings Ball Crowd Pressure	Inappropriate placement of ball Failure to consider ball placement
2. Check goal and goalkeeper	Player checks the goal and goalkeeper in an attempt to read the goal keeper and to make initial selection of where to put penalty	Visual check of goal and goalkeeper's positioning and actions	To gather information regarding the goalkeeper in an attempt to read which way he will dive	Placement of ball	Goalkeeper position Goalkeeper stance Goalkeeper expressions Goal Ball	Ability to visually check the goal and goalkeeper Ability to read goalkeeper	Which way is the goalkeeper likely to dive? Where should I place my kick?	Visual check of goal and goalkeeper	Penalty kick training Training in reading goalkeepers Revision of goalkeepers' typical penalty responses	Goalkeeper Crowd Pressure	Misread goalkeeper's intentions

Task Analysis Methods

Step	Sub-task	Description	Goal	Cues/Information	Input	Ability Required	Decision	Action	Training	Influencing Factors	Errors
3. Select kick strategy	Player makes selection regarding placement and power of penalty	Cognitive decision-making activity of selecting penalty kick placement and power	To select optimum penalty kick strategy with regard to the goalkeeper faced and conditions	Placement of ball or whistle from referee to take penalty	Goalkeeper position; Goalkeeper stance; Goalkeeper expressions; Goal; Ball; Performance in training; Previous performance; Personal preference; Knowledge of this goalkeeper's previous performance; Instructions from teammates/manager/crowd; Conditions (pitch, weather, etc.)	Ability to select optimum penalty strategy; Abilty to block out pressure and maintain optimum strategy selection; Ability to read goalkeeper; Ability to read conditions and comprehend impact on penalty kick	What is the optimum penalty kick strategy (i.e., power and placement) for this goalkeeper and these conditions?	Check the goal and goalkeeper; Make decision regarding optimum penalty kick strategy	Penalty kick training; Training in reading goalkeepers; Revision of goalkeeper's typical penalty responses	Goalkeeper; Score; Teammates/manager; Crowd; Pressure; Own performance in game/over season; Mental state; Goalkeeper performance in game; Criticality of penalty; Overconfidence	Selection of inappropriate penalty kick strategy, e.g., dink. Change selection due to pressure
4. Visualise shot	Player visualises penalty kick	Cognitive simulation of how penalty will unfold	To rehearse chosen penalty kick strategy and generate view of positive outcome	Selection of penalty kick strategy	Simulation of penalty kick	Ability to mentally simulate chosen penalty kick strategy	N/A	Mental simulation/rehearsal of penalty kick	Mental simulation/rehearsal of penalty kicks	Pressure; Referee whistle; Crowd	Failure to perform mental rehearsal

TABLE 3.3 (CONTINUED)
Score Goal from Penalty Kick Task Decomposition Output

Task Step	Task Description	Activity/Behaviour Type	Function	Initiating Event/Cue	Information	Skills Required	Decisions Required	Actions Required	Training Required	Performance Shaping Factors	Typical Errors
5. Make run up	Player makes run up toward ball prior to striking ball	Physical act of running towards ball	To move towards ball and position oneself optimally to strike the ball	Whistle from referee indicating that penalty is to be taken	Whistle Position of ball Chosen penalty strategy	Ability to make run up towards ball and position body optimally for ball strike	How long should the run up be? How quick should the run up be? Should a feign be used?	Physical movement towards ball Look in direction of intended/feigned strike	Training in the use of run and signal to deceive goalkeeper	Weather Pitch Pressure Crowd Current score Criticality of penalty	Mis-time run up Position body inappropriately for striking ball optimally
6. Strike ball	Player strikes ball towards goal based on strategy selected in step 3	Physical act of striking the ball into the goal	To score goal	Whistle from referee indicating that penalty is to be taken	Whistle Ball Chosen penalty strategy Goalkeeper Condition	Ability to strike dead ball Ability to deceive goalkeeper	Do I stick with selected strategy? How hard should I strike the ball? Which way is the goalkeeper likely to dive?	Strike ball based on chosen penalty kick strategy	Penalty kick training Shooting training	Goalkeeper Score Teammates/manager Crowd Pressure Own performance in game/over season Physical state Mental state Goalkeeper performance in game Criticality of penalty Overconfidence Footwear Condition (e.g pitch, weather) Placement of ball	Mis-hit ball Miss target

Task Analysis Methods

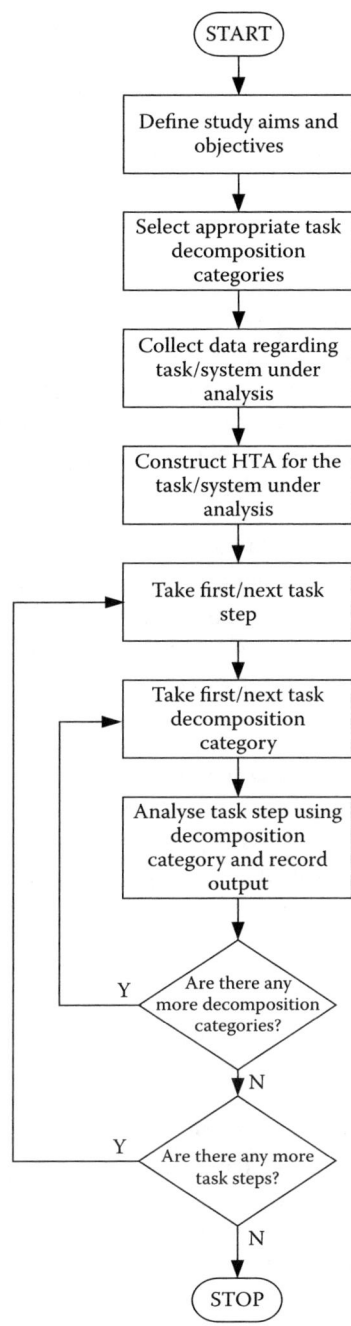

FLOWCHART 3.2 Task decomposition flowchart.

Application in Sport

VPA offers a simplistic means of gaining insight into the processes used during task performance and is clearly suited to the analysis of sports performance; in particular, it appears to be suited to identifying and analysing the differences between the processes and strategies adopted by sports performers of differing abilities and expertise levels (e.g., elite versus amateur sports performers). For these reasons, the VPA method has already received attention from sports scientists.

McPherson and Kernodle (2007), for example, used verbal reports in between points to investigate the differences between problem representation by tennis players of differing levels of expertise during singles games. Nicholls and Polman (2008) used think aloud trials to measure stress and the coping strategies used by high performance adolescent golfers during play. James and Patrick (2004) also report the use of verbal reports to identify the cues used by expert sports performers.

Procedure and Advice

Step 1: Clearly Define Aims of the Analysis

It is important to first clearly define what the overall aims of the VPA are. Without clearly defined aims and objectives, the analysis can lack focus and inadequate or inappropriate data may be gathered. Clear specification of the aims and objectives ensures that the VPA is well designed, that appropriate participants are recruited, and that the appropriate data are collected and analysed.

Step 2: Define Task/Scenario under Analysis

Next, the task or scenario under analysis should be clearly defined and described in accordance with the analysis requirements. It is important that the analysts involved have a clear and detailed understanding of the task or scenario under analysis. For this purpose it is recommended that an HTA of the task or scenario under analysis be constructed.

Step 3: Brief Participant(s)

Participants should be briefed regarding the VPA method, the analysis, and what is required of them during the data collection phase. What they should report verbally should be clarified at this stage and it is important that participants are encouraged to continue verbalising even when they feel that it may not make much sense or that what they are saying may not be of use to the analysts. It is useful at this stage to give the participant(s) a short demonstration of the VPA data collection phase.

Step 4: Conduct Pilot Run

Next, a pilot run of the data collection procedure should be undertaken. This process is important, as it allows any problems with the data collection procedure to be resolved, and gives the participant a chance to ask any questions. A simple small-scale task usually suffices for the pilot run.

Step 5: Undertake Scenario and Record Data

Once the participant fully understands the data collection procedure, and what is required of them as a participant, the VPA data collection phase can begin. This involves getting the participant to perform the task under analysis and instructing them to "think aloud" as they perform it. All verbalisations made by the participant should be recorded using an audio recording device. Analysts should also make pertinent notes throughout. It is also recommended that, if possible, a video recording be made of the task being performed. This allows analysts to compare physical aspects of task performance with the verbalisations made by participants.

Step 6: Transcribe Data

Once the trial is finished, the data collected should be transcribed into a written form. Microsoft Excel or Word is normally used for this purpose.

Step 7: Encode Verbalisations

Next, the verbal transcript needs to be categorised or coded. Depending upon the analysis requirements, the data are coded into pertinent categories. Typically, the data are coded into one of the following five categories: words, word senses, phrases, sentences, or themes. The encoding scheme chosen should then be encoded according to a rationale determined by the aims of the analysis. Walker (2004) recommends that this involve grounding the encoding scheme according to some established concept or theory, such as mental workload or situation awareness. Written instructions should also be developed for the encoding scheme and should be strictly adhered to and referred to constantly throughout the encoding process. Once the encoding type, framework, and instructions are completed, the analyst should proceed to encode the data. Various software packages are available to aid the analyst in this process, such as General Enquirer.

Step 8: Devise Other Data Columns

Once the encoding is complete, any other pertinent data columns should be devised. This allows the analyst to note any mitigating circumstances that may have affected the verbal transcript data obtained.

Step 9: Establish Inter- and Intra-Rater Reliability

It is important that the reliability of the encoding scheme is established (Walker, 2004). Reliability is typically established through reproducibility, i.e., independent raters need to encode previous analyses.

Step 10: Perform Pilot Study

The protocol analysis procedure should now be tested. This is normally done via the conduct of a small pilot study, the output of which can be used to determine whether the verbal data collected are useful, whether the encoding system works, and whether inter- and intra-rater reliability levels are satisfactory. Any problems identified at this stage should be refined before any further analysis of the data is undertaken.

Step 11: Analyse Structure of Encoding

Finally, the data are ready to be analysed. For this purpose, the responses given in each category should be summed by adding the frequency of occurrence noted in each category.

ADVANTAGES

1. VPA provides a rich data source.
2. The data collection phase is quick and typically does not take significantly longer than the task under analysis itself.
3. VPA is particularly effective when used to analyse sequences of activities.
4. All manner of Human Factors concepts can be analysed, including situation awareness, cognition, decision making, teamwork, distraction, human error, and workload.
5. VPA can be used to compare the processes adopted by sports performers of differing expertise and skill levels (e.g., elite versus amateur performers).
6. When done properly, the verbalisations generated provide a genuine insight into the cognitive and physical processes involved in task performance.
7. VPA has been used extensively in a wide range of domains, including sports domains such as tennis and golf.
8. VPA is simple to conduct provided the appropriate equipment is available.
9. VPA data can be used as input data for other Human Factors methods, such as HTA, propositional networks, and cognitive task analysis methods.

DISADVANTAGES

1. Even for simplistic tasks VPA is a highly time-consuming method to apply.
2. The data analysis (encoding) phase is often extremely laborious and time consuming.

3. The difficulties associated with verbalising cognitive behaviour may raise questions over the data collected. Militello and Hutton (2000), for example, point out that researchers have been cautioned about relying on verbal protocol data. Further, Noyes (2006) reports that some mental processes cannot be verbalised accurately.
4. Participants may find it difficult to verbalise certain aspects of task performance.
5. Providing concurrent verbal commentary can sometimes affect task performance.
6. More complex tasks that incur high demand on participants often lead to a reduced quantity of verbalizations.
7. The strict procedure involved is often not adhered to.
8. VPA is often not appropriate during "real-world" task performance, and so it may be difficult to apply during professional sports games.
9. Verbalisations can often be prone to bias on behalf of the participants making them, i.e., participants merely verbalising what they think the analysts want to hear (Bainbridge, 1995; Noyes, 2006).

RELATED METHODS

Most of the time, VPA involves some form of direct observation of the task or scenario under analysis. The data collected during VPA is also often used to inform the conduct of other Human Factors analyses, such as task analysis (e.g. HTA) and situation awareness modelling (e.g. propositional networks). Content analysis is also often undertaken on VPA data for data analysis purposes.

APPROXIMATE TRAINING AND APPLICATION TIMES

The method itself is easy to train and typically incurs only a minimal training time. Applying the VPA method, however, is highly time consuming, with the data analysis phase in particular being hugely time consuming. Walker (2004) points out that, if transcribed and encoded by hand, 20 minutes of verbal transcript data, transcribed at around 130 words per minute, would take between 6 to 8 hours to transcribe and encode.

RELIABILITY AND VALIDITY

According to Walker (2004), the reliability of the VPA method is reassuringly good. For example, Walker, Stanton, and Young (2001) used two independent raters and established inter-rater reliability at Rho = 0.9 for rater 1 and Rho = 0.7 for rater 2. Intra-rater reliability during the same study was also found to be high, being in the region of Rho = 0.95.

TOOLS NEEDED

Collecting the data for VPA requires pen and paper, and audio and video recording devices. For the data analysis phase, Microsoft Excel is normally used. A number of software packages can also be used to support the VPA process, including Observer Pro, General Enquirer, TextQuest, and Wordstation.

EXAMPLE

VPA was used to collect data regarding the cognitive processes involved during running. The example presented is taken from one participant who completed a 7-mile running course (the runner was familiar with the course having ran it previously on numerous occasions during event training). The runner took a portable audio recording device (Philips Voicetracer) on the run and was asked to verbalise all elements of the task. Once the run was completed and the data transcribed, the runner

Task Analysis Methods

and one analyst went through the transcript in order to identify the different events that occurred throughout the run. Four encoding groups were defined: behaviour, cognitive processes, physical condition, and route. The behaviour group consisted of verbalisations relating to the runner's *own behaviour*, *performance*, and the *running environment*. The cognitive processes group consisted of *perception*, *comprehension*, and *projection*. The physical condition group included *condition*, *pain*, and *breathing rate*, and finally the route group was used to cover verbalisations relating to the course *route*. A compressed (i.e., non-salient verbalisations removed) extract from the VPA is presented in Figure 3.4.

Flowchart

(See Flowchart 3.3.)

Recommended Texts

Bainbridge, L. (1995). Verbal protocol analysis. In *Evaluation of human work: A practical ergonomics methodology*, eds. J. R. Wilson and E. N. Corlett, 161–79). London: Taylor & Francis.

Noyes, J. M. (2006). Verbal protocol analysis. In *International encyclopaedia of ergonomics and Human Factors*, 2nd ed., ed. W. Karwowski, 3390–92. London: Taylor & Francis.

Walker, G. H. (2004). Verbal protocol analysis. In *Handbook of Human Factors and ergonomics methods*, eds. N. A. Stanton, A. Hedge, K. Brookhuis, E. Salas, and H. Hendrick, 30.1–30.8). Boca Raton, FL: CRC Press.

OPERATION SEQUENCE DIAGRAMS

Operation Sequence Diagrams (OSDs) are used to graphically describe the activity and interactions between team members (and technology, if required) during collaborative activities. The output of an OSD graphically depicts the activity, including the tasks performed and the interaction between operators over time, using standardised symbols. There are various forms of OSDs, ranging from a simple flow diagram representing task order, to more complex OSDs that account for team interaction and communication. OSDs have recently been used by the authors for the analysis of command and control activities in a number of domains, including the military, emergency services, naval warfare, aviation, energy distribution, air traffic control, and rail domains.

Domain of Application

The OSD method was originally used in the nuclear power and chemical process industries (Kirwan and Ainsworth, 1992); however, the method is generic and can be applied in any domain.

Application in Sport

OSDs can be used to depict sporting activity and are particularly useful for representing complex team tasks, such as those seen in team sports such as rugby league and union, soccer, and American football.

Procedure and Advice

Step 1: Clearly Define Aims of the Analysis

It is important to first clearly define what the overall aims of the analysis are. Without clearly defined aims and objectives, the analysis can lack focus and inadequate or inappropriate data may be gathered. Clear specification of the aims and objectives ensures that appropriate data are collected and analysed.

				Encoding									
				Behavioural		Cognitive				Physical		Rou	Events
Time		Verbalisations		OB	PERF	ENV	PERC	COM	PRO	CONC	PAIN	BR	Rou
Mins	Secs												
1	44	Im just checking my pace as im trying to get to as quick a starting pace as possible		1	1		1	1	1				Start
3	44	Just feeling for any niggles or pain that Im getting...left knee is a bit sore		1			1		1		1	1	
4	36	Im just checking the road ahead for cars as I know that I need to cross at the intersection		1		1	1		1				1 Crossing road at inte
6	14	I am having to move over onto the grass to avoid three pedestrians on the path		1		1	1		1				1 Runner has to move
6	20	Im now checking distance and time to see how quick my first mile was		1			1		1				Completed 1st mile
7	28	Im making a conscious effort to up my pace and lengthen my stride length as quite a big incline is coming up		1	1	1	1		1				1 Approaching steep in
7	52	Im checking my pace as I go up the hill			1	1	1		1				1 Running up steep inc
12	14	Im aware that im struggling a bit and my paced has slowed...im making an effort to increase pace and stride length			1		1		1				
13	17	My mind has wandered and im thinking about work...its because of low workload...just checked my pace and ive slowe		1	1		1		1				
14	19	Im checking the traffic lights at the upcoming intersection as I have to cross here		1		1	1		1				1 Approaching cross ro
15	22	Im increasing my pace and stride length as we are on a long flat section			1	1	1		1				1 Long flat, fast section
16	22	Im trying to make myself focus on stride length and pace as this section is a bit easier and I know that my mind often		1	1		1		1				
17	24	Just noticing a bit of pain in my left knee at the moment		1			1				1	1	
18	26	Im just checking how far ive ran so far on my watch...also check my pace when I glance		1	1		1		1				
19	28	Ive just turned into a strong headwind and im trying to maintain some sort of pace but its very difficult		1	1	1	1		1				1 Strong headwind
20	30	Again im trying to increase my pace as I know the next mile or so is flat and fast		1	1	1	1		1				1 Approaching long fla
21	31	Im wondering how long I can keep up at my current fast which is quicker than normal		1	1		1		1				1 Long flat, fast section
22	48	Just checking distance ran to see how far ive ran and how far is remaining		1			1		1				
23	52	7 Im checking time and working out how long is left...im now deciding when to make a sprint for home		1	1		1		1				1 Approaching sprint fi
24	54	25 Im conscious that my breathing rate has increased significantly		1			1		1			1	Sprint finish

FIGURE 3.4 Runner VPA extract.

Task Analysis Methods

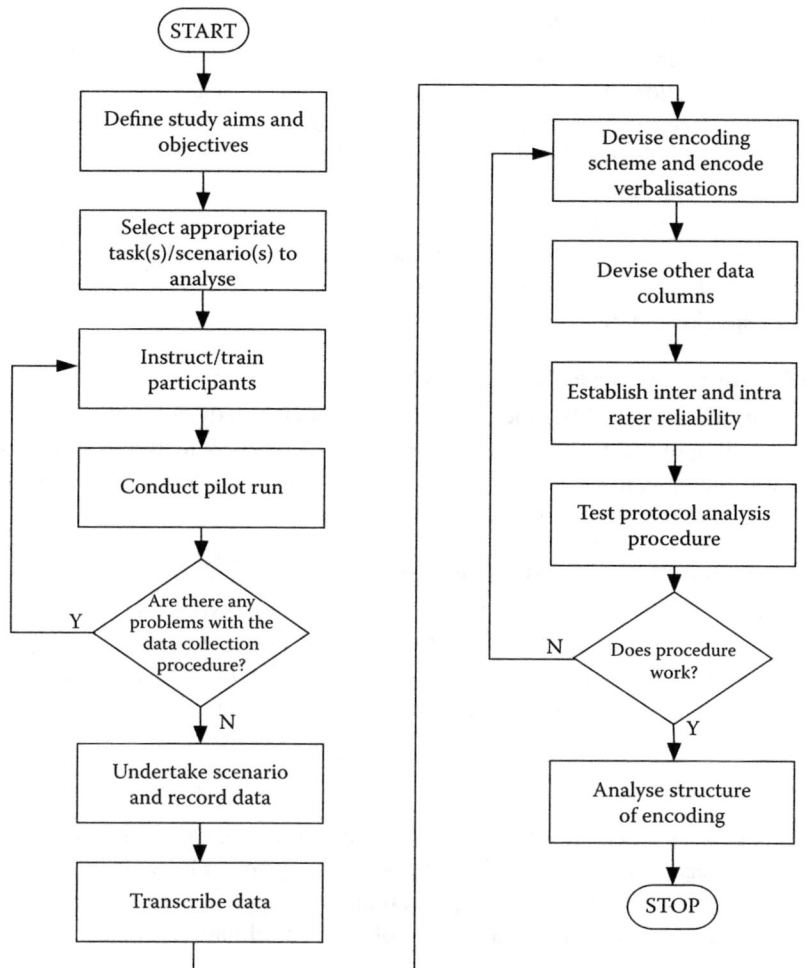

FLOWCHART 3.3 Verbal Protocol Analysis flowchart.

Step 2: Define Task/Scenario under Analysis

Next, the task or scenario under analysis should be clearly defined and described in accordance with the analysis requirements. It is important that the analysts involved have a clear and detailed understanding of the task or scenario under analysis.

Step 3: Collect Data via Observational Study

The next step involves collecting data regarding the task under analysis via observational study. An observational transcript should be created that includes a description of the activities performed and who they are performed by, the communications between team members, and the technology used. A timeline of the task or scenario should also be recorded.

Step 4: Describe the Task or Scenario Using HTA

OSD diagrams are normally built using task analysis outputs. Once the data collection phase is completed, a detailed task analysis should be conducted for the scenario under analysis. The type of task analysis used is determined by the analyst(s), and in some cases, a task list will suffice. However, it is recommended that an HTA be conducted for the task under analysis.

Step 5: Construct the OSD Diagram

Once the task has been described adequately, the construction of the OSD can begin. The process begins with the construction of an OSD template. The template should include the title of the task or scenario under analysis, a timeline, and a row for each agent involved in the task. In order to construct the OSD, it is recommended that the analyst walk through the HTA of the task under analysis, creating the OSD in conjunction. OSD symbols that we have used in the past are presented in Figure 3.5. The symbols involved in a particular task step should be linked by directional arrows in order to represent the flow of activity during the scenario. Each symbol in the OSD should contain the corresponding task step number from the HTA of the scenario. The artefacts used during the communications should also be annotated onto the OSD.

Step 6: Add Additional Analyses Results to OSD

One of the endearing features of the method is that additional analysis results can easily be added to the OSD. According to the analysis requirements, additional task features can also be annotated onto the OSD. For example, in the past we have added coordination demand analysis results to each task step represented in the OSD (e.g., Walker et al., 2006).

Step 7: Calculate Operational Loading Figures

Operational loading figures are normally calculated for each actor involved in the scenario under analysis. Operational loading figures are calculated for each OSD operator or symbol used, e.g., operation, receive, delay, decision, transport, and combined operations. The operational loading figures refer to the frequency in which each agent was involved in the operator in question during the scenario.

ADVANTAGES

1. The OSD provides an exhaustive analysis of the task in question. The flow of the task is represented in terms of activity and information; the type of activity and the actors involved are specified; a timeline of the activity, the communications between actors involved in the task, the technology used, and also a rating of total coordination for each teamwork activity is also provided.
2. The OSD method is particularly useful for analysing and representing distributed teamwork or collaborated activity.
3. OSDs are useful for demonstrating the relationship between tasks, technology, and team members.
4. The method is generic and can be applied to the analysis of sports team performance.
5. The method has high face validity (Kirwan and Ainsworth, 1992).
6. OSDs have been used extensively and have been applied in a variety of domains.
7. A number of different analyses can be overlaid onto the OSD.
8. The OSD method is very flexible and can be modified to suit the analysis needs.
9. The WESTT software package (Houghton, Baber, Cowton, Stanton, and Walker, 2008) can be used to automate a large portion of the OSD procedure.
10. Despite its exhaustive nature, the OSD method requires only minimal training.

DISADVANTAGES

1. The OSD method can be hugely time consuming to apply, particularly for large, complex tasks involving multiple team members.
2. Without any supporting software, the construction of OSDs is overly time consuming and laborious.
3. OSDs can become cluttered and confusing (Kirwan and Ainsworth, 1992).

Task Analysis Methods

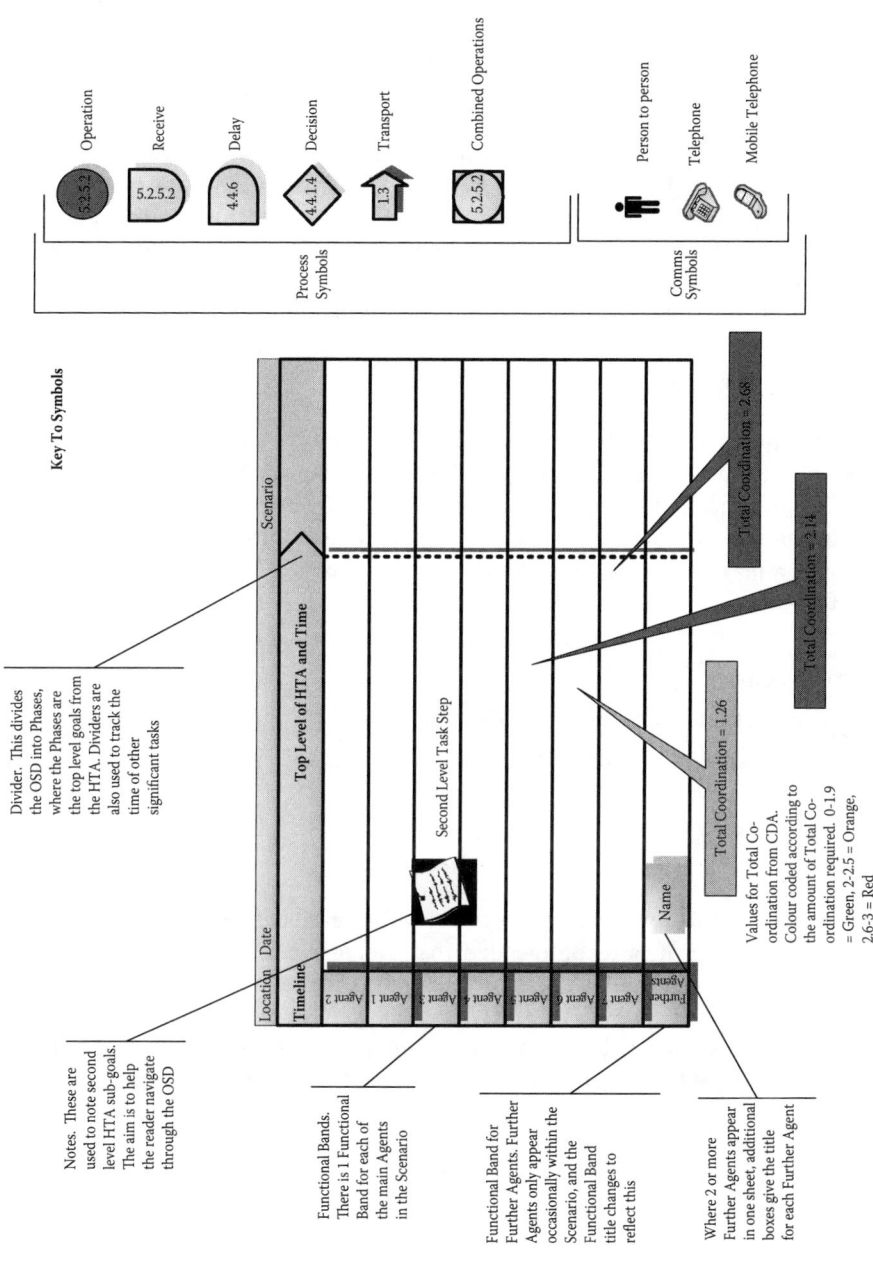

FIGURE 3.5 Standard OSD template.

4. The output of OSDs can become large and unwieldy, and it can be difficult to present outputs in full in reports, presentations, and scientific journal articles.
5. The reliability of the method is questionable. Different analysts may interpret the OSD symbols differently.

RELATED METHODS

Various types of OSDs exist, including temporal operational sequence diagrams, partitioned operational sequence diagrams, decision action diagrams, and spatial operational sequence diagrams (Kirwan and Ainsworth, 1992). During the OSD data collection phase, various data collection procedures, such as observational study and interviews, are typically employed. Task analysis methods such as HTA are also used to provide the input for the OSD. Timeline analysis may also be used in order to construct an appropriate timeline for the task or scenario under analysis. Additional analyses results can also be annotated onto an OSD, such as coordination analysis and communications analysis outputs. The OSD method is also used as part of the Event Analysis of Systemic Teamwork framework (Stanton et al., 2005) for analysing collaborative activities.

APPROXIMATE TRAINING AND APPLICATION TIMES

The OSD method requires only minimal training; however, analysts should be proficient in observational study and task analysis methods, such as HTA, before they apply the approach. The application time for OSDs is typically high. The construction of the OSD in particular can be highly time consuming, although this is dependent upon the complexity and duration of the task under analysis. From our own experiences with the method, constructing OSD diagrams can take anywhere between 1 and 5 days.

RELIABILITY AND VALIDITY

According to Kirwan and Ainsworth (1992), OSD methods possess a high degree of face validity. The intra-analyst reliability of the method may be suspect, as different analysts may interpret the OSD symbols differently.

TOOLS NEEDED

When conducting an OSD analysis, pen and paper can be sufficient. However, to ensure that data collection is comprehensive, it is recommended that video or audio recording devices are also used. For the construction of the OSD, it is recommended that a suitable drawing package, such as Microsoft Visio, is used. The WESTT software package (Houghton, Baber, Cowton, Stanton, and Walker, 2008) can also be used to automate a large portion of the OSD construction process (WESTT constructs the OSD based upon an input of the HTA for the scenario under analysis).

EXAMPLE

An OSD was developed based on an HTA description of a rugby union scrum. A scrum or "scrummage" is used in rugby union to restart play when the ball has either been knocked on, gone forward, has not emerged from a ruck or maul, or when there has been an accidental offside (BBC sport, 2008). Nine players from either side are involved in the scrum, including the hooker, loosehead prop, tight-head prop, two second rows, blind-side flanker, open-side flanker, and number 8, who form the scrum, and the scrum half, who feeds the ball into the scrum and/or retrieves the ball from the scrum. For the task definition part of the analysis, an HTA was conducted. The HTA is presented in Figure 3.6. The OSD for the scrum task is presented in Figure 3.7.

Task Analysis Methods

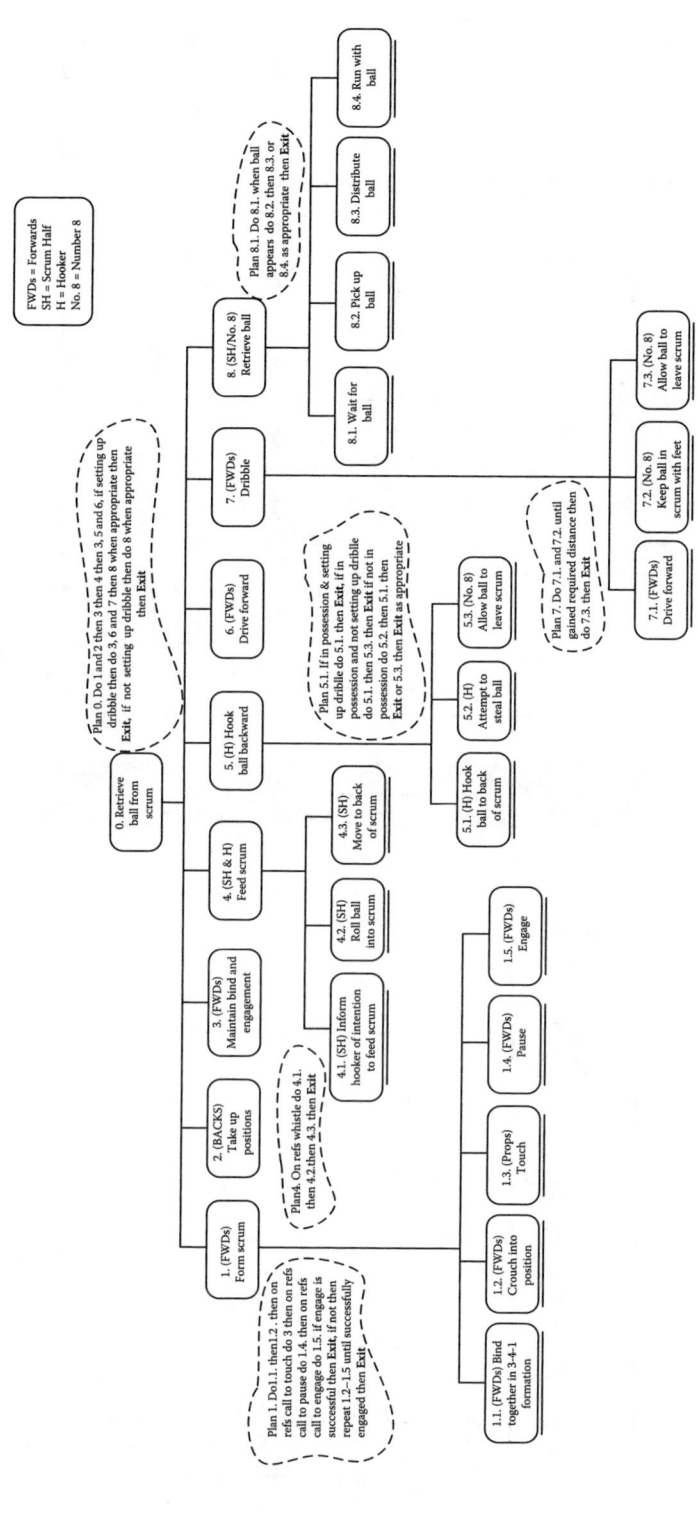

FIGURE 3.6 Scrum HTA description.

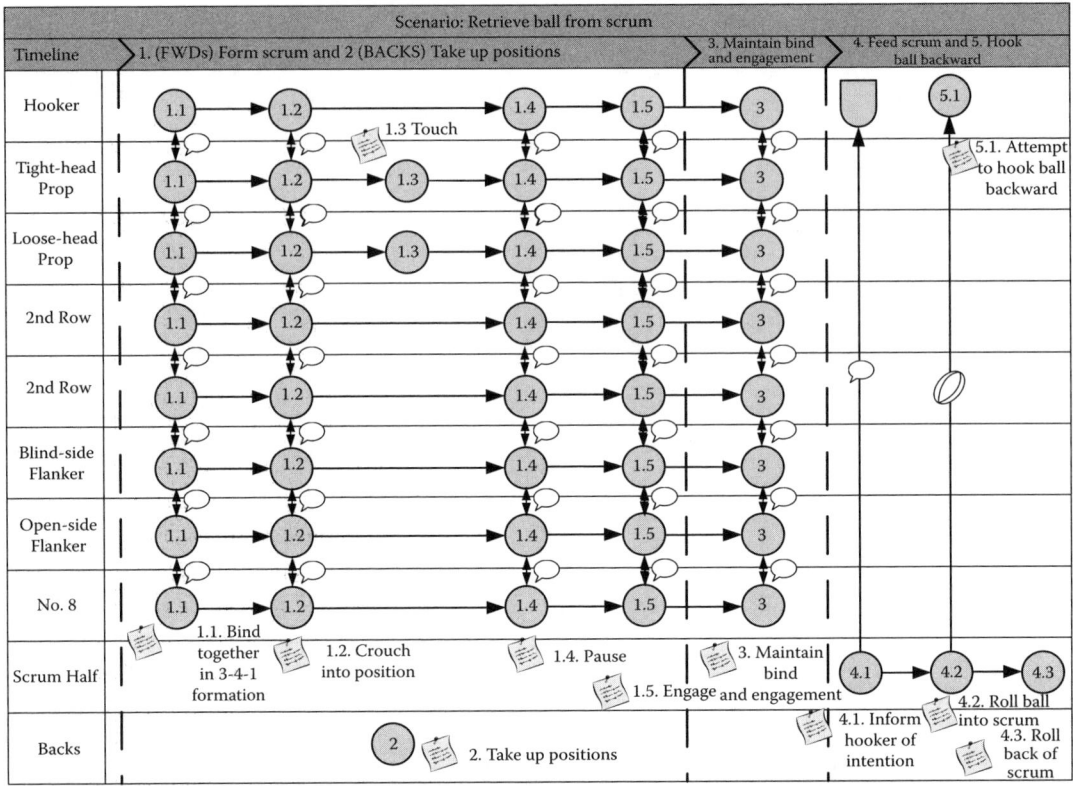
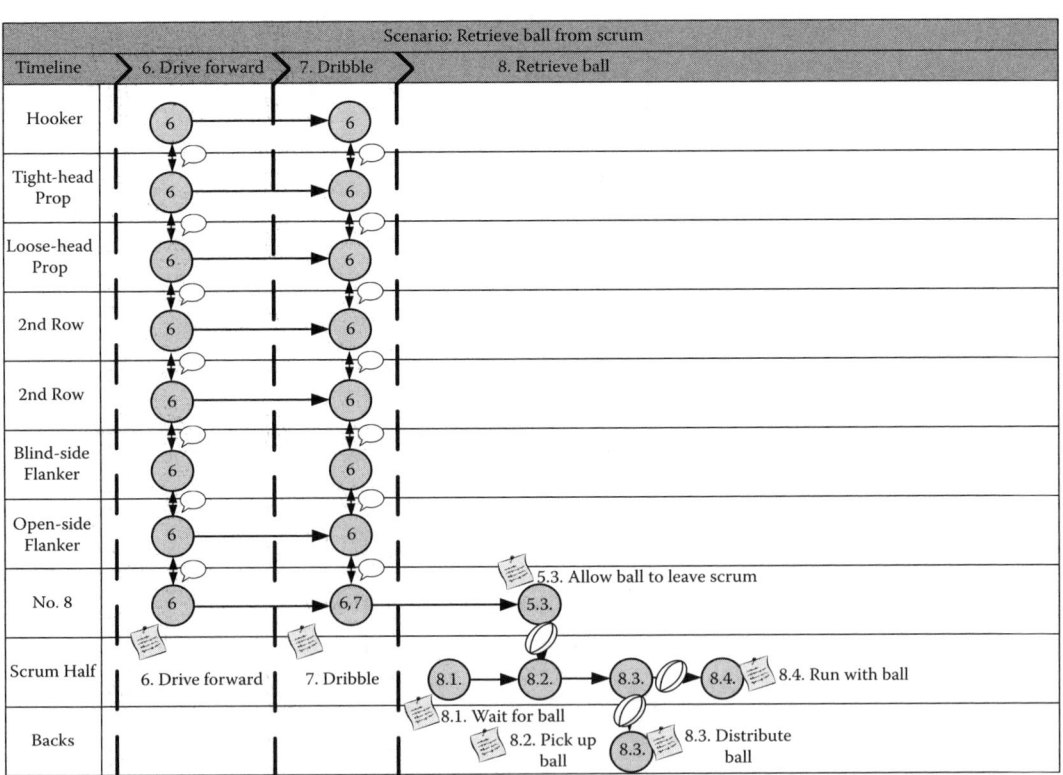

FIGURE 3.7 Scrum OSD diagram.

Flowchart

(See Flowchart 3.4.)

Recommended Texts

Kirwan, B., and Ainsworth, L. K. (1992). *A guide to task analysis*. London, UK: Taylor & Francis.

Stanton, N. A., Salmon, P. M., Walker, G., Baber, C., and Jenkins, D. P. (2005). *Human Factors methods: A practical guide for engineering and design*. Aldershot, UK: Ashgate.

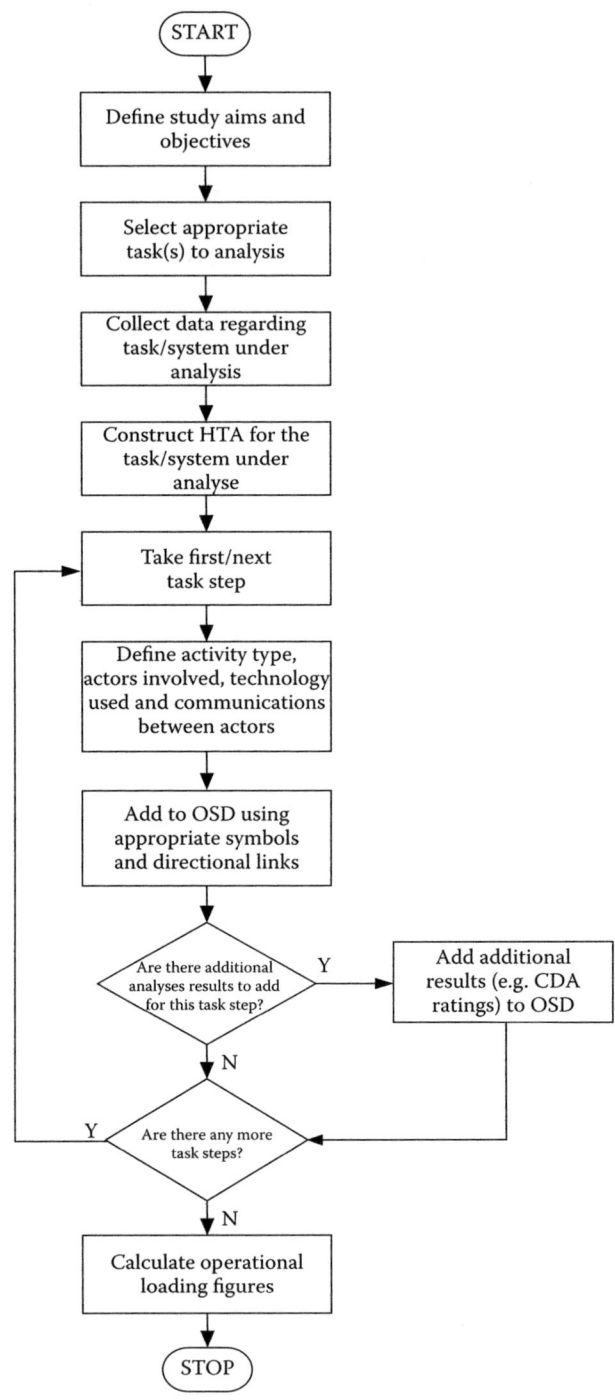

FLOWCHART 3.4 Operation sequence diagram flowchart.

4 Cognitive Task Analysis

INTRODUCTION

Cognitive Task Analysis (CTA) methods focus exclusively on the cognitive processes and skills required during task performance. CTA methods are used for "identifying the cognitive skills, or mental demands, needed to perform a task proficiently" (Militello and Hutton, 2000, p. 90), and to identify and describe the knowledge, thought processes, and goal structures underlying task performance (Schraagen, Chipman, and Shalin, 2000). As tasks have become more cognitive in nature, and as the cognitive demands imposed on operators have increased (due to increases in the use of technology and complex procedures), the need for system, procedure, and training programme designers to understand the cognitive aspects of task performance has become paramount. Accordingly, CTA methods provide a structured means for describing and representing the cognitive elements that underlie goal generation, decision making, and judgements (Militello and Hutton, 2000). The outputs derived from CTA efforts are used for a variety of different purposes including, amongst other things, to inform the design of new technology, systems, procedures and processes; for the development of training procedures and interventions; for allocation of functions analysis; and for the evaluation of individual and team performance within complex sociotechnical systems.

Flanagan (1954) first probed the cognitive processes associated with the decisions and actions made by pilots during near-miss incidents using the Critical Incident Technique (CIT; Flanagan, 1954). The foundations for cognitive task analysis methods were laid, however, in the 1960s when, in response to the need to better understand tasks that were becoming more complex and cognitive in nature, the HTA method was developed (Annett and Stanton, 2000). HTA was unique at the time in that, in addition to the physical tasks being performed, it also attempted to describe the cognitive elements of task performance. The actual term "Cognitive Task Analysis," however, did not appear until the early 1980s when it began to be used in academic research texts. Hollnagel (2003) points out that the term was first used in 1981 to describe approaches used to understand the cognitive activities required in man-machine systems. Since then, CTA methods have enjoyed widespread application, particularly in the defence, process control, and emergency services domains. For example, domains where CTA has recently been applied include the military (e.g., Jenkins, Stanton, Walker, and Salmon, 2008; Riley et al., 2006), emergency services (Militello and Hutton 2000), aviation (O'Hare et al., 2000), energy distribution (Salmon et al., 2008), rail transport (Walker et al., 2006), naval maintenance (Schaafstal and Schraagen, 2000), and even white-water rafting (O'Hare et al., 2000).

There are many CTA methods available; the CTA Resource Web site (www.ctaresource.com), for example, lists over 100 methods designed to evaluate and describe the cognitive aspects of task performance. Most approaches involve some sort of observation of the task under analysis and then interviews with SMEs (Militello and Hutton, 2000); however, Roth, Patterson, and Mumaw (2002) describe three different kinds of approach. The first involves analysing the domain in question in terms of goals and functions in order to determine the cognitive demands imposed by the tasks performed. The second involves the use of empirical methods, such as observation and interview methods, in order to determine how the users perform the task(s) under analysis, allowing a specification of the knowledge requirements and strategies involved. The third and more recent approach involves developing computer models that can simulate the cognitive activities required during the task under analysis. Roth (2008), on the other hand, describes five different forms of the CTA method, including structured interview approaches; critical incident analysis approaches; concept mapping methods, which represent knowledge concepts and the relationships between them; psychological scaling methods, which look at knowledge organisation; and cognitive field observation methods, which look at actual performance in the real world or simulated settings.

CTA methods offer a relatively simplistic approach to understanding the cognitive elements of task performance. In addition to looking at knowledge, goals, and thought processes, this approach allows various concepts to be studied, such as expertise, decision making, problem-solving strategies, situation awareness, and mental workload. The utility of CTA approaches is reflected in their current popularity; however, these methods are not without their flaws. The resources invested during CTA applications are typically high, and CTA methods are often criticised due to the high financial and time costs associated with them (Seamster, Redding, and Kaempf, 2000; Shute, Torreano and Willis, 2000) and the requirement for highly skilled analysts (Seamster, Redding, and Kaempf, 2000; Shute, Torreano, and Willis, 2000; Stanton et al., 2005). CTA methods also rely heavily on analysts having high levels of access to SMEs—something that is often difficult to organise and is often not available. Interpretation and application of CTA outputs has also been problematic. Once a CTA has been conducted, exactly what the results mean in relation to the problem and analysis goals can also be difficult to understand or misinterpreted. Shute, Torreano, and Willis (2000), for example, highlight the imprecise and vague nature of CTA outputs, whereas Potter, Roth, Woods, and Elm (2000) describe the bottleneck that occurs during the transition from CTA to system design, and suggest that the information gained from CTA must be effectively translated into design requirements and specifications. In conclusion to a review of CTA methods and computer-based CTA tools, Schraagen, Chipman, and Shalin (2000) reported that, although there were a large number of CTA methods available, they were generally limited. It was also concluded that there is limited guidance available in assisting researchers in the selection of the most appropriate CTA methods, in how to use CTA methods, and also in how CTA outputs can be utilised.

Roth's (2008) description of previous CTA method applications highlights their potential utility within the sporting domains. According to Roth (2008), CTA approaches have been used to inform the design of new technology and performance aids, in the development of training programmes, for performance evaluation, to investigate human error, to evaluate competing design proposals, for allocation of functions analysis, and to identify experts' cognitive strategies. All of these applications would be of use to sports scientists. The design of technology and performance aids for professional sports requires an in-depth understanding of the cognitive elements of task performance, and yet, to the authors' knowledge at least, CTA methods have not previously been used for this purpose. The identification of experts' or elite performers' cognitive strategies is also of interest to sports scientists, in particular how these differ from novice and amateur performer strategies.

CTA approaches have had some exposure in a sporting context. O'Hare et al. (2000), for example, describe an application of a semi-structured interview CTA method in the white-water rafting domain. In doing so, they wished to demonstrate the benefits of theoretically driven CTA in supporting enhanced system design effectiveness through improved design and training. Their findings had a number of implications for training, both at the novice and guide level. Due to the requirement during non-routine incidents for rapid decision making, and commitment to a course of action and action selection based on experience, O'Hare et al. (2000) recommended that guide training should be tailored to enable them to recognise critical environmental cues that are indicative of potential trouble, while avoiding distracting or irrelevant information. Potential training measures suggested by O'Hare et al. (2000) were to provide training on mental simulation in order to select appropriate routes and to demonstrate the cue recognition strategies of expert white-water rafters through multi-media packages. Hilliard and Jamieson (2008) describe an application of the Cognitive Work Analysis (CWA; Vicente, 1999) framework in the design of a graphical interface for driver cognitive support tools in solar car racing. By modelling the solar racing domain using CWA methods, the authors were able to develop a prototype in-vehicle display designed to decrease monitoring latency,

Cognitive Task Analysis

increase the accuracy of fault diagnosis, and free up driver cognitive resources to enable more efficient strategy planning (Hilliard and Jamieson, 2008). Hilliard and Jamieson (2008) report that the CWA analysis informed the content, form, and structure of the interface developed, and also that the racing team involved adopted the design and were, at the time of writing, planning to implement it during the upcoming race season. Burns and Hajdukiewicz (2004) also describe an application of CWA in the design of responsible casino gambling interfaces.

This chapter focuses on the following four CTA methods: CWA (Vicente, 1999), the Critical Decision Method (CDM; Klein, Calderwood, and McGregor, 1989), concept mapping (Crandall, Klein, and Hoffman, 2006), and Applied Cognitive Task Analysis (ACTA; Militello and Hutton, 2000). Alongside HTA, the CWA framework is arguably currently the most popular Human Factors method. Developed at the Risø National Laboratory in Denmark (Rasmussen, 1986), CWA offers a framework for the evaluation and design of complex sociotechnical systems. The framework focuses on constraints, which is based on the notion that making constraints explicit in an interface can potentially enhance human performance (Hajdukiewicz and Vicente, 2004), and comprises the following five phases, each of which model different constraint sets: Work Domain Analysis (WDA), control task analysis (ConTA), strategies analysis, Social Organisation and Cooperation Analysis (SOCA), and Worker Competencies Analysis (WCA). Probably the most well-known CTA approach is the CDM (Klein, Calderwood, and McGregor, 1989), which uses semi-structured interviews and cognitive probes in order to analyse the cognitive processes underlying decision making and performance in complex environments. Using this approach, tasks are decomposed into critical decision points, and cognitive probes are used to elicit information regarding the cognitive processes underlying decision making at each decision point. Concept mapping (Crandall, Klein, and Hoffman, 2006) is used to represent knowledge and does this via the use of networks depicting knowledge-related concepts and the relationships between them. ACTA (Militello and Hutton, 2000) offers a toolkit of three semi-structured interview methods that can be used to investigate the cognitive demands and skills associated with a particular task or scenario. The ACTA approach was developed so that the information generated could be represented in a manner that could easily be used to improve training programmes or interface design (Militello and Hutton, 2000). A summary of the CTA methods described is presented in Table 4.1.

COGNITIVE WORK ANALYSIS

More of a framework than a rigid methodology, CWA (Vicente, 1999) was originally developed at the Risø National Laboratory in Denmark (Rasmussen, 1986) for use within nuclear power process control applications. Underlying the approach was a specific need to design for new or unexpected situations; in a study of industrial accidents and incidents, Risø researchers found that most accidents began with non-routine operations. CWA's theoretical roots lie in general and adaptive control system theory and also Gibson's Ecological psychology theory (Fidel and Pejtersen, 2005). The approach itself is concerned with constraints, based on the notion that making constraints explicit in an interface can potentially enhance human performance (Hajdukiewicz and Vicente, 2004).

The CWA framework comprises the following five phases, each of which model different constraint sets:

1. *Work Domain Analysis (WDA)*. The WDA phase involves modelling the system in question based on its purposes and the constraints imposed by the environment. The Abstraction Hierarchy (AH) and Abstraction Decomposition Space (ADS) methods are used for this purpose. According to Jenkins et al. (2008), WDA identifies a fundamental set of constraints that are imposed on the actions of any actor. In modelling a system in this way, the systemic constraints that shape activity are specified. This formative approach leads to an event, actor, and time-independent description of the system (Sanderson, 2003; Vicente, 1999).

TABLE 4.1
Cognitive Task Analysis Methods Summary Table

Name	Domain	Application in Sport	Training Time	App. Time	Input Methods	Tools Needed	Main Advantages	Main Disadvantages	Outputs
Cognitive Work Analysis (CWA; Vicente, 1999)	Generic	Task description; Task analysis; Coaching design and evaluation; Tactics design, evaluation, and selection; Sports product design and evaluation; Performance evaluation; Error prediction; Allocation of functions	High	High	Observational study; Interviews; Questionnaires; Walkthrough	Pen and paper; Microsoft Visio; CWA software tool	1. A flexible framework that can be applied for a variety of different purposes 2. The approach is formative, and goes beyond the normative and descriptive analyses provided by traditional task analysis approaches 3. The five different phases allow the approach to be highly comprehensive	1. CWA is more complex than other methods and may require considerable training and practice 2. Can be extremely time consuming to apply 3. Some of the methods and phases have received only limited attention and are still in their infancy	Abstraction Hierarchy model of system; Control task analysis; Strategies analysis; Social organisation and cooperation analysis; Worker competencies analysis
Critical Decision Method (CDM; Klein, Calderwood, and McGregor, 1989)	Generic	Cognitive task analysis; Coaching design and evaluation; Tactics design, evaluation, and selection; Sports product design and evaluation; Performance evaluation; Identification of strategies	High	High	Interviews; Observational study	Pen and paper; Audio recording device; Word processing software	1. Can be used to elicit information regarding the cognitive processes and skills employed during task performance 2. Various sets of CDM probes are presented in the academic literature 3. Particularly suited to comparing the decision-making strategies used by different performers	1. Quality of data obtained is highly dependent upon the skill of the interviewer 2. Extent to which verbal interview responses reflect exactly the cognitive processes employed by decision makers during task performance is questionable 3. Highly time consuming	Transcript of decision-making process

Cognitive Task Analysis

Method	Domain	Application	Training time	Related methods	Tools needed	Advantages	Disadvantages	Output
Concept maps (Crandall, Klein, and Hoffman, 2006)	Generic	Cognitive task analysis; Coaching design and evaluation; Tactics design, evaluation, and selection; Sports product design and evaluation; Performance evaluation; Identification of strategies	Low	Observational study; Interviews; Questionnaires; HTA	Pen and paper; Microsoft Visio; Whiteboard; Flipchart	1. Can be used to elicit information regarding the knowledge used during task performance. 2. Relatively easy to learn and quick to apply. 3. Particularly suited to comparing the knowledge used by different performers	1. A high level of access to SMEs is required. 2. Can be difficult and time consuming, and outputs can become complex and unwieldy. 3. A high level of skill and expertise is required in order to use the method to its maximum effect	Concept map of linked concepts underpinning knowledge required during task performance
Applied Cognitive Task Analysis (ACTA; Militello and Hutton, 2000)	Generic	Cognitive task analysis; Coaching design and evaluation; Tactics design, evaluation, and selection; Sports product design and evaluation; Performance evaluation; Identification of strategies	Low	High	Observational study; Interviews	Pen and paper; Audio recording device; Word processing software	1. Offers a structured approach to CTA and aids the analyst through the provision of probes and questions. 2. Comprehensive, covering various aspects of the cognitive elements of task performance. 3. The approach is generic and seems suited to analysing the cognitive requirements associated with sporting tasks	1. Few applications of the method reported in the literature. 2. Highly time consuming. 3. The process may be laborious, for both analysts and interviewees, since there appears to be some repetition between the three interviews used	Interview transcripts detailing the cognitive elements of task performance; Cognitive demands table highlighting difficult tasks, common errors, and the cues and strategies used.

2. *Control Task Analysis (ConTA)*. The second phase, ConTA, is used to identify the tasks that are undertaken within the system and the constraints imposed on these activities during different situations. ConTA focuses on the activity necessary to achieve the purposes, priorities and values, and functions of a work domain (Naikar, Moylan, and Pearce, 2006). Rasmussen's decision ladder (Rasmussen, 1976; cited in Vicente, 1999) and the contextual activity template of Naikar, Moylan, and Pearce (2006) are used for the ConTA phase.
3. *Strategies Analysis*. The strategies analysis phase is used to identify how the different functions can be achieved. Vicente (1999) points out that whereas the ConTA phase provides a *product* description of *what* needs to be done, strategies analysis provides a *process* description of *how* it can be done; Jenkins et al. (2008) point out that the strategies analysis phase fills the "black-box" that is left by the ConTA phase. It involves the identification of the different strategies that actors might employ when performing control tasks.
4. *Social Organisation and Cooperation Analysis (SOCA)*. The fourth phase is used to identify how the activity and associated strategies are distributed amongst human operators and technological artefacts within the system in question, and also how these actors could potentially communicate and cooperate (Vicente, 1999). The objective of this phase is to determine how social and technical factors can work together in a way that enhances system performance (Vicente, 1999).
5. *Worker Competencies Analysis (WCA)*. The fifth and final stage attempts to identify the competencies that an ideal worker should exhibit (Vicente, 1999); it focuses on the cognitive skills that are required during task performance. Worker competencies analysis uses Rasmussen's Skill, Rule, and Knowledge (SRK) framework in order to classify the cognitive activities employed by actors during task performance.

The different CWA phases therefore allow researchers to specify the constraints related to why the system exists (WDA), as well as what activity is conducted (ConTA), how the activity is conducted (strategies analysis and WCA), with what the activity under analysis is conducted (WDA), and also who the activity is conducted by (SOCA).

DOMAIN OF APPLICATION

The CWA framework was originally developed for applications within the nuclear power domain; however, its generic nature allows it to be applied within any sociotechnical (i.e., consisting of social, technical, and psychological elements) system. Since its conception, CWA has been applied in various domains, including air traffic control (e.g., Ahlstrom, 2005), road transport (e.g., Salmon, Stanton, Regan, Lenne, and Young, 2007), aviation (e.g., Naikar and Sanderson, 2001), healthcare (e.g., Watson and Sanderson, 2007), maritime (e.g., Bisantz, Roth, Brickman, Gosbee, Hettinger, and McKinney, 2003), manufacturing (e.g., Higgins, 1998), the military (e.g., Jenkins et al., 2008), petrochemical (e.g., Jamieson and Vicente, 2001), process control (e.g., Vicente, 1999), and rail transport (e.g., Jansson, Olsson, and Erlandsson, 2006).

APPLICATION IN SPORT

There is great potential for applying the CWA framework in a sporting context, both for system evaluation and design purposes. The flexibility of the framework and the comprehensiveness of its five phases has led to it being applied in a variety of domains for various purposes, including system modelling (e.g., Hajdukiewicz, 1998), system design (e.g., Bisantz et al., 2003), training needs analysis (e.g., Naikar and Sanderson, 1999), training programme evaluation and design (e.g., Naikar and Sanderson, 1999), interface design and evaluation (Vicente, 1999), information requirements specification (e.g., Ahlstrom, 2005), tender evaluation (Naikar and Sanderson, 2001), team design (Naikar, Pearce, Drumm, and Sanderson, 2003), allocation of functions (e.g., Jenkins et

Cognitive Task Analysis

al., 2008), the development of human performance measures (e.g., Yu, Lau, Vicente, and Carter, 2002), and error management strategy design (Naikar and Saunders, 2003). Potential applications in sport include for modelling and evaluating systems, designing and evaluating training and coaching methods, designing and evaluating sports technology, performance aids and tactics, identifying the cognitive requirements of sports tasks, and for allocations of functions analysis. To date, the approach has been applied in a sporting context for interface design and performance optimisation purposes. Hilliard and Jamieson (2008), for example, describe an application of CWA in the design of a graphical interface for driver cognitive support tools for a solar car racing team. In addition, Burns and Hajdukiewicz (2004) describe an application of the WDA component of CWA in the design of casino gambling interfaces.

Procedure and Advice

Step 1: Clearly Define Aims of the Analysis

It is first important to clearly define the aims of the analysis. Exactly what the aims of the analysis to be undertaken are should be clearly defined so that the appropriate CWA phases are undertaken. It is important to note that it may not be necessary to complete a full five phase CWA; rather, the appropriate phases should be undertaken with regard to overall aims of the analysis.

Step 2: Select Appropriate CWA Phase(s)

Once the aims of the analysis are clearly defined, along with the required outputs, the analyst(s) should select the appropriate CWA phases to undertake. The combination of phases used is entirely dependent on the analysis requirements.

Step 3: Construct Work Domain Analysis

The WDA phase involves modelling the system in question based on a functional description of the constraints present within the system. The Abstraction Hierarchy (AH) and Abstraction Decomposition Space (ADS) methods are normally used for this phase. It is recommended that an AH is constructed first and then the ADS is constructed based on the AH. The AH consists of five levels of abstraction, ranging from the most abstract level of purposes to the most concrete level of form (Vicente, 1999). A description of each of the five AH levels is given below (Naikar, Hopcroft, and Moylan, 2005). Most of the levels have two names; the first set is typically more appropriate for causal systems, whereas the second set is more appropriate for intentional systems.

- *Functional purpose*—the overall purposes of the system and the external constraints on its operation
- *Abstract function/values and priority measures*—the criteria that the system uses for measuring progress towards the functional purposes
- *Generalised function/purpose-related functions*—the general functions of the work system that are necessary for achieving the functional purposes
- *Physical function/object-related processes*—the functional capabilities and limitations of the physical objects within the system that enable the purpose-related functions
- *Physical form/physical objects*—the physical objects within the work system that affords the physical functions

The ADS comprises a combination of the AH and a decomposition hierarchy and thus provides a two-dimensional representation of the system under analysis (Vicente, 1999). Each of the cells within the ADS presents a different representation of, or "way of thinking about," the same system. For example, the top left cell describes the functional purposes of the entire system (e.g., why the system exists and what its primary purposes are), while the bottom right cell describes the physical components that comprise the system (e.g., the people, objects, tools, and technological artefacts used within the system).

The decomposition hierarchy uses a number of levels of resolution (typically five), ranging from the broadest level of total system to the finest level of component, comprising total system, sub-system, function unit, sub-assembly, and component (Vicente, 1999). WDA is flexible in that all of the cells within the ADS do not need to be populated during the course of an analysis. The ADS also employs structural means-ends relationships in order to link the different representations of the system. This means that every node in the ADS should be the end that is achieved by all of the linked nodes below it, and also the means that (either on its own or in combination with other nodes) can be used to achieve all of the linked nodes above it. The means-ends relationships capture the affordances of the system (Jenkins et al., 2008) in that they examine what needs to be done and what the means available for achieving these ends are. The ADS provides an activity independent description of the work domain in question. It focuses on the purpose and constraints of the domain, which can be used to understand how and why structures work together to enact the desired purpose.

Naikar, Hopcroft, and Moylan (2005) and Burns and Hadjukiewicz (2004) both present guidance on how to undertake the WDA phase. Based on both sources, Jenkins et al. (2008) present the following guidance:

1. *Establish the purpose of the analysis.* At this point it is important to consider what is sought, and expected, from the analysis. This will influence which of the five phases are used and in what ratio.
2. *Identify the project constraints.* Project constraints such as time and resources will all influence the fidelity of the analysis.
3. *Identify the boundaries of the analysis.* The boundary needs to be broad enough to capture the system in detail; however, the analysis needs to remain manageable. It is difficult to supply heuristics, as the size of the analysis will in turn define where the boundary is set.
4. *Identify the nature of the constraints in the work domain.* According to Naikar, Hopcroft, and Moylan (2005), the next step involves identifying the point on the causal-intentional continuum on which the focus system falls. The purpose of this is to gain an insight into the nature of the constraints that should be modelled in the WDA (Hajdukiewicz et al., 1999; cited in Naikar, Hopcroft, and Moylan, 2005). Naikar, Hopcroft, and Moylan (2005) describe five categories of work systems that can be used to identify where a particular system lies on the causal-intentional continuum.
5. *Identify the sources of information for the analysis.* Information sources are likely to include, among others, engineering documents (if the system exists), the output of structured interviews with SMEs, and the outputs of interviews with stakeholders. The lower levels of the hierarchy are most likely to be informed from engineering documents and manuals capturing the physical aspects of the system. The more abstract functional elements of the system are more likely to be elicited from stakeholders or literature discussing the aims and objectives of the system.
6. *Construct the AH/ADS with readily available sources of information.* Often the easiest and most practical way of approaching the construction of the AH is to start at the top, then at the bottom and meet in the middle. In most cases, the functional purpose for the system should be clear—it is the reason that the system exists. By considering ways of determining how to measure the success of the functional purpose(s), the values and priority measures can be set. In many cases, at least some of the physical objects are known at the start of the analysis. By creating a list of each of the physical objects and their affordances, the bottom two levels can be partially created. Often, the most challenging part of an AH is creating the link in the middle between the physical description of the system's components and a functional description of what the system should do. The purpose-related functions level involves considering how the object-related processes can be used to have an effect on the

identified values and priority measures. The purpose-related functions should link upward to the values and priority measures to explain why they're required; they should also link down to the object-related processes, explaining how they can be achieved. The next stage is to complete the means-ends links, checking for unconnected nodes and validating the links. The why-what-how triad should be checked at this stage.

7. *Construct the AH/ADS by conducting special data collection exercises.* After the first stage of constructing the AH/ADS there are likely to be a number of nodes that are poorly linked, indicating an improper understanding of the system. At this stage, it is often necessary to seek further information on the domain from literature or from SMEs.
8. *Review the AH/ADS with domain experts.* The validation of the AH/ADS is a very important stage. Although this has been listed as stage 8 it should be considered as an iterative process. If access to SMEs is available, it is advantageous to consult with them throughout the process. Often the most systematic process for validating the AH is to step through, node-by-node checking the language. Each of the links should be validated along with each of the correct rejections.
9. *Validate the AH/ADS.* Often the best way to validate the ADS is to consider known recurring activity, checking to see that the AH contains the required physical objects, and that the values and priority measures captured cover the function aims of the modelled activity.

Naikar, Hopcroft, and Moylan (2005) also present a series of prompts designed to aid analysts wishing to conduct WDAs. Examples of their prompts are presented in Table 4.2.

Step 4: Conduct Control Task Analysis

The control task analysis phase is used to investigate what needs to be done independently of how it is done or by whom; it allows the requirements associated with known, recurring classes of situations to be identified (Jenkins et al., 2008). The decision ladder (Figure 4.1) is normally used for the control task analysis phase; however, Naikar, Moylan, and Pearce (2006) have more recently proposed the Contextual Activity Template (CAT) for this phase.

The decision ladder contains two types of nodes: data processing activities, as represented by the rectangular boxes, and states of knowledge that arise from the data processing activities, as represented by the circular nodes. Vicente (1999) points out that the ladder represents a linear sequence of information processing steps but is bent in half; novices are expected to follow the ladder in a linear fashion, whereas experts use short cuts (known as leaps and shunts) to link the two halves of the ladder. "Shunts" connect an information processing activity to a state of knowledge, and "leaps" connect two states of knowledge. Undertaking decision ladder analyses typically involves interviewing SMEs in order to determine how they undertake the control tasks identified in the generalised functions component of the WDA.

Step 5: Conduct Strategies Analysis

The strategies analysis phase involves identifying the different strategies that can potentially be employed to achieve control tasks; as Jenkins et al. (2008) point out, the strategy adopted in a given situation may vary significantly depending upon the constraints imposed. Naikar (2006) suggests that strategies analysis is concerned with identifying general categories of cognitive procedures. The strategy adopted is dependent upon various factors, including workload and task demands, training, time pressure, experience, and familiarity with the current situation. Ahlstrom (2005) proposes a modified form of Vicente's (1999) information flow map for the strategies analysis phase; here the situation is broken down into a "start state" and a desired "end state." The strategies available for achieving the end state connect the two states. An example of a simplified flow map is presented in Figure 4.2. The strategies available are typically identified via interviews with appropriate SMEs.

TABLE 4.2
Example WDA Prompts

	Prompts	**Keywords**
Functional purposes	Purposes: • For what reasons does the work system exist? • What are the highest-level objectives or ultimate purposes of the work system? • What role does the work system play in the environment? • What has the work system been designed to achieve? External constraints: • What kinds of constraints does the environment impose on the work system? • What values does the environment impose on the work system? • What laws and regulations does the environment impose on the work system?	Purposes: Reasons, goals, objectives, aims, intentions, mission, ambitions, plans, services, products, roles, targets, aspirations, desires, motives, values, beliefs, views, rationale, philosophy, policies, norms, conventions, attitudes, customs, ethics, morals, principles External constraints: Laws, regulations, guidance, standards, directives, requirements, rules, limits, public opinion, policies, values, beliefs, views, rationale, philosophy, norms, conventions, attitudes, customs, ethics, morals, principles
Values and priority measures	• What criteria can be used to judge whether the work system is achieving its purposes? • What criteria can be used to judge whether the work system is satisfying its external constraints? • What criteria can be used to compare the results or effects of the purpose-related functions on the functional purposes? • What are the performance requirements of various functions in the work system? • How is the performance of various functions in the work system measured or evaluated and compared?	Criteria, measures, benchmarks, tests, assessments, appraisals, calculations, evaluations, estimations, judgements, scales, yardsticks, budgets, schedules, outcomes, results, targets, figures, limits Measures of: effectiveness, efficiency, reliability, risk, resources, time, quality, quantity, probability, economy, consistency, frequency, success Values: laws, regulations, guidance, standards, directives, requirements, rules, limits, public opinion, policies, values, beliefs, views, rationale, philosophy, norms, conventions, attitudes, customs, ethics, morals, principles
Purpose-related functions	• What functions are required to achieve the purposes of the work system? • What functions are required to satisfy the external constraints on the work system? • What functions are performed in the work system? • What are the functions of individuals, teams, and departments in the work system?	Functions, roles, responsibilities, purposes, tasks, jobs, duties, occupations, positions, activities, operations
Object-related processes	• What can the physical objects in the work system do or afford? • What processes are the physical objects in the work system used for? • What are the functional capabilities and limitations of physical objects in the work system?	Processes, functions, purposes, utility, role, uses, applications, functionality, characteristics, capabilities, limitations, capacity, physical processes, mechanical processes, electrical processes, chemical processes
Physical objects	• What are the physical objects or physical resources in the work system—both man-made and natural? • What physical objects or physical resources are necessary to enable the processes and functions of the work system?	Man-made and natural objects: tools, equipment, devices, apparatus, machinery, items, instruments, accessories, appliances, implements, technology, supplies, kit, gear, buildings, facilities, premises, infrastructure, fixtures, fittings, assets, resources, staff, people, personnel, terrain, land, meteorological features

Source: Adapted from Naikar N., Hopcroft, R., and Moylan A. (2005). Work domain analysis: Theoretical concepts and methodology. Defence Science and Technology Organisation Report, DSTO-TR-1665, Melbourne, Australia.

Cognitive Task Analysis

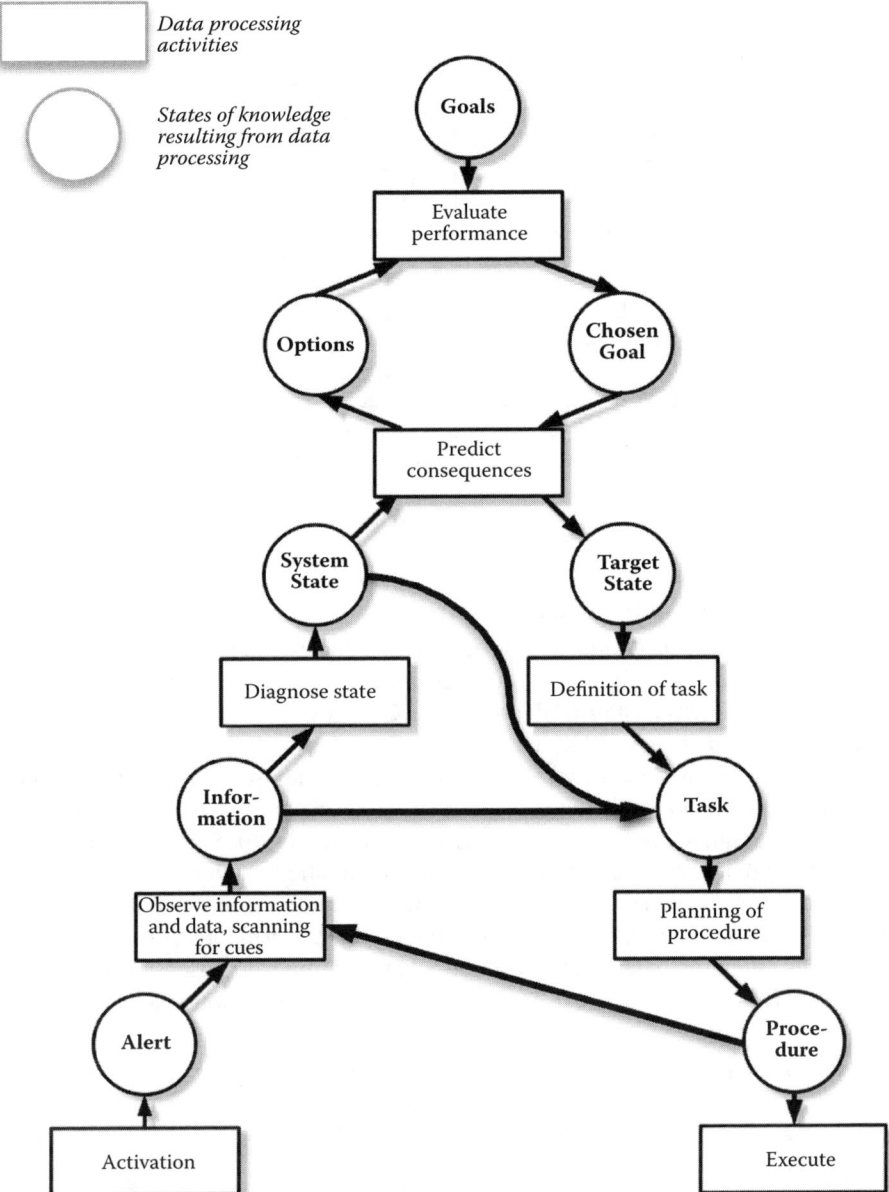

FIGURE 4.1 Decision ladder (showing leaps and shunts).

Step 6: Conduct Social Organisation and Cooperation Analysis

The SOCA phase looks at how tasks can potentially be allocated across a system (comprising human and technological operators). The objective of this phase is to determine how social and technical factors can work together in a way that maximises system performance. SOCA typically uses the outputs generated from the preceding three phases as its input. For example, simple shading can be used to map actors onto the ADS, decision ladder, and strategies analysis outputs. An example SOCA, based on the strategies analysis flow map presented in Figure 4.2, is depicted in Figure 4.3. SOCA is best informed through SME interviews, observation, and review of relevant documentation such as training manuals and standard operating procedures.

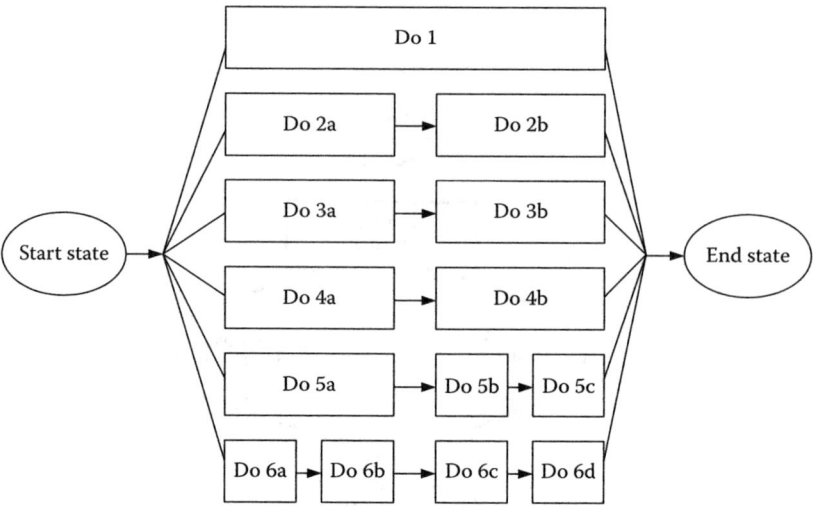

FIGURE 4.2 Strategies analysis simplified flow map.

Step 7: Conduct Worker Competencies Analysis

The final phase, WCA, involves identifying the competencies required for undertaking activity within the system in question. This phase is concerned with identifying the psychological constraints that are applicable to system design (Kilgore and St-Cyr, 2006). Vicente (1999) recommends that the Skill, Rule, and Knowledge (SRK) framework (Rasmussen, 1983; cited in Vicente, 1999) be used for this phase. The SRK framework describes three hierarchical levels of human behaviour: skill, rule, and knowledge-based behaviour. Each of the levels within the SRK framework defines a different level of cognitive control or human action (Vicente, 1999). Skill-based behaviour occurs in routine situations that require highly practised and automatic behaviour and where there is only small conscious control on behalf of the operator. According to Vicente (1999), skill-based behaviour consists of smooth, automated, and highly integrated patterns of action that are performed without conscious attention. The second level of behaviour, the rule-based level, occurs when the situation deviates from the normal but can be dealt with by the operator applying rules that are

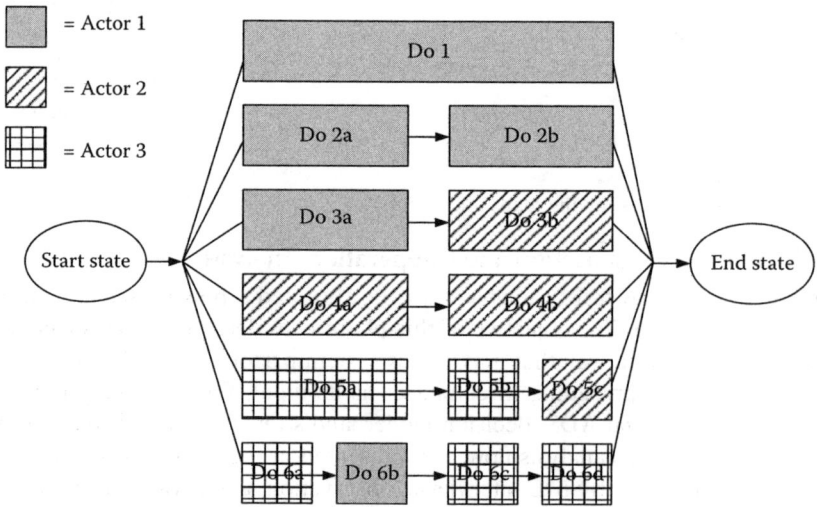

FIGURE 4.3 Mapping actors onto strategies analysis output for SOCA.

Cognitive Task Analysis

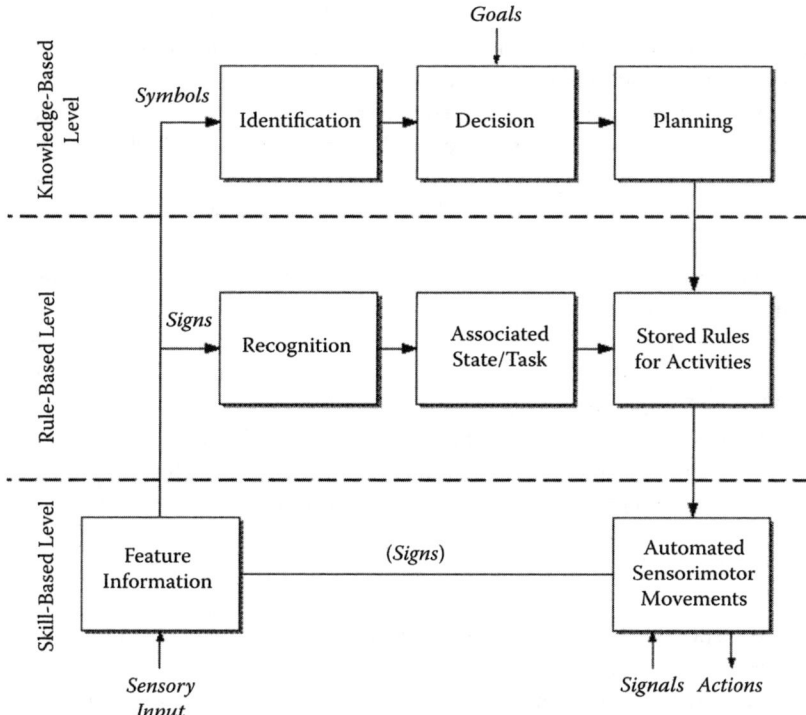

FIGURE 4.4 SRK behavioural classification scheme.

either stored in memory or are readily available, for example, emergency procedures. According to Vicente (1999), rule-based behaviour consists of stored rules derived from procedures, experience, instruction, or previous problem-solving activities. The third and highest level of behaviour is knowledge-based behaviour, which typically occurs in non-routine situations (i.e., emergency scenarios) where the operator has no known rules to apply and has to use problem-solving skills and knowledge of the system characteristics and mechanics in order to achieve task performance. According to Vicente (1999), knowledge-based behaviour consists of deliberate, serial search based on an explicit representation of the goal and a mental model of the functional properties of the environment. Further, knowledge-based behaviour is slow, serial, and effortful, as it requires conscious, focused attention (Vicente, 1999). The SRK framework is presented in Figure 4.4.

Worker competencies are expressed in terms of the skill, rule, and knowledge-based behaviours required. It is recommended that interviews with SMEs and relevant documentation review (e.g., procedures, standard operating instructions, training manuals) are used for the WCA phase.

Step 8: Review and Refine Outputs with SMEs

Once the appropriate phases are completed, it is important that they are reviewed and refined on the basis of input from SMEs. Undertaking CWA is a highly iterative process and the original outputs are likely to change significantly over the course of an analysis.

ADVANTAGES

1. CWA provides a flexible framework that can be applied for a variety of different purposes.
2. There is great potential for applying the framework in a sporting context, particularly for the design of sports technology.

3. The approach is formative, and so goes beyond the normative and descriptive analyses provided by traditional cognitive task analysis and task analysis approaches.
4. The approach has sound underpinning theory.
5. The five different phases allow the approach to be highly comprehensive. Researchers can potentially identify the constraints related to the environment in which activity takes place, why the system exists, what activity is conducted and what with, how the activity can be conducted, who the activity can be conducted by, and what skills are required to conduct the activity.
6. The framework has received considerable attention of late and has been applied in a range of domains for a range of purposes.
7. Due to its formative nature the framework can deal with non-routine situations.
8. The approach goes further than most Human Factors approaches in bridging the gap between Human Factors analysis and system design.
9. CWA can be used for both design and evaluation purposes.
10. Jenkins et al. (2008) recently developed a CWA software tool that supports both experts and novices in applying the method and significantly expedites the process.

Disadvantages

1. CWA is more complex than other Human Factors methods and researchers wishing to use the framework may require considerable training and practice.
2. CWA can be extremely time consuming to apply, although this can be circumvented somewhat by using appropriate analysis boundaries.
3. Some of the methods and phases within the framework have received only limited attention and are still in their infancy. It is notable that there are only limited applications using the latter three phases published in the academic literature. As a corollary, guidance on the latter three phases is also scarce.
4. Reliability of the approach is questionable and validity is difficult to assess.
5. CWA outputs can often be large, complex, and unwieldy and can be difficult to present satisfactorily.
6. It is often difficult for analysts to conduct the analysis independent of activity.

Related Methods

In terms of other similar Human Factors approaches, CWA remains the only formative system design and evaluation framework available. CWA is often compared with HTA (e.g., Hajdukiewicz and Vicente, 2004; Miller and Vicente, 2001; Stanton, 2006; Salmon, Jenkins, Stanton, and Walker, in press); however, this is more due to popularity and theoretical differences rather than any similarities between the two. As described above, a number of different methods have previously been used to support CWA efforts, including the ADS, AH, decision ladder, contextual activity template, information flow maps, and the SRK framework.

Approximate Training and Application Times

The training time associated with the CWA framework is high, particularly if all phases and methods are to be trained. The application time is also typically high, although the setting of appropriate analysis boundaries goes a long way to reducing it to a more reasonable figure. The application time therefore varies depending on the analysis aims and the boundaries set. For example, Naikar and Sanderson (2001) report that a WDA of an airborne early warning and control system took around six months to complete, whereas Salmon et al. (in press) report an application of the WDA, control

Cognitive Task Analysis

task analysis, and SOCA phases for the evaluation of an attack helicopter digitised mission planning system that took around 20 hours to complete.

RELIABILITY AND VALIDITY

The reliability and validity of the CWA framework is difficult to assess. The reliability may be questionable, although there are no data presented in the literature.

TOOLS NEEDED

CWA can be applied using pen and paper. As observational study and interviews are normally used during the data collection phase, video and audio recording equipment is typically required. A software drawing package, such as Microsoft Visio, is also helpful for producing the CWA outputs. Jenkins et al. (2008) recently developed a CWA software tool that supports construction of the CWA outputs for each phase and allows files to be saved, copied, edited, and exported into Microsoft Word documents.

EXAMPLE

For this example, selected CWA phases were conducted for the golf domain. An AH (from the WDA phase) for the golf domain is presented in Figure 4.5. The functional purpose of the golf "system" in this case is for the golfer to shoot as low a score as possible. At the bottom of the AH, the physical objects include those physical objects that are necessary to enable the processes and functions of the work system. In this case they include the course, holes, golf clubs, GPS device, golf ball and tee, distance markers, crowd, flags, and hazards (since the AH is actor independent the golfer is not included at this level). The next level up in the AH specifies what the physical objects do or afford. For example, the golf tee affords elevation and support of the golf ball; the clubs afford striking of the ball and the production of spin; and the course card depicts the course and hole layout, provides distance information, and allows the golfer to determine his or her position on the course. The generalised functions level specifies the functions that are required to achieve the purposes of the work system, such as club and shot selection, shot visualisation, setting up swing, and the various types of shots available (e.g., drive, approach, lay up, pitch and run). The abstract functions level specifies the criteria that can be used to determine whether the work system is achieving its purposes or not, and in this case includes overall score in strokes, number of strokes per hole, number of fairways hit, greens in regulation, and number of putts per hole.

An example of means-ends links can be seen in Figure 4.5 by tracing the links up the hierarchy from the golf course card. This shows how the golf course card depicts the course and hole layout, provides distance information, and allows the golfer to determine his or her position on the course. This information then informs club and shot selection and allows the golfer to visualise the forthcoming shot, which all play a key role in the achievement of a low overall score, optimum score per hole, number of fairways hit, and number of greens in regulation achieved. Optimum performance on these priority measures is essential for the golfer to achieve the lowest score possible, which is the functional purpose in this case.

An example decision ladder for an approach shot is presented next. The decision ladder represents the processes and knowledge states involved in making an approach shot, and is presented in Figure 4.6. Constructing the decision ladder involved discussing and walking through the process of making an approach shot with one golf SME.

Next, the strategies analysis phase is used to identify the range of strategies available for getting from a particular start state to a desired end state. For this example, we present an example strategies analysis flowchart for an approach shot to the green, with the start state being "ball on fairway,"

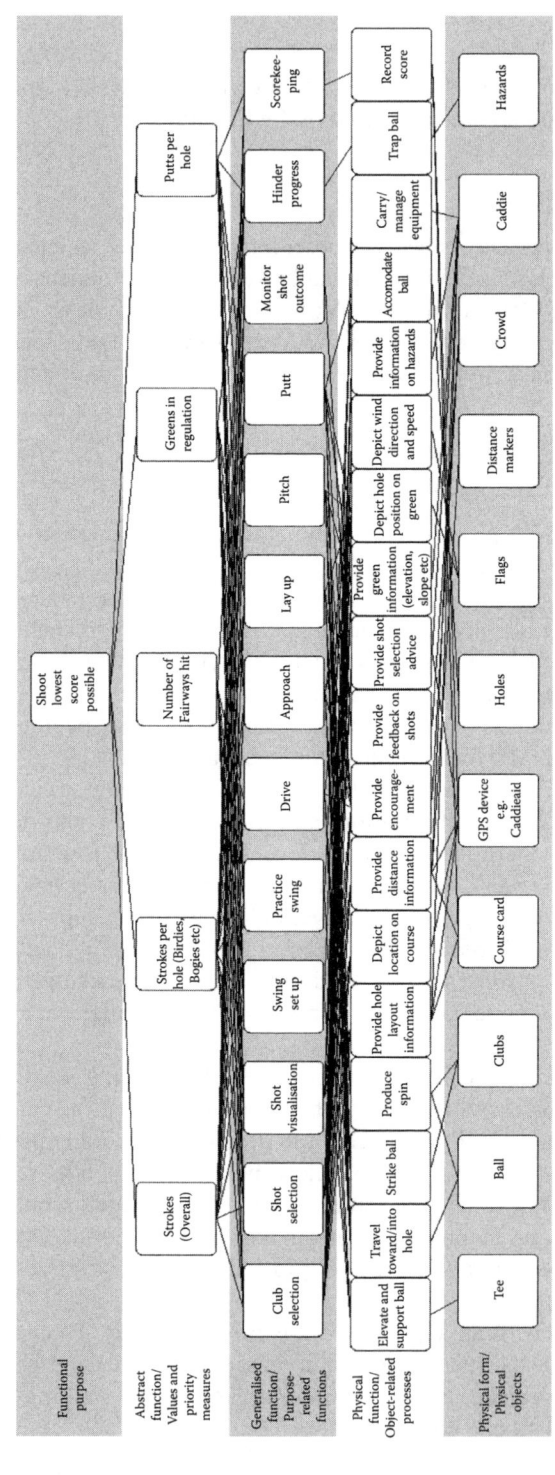

FIGURE 4.5 Golf abstraction hierarchy.

Cognitive Task Analysis

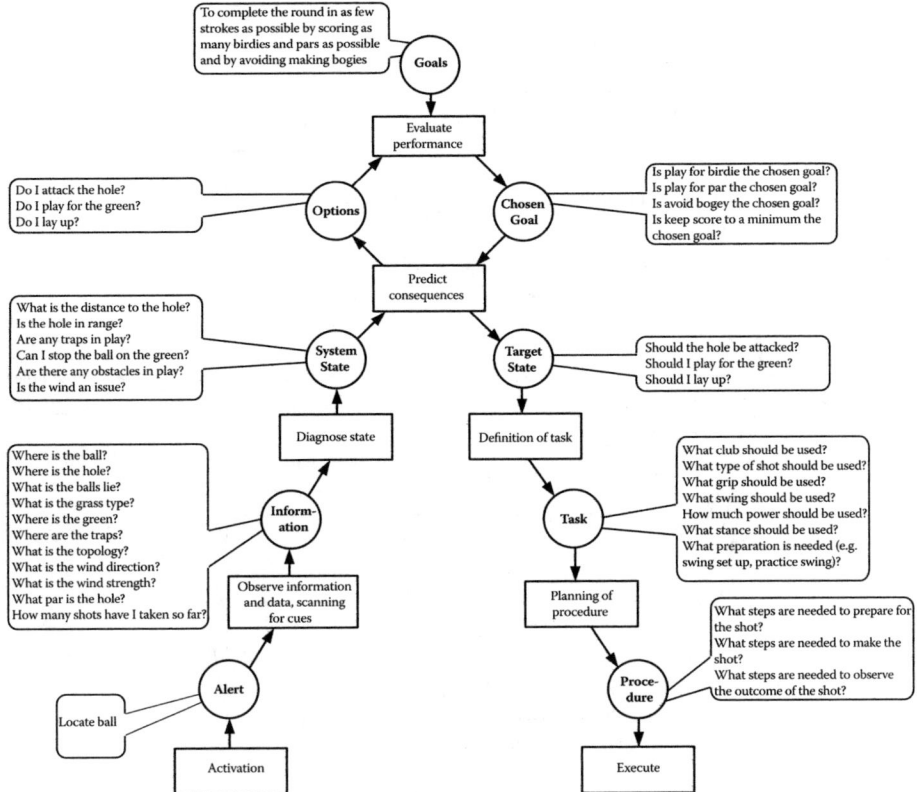

FIGURE 4.6 Golf shot decision ladder.

and the desired end state being "ball in hole." The approach shot strategies analysis flow map is presented in Figure 4.7.

Recommended Texts

Bisantz, A. M., and Burns, C. M. (2008). *Applications of cognitive work analysis*. Boca Raton, FL: CRC Press.

Jenkins, D. P., Stanton, N. A., Walker, G. H., and Salmon, P. M. (2009). *Cognitive work analysis: Coping with complexity*. Aldershot, UK: Ashgate.

Vicente, K. J. (1999). *Cognitive work analysis: Toward safe, productive, and healthy computer-based work*. Mahwah, NJ: Lawrence Erlbaum Associates.

CRITICAL DECISION METHOD

The Critical Decision Method (CDM; Klein, Calderwood, and McGregor, 1989) uses semi-structured interviews and cognitive probes to identify the cognitive processes underlying decision making in complex environments. Typically, scenarios are decomposed into critical decision points and so-called "cognitive probes" (targeted interview probes focussing on cognition and decision making) are used to identify and investigate the cognitive processes underlying operator performance at each decision point.

The CDM approach can be applied for a range of purposes, including for the identification of training requirements, the development of training materials, and for the evaluation of task performance. It is also useful for comparing the decision-making strategies employed by different (e.g., novice and expert) operators. The CDM is a popular approach and has been applied in various domains. For example, recent published applications have taken place in the emergency services

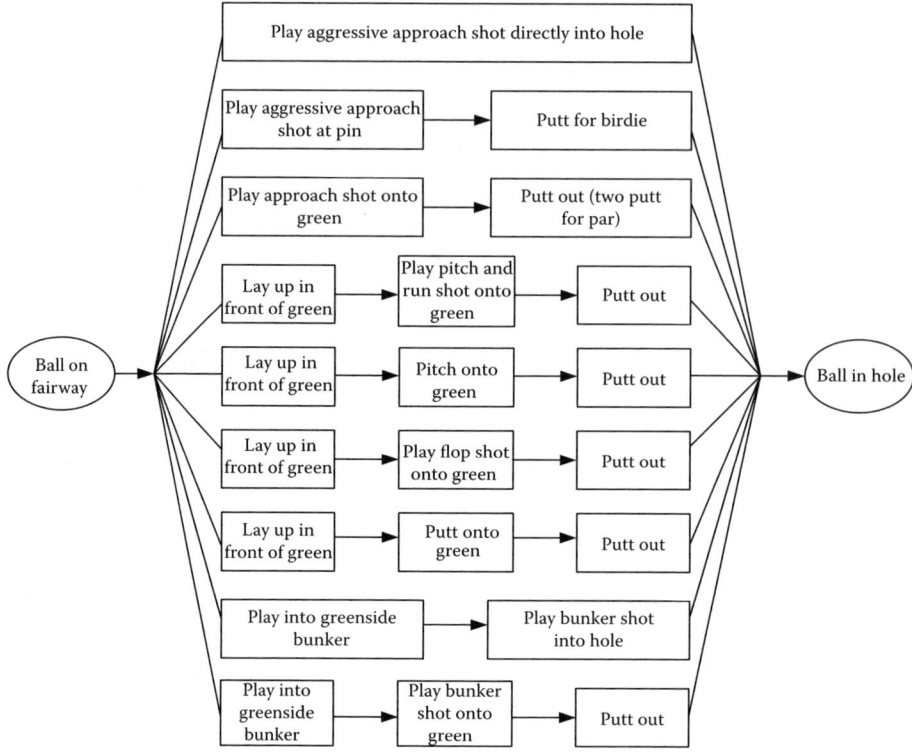

FIGURE 4.7 Example approach shot strategies analysis flow map.

(Blandford and Wong, 2004), the military (Salmon et al., 2009), energy distribution (Salmon et al., 2008), road transport (Stanton, Walker, Young, Kazi, and Salmon, 2007), rail transport (Walker et al., 2006), and white-water rafting (O'Hare et al., 2000) domains.

DOMAIN OF APPLICATION

The procedure is generic and can be applied in any domain.

APPLICATION IN SPORT

There is great potential for applying CDM in a sporting context; the approach offers a simplistic means of gathering data regarding the cognitive processes used during task performance and is clearly suited to the analysis of sports performance. In particular, it appears to be suited to identifying and analysing the differences between the cognitive processes adopted by sports performers of differing abilities and expertise levels (e.g., elite versus amateur sports performers). Further, it may also be useful for identifying training requirements for complex cognitive tasks.

PROCEDURE AND ADVICE

Step 1: Clearly Define Aims of the Analysis

It is first important to clearly define the aims of the analysis. Exactly what the aims of the analysis to be undertaken are should be clearly defined so that appropriate scenarios can be analysed using appropriate participants and an appropriate set of CDM probes.

Step 2: Identify Scenarios to Be Analysed

Once the aims of the analysis are clearly defined, it is next important to identify what scenarios should be analysed. In a sporting context, for example, this may be a particular event or a series of specific training or game scenarios. Most CDM analyses focus on non-routine events or critical incidents. For example, in a game of soccer a critical incident may be a penalty kick or a free kick located on the edge of the oppositions box, or a one-on-one attack situation. In golf, a critical incident may be a bunker shot or, alternatively, an entire hole may be broken down into critical incidents, such as select strategy for hole, make drive, make approach shot, and putting out. Critical incidents can also be identified once the scenario has unfolded by asking participants, post task, to identify the critical incidents that occurred during task performance.

Step 3: Select/Develop Appropriate CDM Interview Probes

The CDM works by probing SMEs using predefined "cognitive probes," which are designed specifically to elicit information regarding the cognitive processes undertaken during task performance. It is therefore highly important that an appropriate set of probes are selected or developed prior to the analysis. There are a number of sets of CDM probes available in the literature (e.g., Crandall, Klein, and Hoffman, 2006; Klein and Armstrong, 2004; O'Hare et al., 2000); however, it may also be appropriate to develop a new set of probes depending on the analysis requirements. A set of CDM probes that we have used in the past (e.g., Salmon et al., 2009; Stanton et al., 2006; Stewart et al., 2008; Walker et al., 2006) is presented in Table 4.3.

TABLE 4.3
Critical Decision Method Probes

Goal Specification	What Were Your Specific Goals at the Various Decision Points?
Cue identification	What features were you looking for when you formulated your decision?
	How did you know that you needed to make the decision?
	How did you know when to make the decision?
Expectancy	Were you expecting to make this sort of decision during the course of the event?
	Describe how this affected your decision-making process.
Conceptual	Are there any situations in which your decision would have turned out differently?
Influence of uncertainty	At any stage, were you uncertain about either the reliability or the relevance of the information that you had available?
Information integration	What was the most important piece of information that you used to formulate the decision?
Situation awareness	What information did you have available to you at the time of the decision?
Situation assessment	Did you use all of the information available to you when formulating the decision?
	Was there any additional information that you might have used to assist in the formulation of the decision?
Options	Were there any other alternatives available to you other than the decision you made?
Decision blocking—stress	Was their any stage during the decision-making process in which you found it difficult to process and integrate the information available?
Basis of choice	Do you think that you could develop a rule, based on your experience, which could assist another person to make the same decision successfully?
Analogy/generalisation	Were you at any time reminded of previous experiences in which a similar/different decision was made?

Source: O'Hare. D., Wiggins, M., Williams, A., and Wong, W. (2000). Cognitive task analysis for decision centered design and training. In *Task analysis*, eds. J. Annett and N. A. Stanton, 170–90. London: Taylor & Francis.

Step 4: Select Appropriate Participant(s)

Once the aims, task, and probes are defined, appropriate participants should be selected; again, this is entirely dependent upon the analysis context and requirements. Normally the participants selected should be SMEs for the domain in question, or the primary decision makers in the chosen task or scenario under analysis.

Step 5: Observe Scenario under Analysis or Gather Description of Incident

The CDM is normally applied based on an observation of the task or scenario under analysis; however, it can also be used retrospectively to analyse incidents that occurred some time ago. If the analysis is based on direct observation, then this step involves the analyst(s) observing the task or scenario. An observational transcript should be completed for the scenario (see Chapter 2). If the analysis is based on an incident that has already occurred then the analyst and SME should work together to develop a description of the incident.

Step 6: Define Timeline and Critical Incidents

Once the observation is complete, the analyst and participant should develop a task model or event timeline for the scenario. This is then used to define a series (normally four or five) of critical incidents or incident phases for further analysis. These normally represent critical decision points for the scenario in question; however, distinct incident phases have also been used in the past.

Step 7: Conduct CDM Interviews

CDM interviews should then be conducted with the participant(s) for each critical incident or key decision point identified during step 6. This involves administering the cognitive probes in a semi-structured interview format in order to elicit information regarding the cognitive processes employed by the decision maker during each critical incident or incident phase. An interview transcript should be recorded by the interviewer and the interview should be recorded using video and/or audio recording equipment.

Step 8: Transcribe Interview Data

Once the interviews are completed, the data should be transcribed using a word processing software package such as Microsoft Word. It is normally useful for data representation purposes to produce CDM tables containing the cognitive probes and interviewee responses for each participant.

Step 9: Analyse Data as Required

The CDM data should then be analysed accordingly based on the analysis requirements. For example, when comparing the decision-making strategies employed by elite and novice sports performers, the analyst should compare and contrast the interviewee responses. Often content analysis is used to pick out key themes or concepts.

ADVANTAGES

1. The CDM can be used to elicit information regarding the cognitive processes and skills employed during task performance.
2. Various sets of CDM probes are available (e.g., Crandall, Klein, and Hoffman, 2006; Klein and Armstrong, 2004; O'Hare et al., 2000).
3. The CDM approach is particularly suited to comparing the decision-making strategies used by different operators and could be applied in a sporting context for this purpose.
4. The method is a popular one and has been applied in a number of different domains.

Cognitive Task Analysis

5. The data obtained are useful and can be used for a number of purposes, including performance evaluation, comparison of decision-making strategies, training programme design and evaluation, situation awareness requirements analysis, and interface and system design.
6. The method offers a good return in terms of data collected in relation to time invested.
7. The flexibility of the approach allows all manner of Human Factors concepts to be studied, including decision making, situation awareness, human error, workload, and distraction.
8. Since it uses interviews, CDM offers a high degree of control over the data collection process. Targeted interview probes can be designed a priori and interviewers can direct interviews as they see fit.
9. The data obtained can be treated both qualitatively and quantitatively.
10. The approach has sound underpinning theory.

Disadvantages

1. The output offers little direct input into design.
2. The data obtained are highly dependent upon the skill of the interviewer and the quality and willingness to participate of the interviewee.
3. Participants may find it difficult to verbalise the cognitive components of task performance, and the extent to which verbal interview responses reflect exactly the cognitive processes employed during task performance is questionable.
4. Designing, conducting, transcribing, and analysing interviews is an extremely time-consuming process and the large amount of time required may limit the number of participants that can be used.
5. The reliability and validity of interview methods is difficult to address and it appears that the reliability of this approach is questionable. Klein and Armstrong (2004), for example, point out that methods that analyse retrospective incidents may have limited reliability due to factors such as memory degradation.
6. A high level of expertise is required in order to use the CDM to its maximum effect (Klein and Armstrong, 2004).
7. Since it involves the conduct of interviews, the CDM approach is susceptible to a range of interviewer and interviewee biases.
8. When analyzing real-world tasks, participants often do not have sufficient free time to fully engage in interviews. Often interviewers are given only a short period of time to collect their data.
9. It is often difficult to gain the levels of access to SMEs that are required for successful completion of CDM interviews.
10. Depending on the subject area, interviewees may be guarded with their responses for fear of reprisals.

Related Methods

The CDM is based on Flanagan's Critical Incident Technique (Flanagan, 1954) and is primarily an interview-based approach. The CDM procedure also utilises observational study methods and timeline analysis methods during the scenario observation and description phases. More recently, CDM outputs have been used to analyse, via content analysis, situation awareness and situation awareness requirements in command and control environments (Salmon et al., 2009).

Approximate Training and Application Times

Although the time taken for analysts to understand the CDM procedure is minimal, the training time is high due to the requirement for experience in interviews and for trainees to grasp cognitive psychology (Klein and Armstrong, 2004). In addition, once trained in the method, analysts are likely to require significant practice until they become proficient in its application. The application time is dependent upon the probes used and the number of participants involved; however, due to the high levels of data generated and the requirement to transcribe the interview data, the application time is typically high. Normally a typical CDM interview, focussing on four to five critical incidents, would take around 1 to 2 hours. The transcription process for a 1 to 2 hour CDM interview normally takes between 2 and 4 hours.

Reliability and Validity

The reliability of the CDM approach is questionable. Klein and Armstrong (2004) suggest that there are concerns over reliability due to factors such as memory degradation. The validity of the approach may also be questionable, due to the difficulties associated with the verbalisation of cognitive processes. Interview approaches also suffer from various forms of bias that affect reliability and validity levels.

Tools Needed

At a simplistic level, CDM can be conducted using only pen and paper; however, it is recommended that the scenario and interviews are recorded using video and audio recording devices. The interviewer also requires a printed set of CDM probes for conducting the CDM interviews.

Example

As part of a study focussing on the decision-making strategies used by novice, amateur, and elite fell runners (see Chapter 10), the CDM approach was used to analyse fell runner decision making during a recent local amateur fell race. Upon receipt of consent to take part, participating runners were asked to complete CDM transcripts upon completion of the race. The race was broken down into two key decision points, the first being the ascent (885 ft) part of the race, and the second being the descent towards the finish. A series of cognitive probes, adapted from those presented in Table 4.3 were used. Example outputs are presented in Table 4.4 and Table 4.5.

Flowchart

(See Flowchart 4.1.)

Recommended Texts

Crandall, B., Klein, G., and Hoffman, R. (2006). *Working minds: A practitioner's guide to cognitive task analysis*. Cambridge, MA: MIT Press.
Klein, G., and Armstrong, A. A. (2004). Critical decision method. In *Handbook of Human Factors and ergonomics methods*, eds. N. A. Stanton, A. Hedge, E. Salas, H. Hendrick, and K. Brookhaus, 35.1–35.8. Boca Raton, FL: CRC Press.
Klein, G., Calderwood, R., and McGregor, D. (1989). Critical decision method for eliciting knowledge. *IEEE Transactions on Systems, Man, and Cybernetics* 19(3):462–72.

CONCEPT MAPS

Background and Applications

Concept maps (Crandall, Klein, and Hoffman, 2006) are used to elicit and represent knowledge via the use of networks depicting concepts and the relationships between them. Representing knowledge in the form of a network is a popular approach that has been used by cognitive psychologists for many years. According to Crandall, Klein, and Hoffman (2006), concept maps were first developed by Novak (1977; cited in Crandall, Klein, and Hoffman, 2006) in order to understand and track changes in his students' knowledge of science. Concept maps are based on Ausubel's theory of learning (Ausubel, 1963; cited in Crandall, Klein, and Hoffman, 2006), which suggests that meaningful learning occurs via the assimilation of new concepts and propositions into existing concepts and propositional frameworks in the mind of the learner. Crandall, Klein, and Hoffman (2006) point out that this occurs via subsumption (realising how new concepts relate to those already known), differentiation (realising how new concepts draw distinctions with those already known), and reconciliation (of contradictions between new concepts and those already known). Crandall, Klein, and Hoffman (2006) cite a range of studies that suggest that building good concept maps leads to longer retention of knowledge and a greater ability to apply knowledge in novel settings. An example concept map of the concept map method (adapted from Crandall, Klein, and Hoffman, 2006) is presented in Figure 4.8.

Domain of Application

Concept maps were originally developed as an educational method for supporting meaningful learning (Ausubel and Novak, 1978; cited in Crandall, Klein, and Hoffman, 2006); however, the approach is generic and can be applied in any domain. Crandall, Klein, and Hoffman (2006) cite a range of domains in which the method has been applied, including education, astrobiology, rocket science, and space exploration.

Application in Sport

Concept maps are used to elicit and represent SME knowledge; therefore, there is great scope to use them in a sporting context in order to determine and represent elite sports performer knowledge during task performance. Concept maps can also be used to compare and contrast the knowledge of sports performers of differing ability, for example, novice versus elite sports performers. Crandall, Klein, and Hoffman (2006) point out that research using the approach has demonstrated that expertise is typically associated not only with more detailed knowledge but also with better organisation of knowledge when compared to novices. The identification of knowledge requirements for different sports is also an important line of enquiry for training and coaching programme development. The output of concept map–type analyses can therefore be used for a range of purposes, including coaching and training programme development, sports product design, and information requirements analysis.

Procedure and Advice

Step 1: Clearly Define Aims of the Analysis
It is first important to clearly define the aims of the analysis. Exactly what the aims of the analysis to be undertaken are should be clearly defined so that appropriate scenarios and participants can be focussed on.

Step 2: Identify Scenarios to Be Analysed
Once the aims of the analysis are clearly defined, it is next important to identify what scenarios should be analysed. In a sporting context, for example, this may be a particular task, event, or a series of specific training or game scenarios. It may also be useful to use different event or game phases.

TABLE 4.4
Example "Ascent" CDM Output

Ascent

Goal specification
What were your specific goals during this part of the race?
- To get to Captain Cook's monument (the top of the ascent) as quickly as possible.
- To stay in touch with the runners around me at the beginning of the ascent (to not let any runners get away from me during the ascent).

Decisions
What decisions did you make during this part of the race?
- Which route to take up the ascent.
- When it was more appropriate to stop and walk, i.e., when the trade-off between speed and energy expired became such that it was more appropriate to walk up the ascent.
- Whether to try and keep up with runners passing me.
- When to get going again.

Cue identification
What features were you looking for when you formulated your decisions?
- Physical condition, heart rate, breathing rate, pain in legs and lungs, incline level, other runners' strategies, pace (minute miles), other runners' pace, terrain, etc.

How did you know that you needed to make the decisions? How did you know when to make the decisions?
- All decisions were part and parcel of competing in the race.

Expectancy
Were you expecting to make these sorts of decisions during the course of the event?
- Yes.

Describe how this affected your decision-making process.
- I knew what information I was looking for and what elements I needed to monitor during the ascent phase.

Conceptual
Are there any situations in which your decisions would have turned out differently?
- Yes, if I felt better or felt that I could maintain a quicker pace, or if pain levels and my physical condition were such that I had to stop earlier on the ascent.

Influence of uncertainty
At any stage, were you uncertain about either the reliability or the relevance of the information that you had available?
- I wasn't sure how far the ascent would go on for at such a steep incline. I wasn't sure if it was going to get worse or better (the incline, that is) and also how far from the end of the ascent I was.

Information integration
What was the most important piece of information that you used to make your decisions?
- My physical condition, e.g., my heart rate, breathing rate, pain levels, legs, fatigue, etc.

Situation awareness
What information did you have available to you at the time of the decisions?
- Physical condition (all subjective, of course, including my heart rate, breathing rate, pain levels, legs, fatigue, etc.).
- Pace (minute miles) taken from Garmin Forerunner watch.
- Distance travelled taken from Garmin Forerunner watch.
- Distance remaining based on calculation using distance travelled information and knowledge of race length.
- Time taken from Garmin Forerunner watch.
- Visual information, including incline, routes available, other runners, etc.
- Knowledge of route taken from map prior to race.
- Previous experience of route and race.

Situation assessment
Did you use all of the information available to you when making decisions?
- Yes, but some information was more prominent than others in my decision making, such as my own physical condition.

Was there any additional information that you might have used to assist you in making decisions?
- Yes, distance remaining on the ascent.
- A real-time display of route depicting my own position, maybe a 3D elevation diagram also. Such information would have allowed me to determine whether I could have maintained my pace (as I would have known that I was towards the end of the ascent or nowhere near the end of it).

Cognitive Task Analysis

TABLE 4.4 (CONTINUED)
Example "Ascent" CDM Output

	Ascent
Options	*Were there any other alternatives available to you other than the decisions you made?* • Yes, to keep a quicker pace or to slow or stop earlier, to take a different route, and whether to try to go with passing runner.
Decision blocking—stress	*Was there any stage during the decision-making process in which you found it difficult to process and integrate the information available?* • No.
Basis of choice	*Do you think that you could develop a rule, based on your experience, which could assist another person to make the same decisions successfully?* • Well, I could certainly develop a personal rule for when to slow down/speed up on severe ascents. • I guess I could identify what runners need to be focussing on in order to make such decisions, yes.
Analogy/ generalisation	*Were you at any time reminded of previous experiences in which similar/different decisions were made?* • Yes, previous fell racing events.

Step 3: Select Appropriate Participant(s)

Once the scenario under analysis is defined, the analyst(s) should proceed to identify an appropriate SME or set of SMEs. Typically, experts for the domain and system under analysis are used; however, if the analysis is focussing on a comparison of novices versus elite performers, then a selection of participants from each group should be used.

Step 4: Observe the Task or Scenario under Analysis

It is important that the analyst(s) involved familiarises him- or herself with the task or scenario under analysis. This normally involves observing the task or scenario, but might also involve reviewing any relevant documentation (e.g., training manuals, standard operating procedures, existing task analyses) and holding discussions with SMEs. If an observation is not possible, a walkthrough of the task may suffice. This allows the analyst to understand the task and the participant's role during task performance.

Step 5: Introduce Participants to Concept Map Method

Crandall, Klein, and Hoffman (2006) suggest that it is important to give an introductory presentation about the concept map method to the participants involved. The presentation should include an introduction to the method, its background, an overview of the procedure, and some example applications, including a description of the methodology employed and the outputs derived.

Step 6: Identify Focus Question

Next, the knowledge elicitation and concept map construction phase can begin. Crandall, Klein, and Hoffman (2006) recommend that one analyst should act as the interviewer and one analyst act as the mapper, constructing the map on-line during the knowledge elicitation phase. They stress that the interviewing analyst should act as a facilitator, effectively supporting the participant in describing their knowledge during the task or scenario under analysis. This involves the use of suggestions such as "leads to?" "comes before?" and "is a precondition for?" (Crandall, Klein, and Hoffman, 2006). Crandall, Klein, and Hoffman (2006) recommend that the facilitator and participant(s) should first identify a focus question that addresses the problem or concept that is to be the focus of the analysis. Examples in a sporting context could be "How does a golfer prepare for a drive?" or "What is a scrum/rook/mall?"

TABLE 4.5
Example "Descent" CDM Output

Descent

Goal specification	*What were your specific goals during this part of the race?* • To maintain a sub 6-minute, 30-second–mile pace. • To not let any runners pass me. • To pass all runners ahead of me in my visual field of view.
Decisions	*What decisions did you make during this part of the race?* • What routes to take down the various sharp descents. • When to quicken pace. • When to reign pace in.
Cue identification	*What features were you looking for when you formulated your decisions?* • For routes, I was looking at the terrain, obstacles, and other runners' routes down. • For pace decisions, I was looking at my own physical condition, heart rate, breathing rate, pain in legs and lungs, incline level, pace (minute miles), and other runners' pace. *How did you know that you needed to make the decisions? How did you know when to make the decisions?* • Based on my knowledge of the race route I knew that the last 2 and a half miles were pretty much a constant descent all the way to the finish. I also knew that gains were to be made by not going down the typical descent routes.
Expectancy	*Were you expecting to make these sorts of decisions during the course of the event?* • Yes. *Describe how this affected your decision-making process.* • I knew what information I was looking for and what elements I needed to monitor during the descent phase.
Conceptual	*Are there any situations in which your decisions would have turned out differently?* • Yes, if I knew for sure that I could maintain it without breaking down, I would have chosen to run at an even quicker pace on the descent phase. Also, if I knew what the best descent routes were for definite, I would have taken them.
Influence of uncertainty	*At any stage, were you uncertain about either the reliability or the relevance of the information that you had available?* • Yes, I was very uncertain about the distance remaining, which I was trying to calculate mentally but did not know for sure as I did not know for definite how long the race was in duration.
Information integration	*What was the most important piece of information that you used to make your decisions?* • My physical condition, e.g., my heart rate, breathing rate, pain levels, legs, fatigue, etc. Basically whether or not I felt I could maintain a quicker pace all the way to the finish.
Situation awareness	*What information did you have available to you at the time of the decisions?* • Physical condition (all subjective, of course, including my heart rate, breathing rate, pain levels, legs, fatigue, etc.). • Pace (minute miles) taken from Garmin Forerunner watch. • Distance travelled taken from Garmin Forerunner watch. • Distance remaining based on calculation using distance travelled information and knowledge of race length. • Time taken from Garmin Forerunner watch. • Visual information, including terrain, routes available, other runners, etc. • Knowledge of route taken from map prior to race. • Previous experience of route and race.
Situation assessment	*Did you use all of the information available to you when making decisions?* • Yes, but physical condition, distance remaining, terrain, and other runners' strategies were the most important for the decisions described. *Was there any additional information that you might have used to assist you in making decisions?* • Exact distance remaining, for example, a countdown facility on my watch. • Knowledge (based on training) of my own ability to maintain a quicker pace.

TABLE 4.5 (CONTINUED)
Example "Descent" CDM Output

	Descent
Options	*Were there any other alternatives available to you other than the decisions you made?*
	• Yes, different descent routes, and a quicker pace taken on earlier or later.
Decision blocking— stress	*Was their any stage during the decision-making process in which you found it difficult to process and integrate the information available?*
	• I did find it difficult to estimate the distance remaining as the figures weren't exact and I had other runners on my tail and felt under pressure and exhausted.
Basis of choice	*Do you think that you could develop a rule, based on your experience, which could assist another person to make the same decisions successfully?*
	• Yes, see ascent response.
Analogy/ generalisation	*Were you at any time reminded of previous experiences in which similar/different decisions were made?*
	• Yes, previous fell races.

Step 7: Identify Overarching Concepts

Following the focus question, the participant should be asked to identify between 5 and 10 of the most important concepts underlying the concept of interest (Crandall, Klein, and Hoffman, 2006). These concepts should be organised in a so-called *step 1* concept map. Crandall, Klein, and Hoffman (2006) suggest that the most important or most closely related concepts should be located toward the top of the concept map.

Step 8: Link Concepts

Once the concepts are defined, the next phase involves linking them based on the relationships between them. Directional arrows and linking words are used on the concept map for this purpose. According to Crandall, Klein, and Hoffman (2006), the links between concepts can express causal relations (e.g., is caused by, results in, leads to), classificational relations (e.g., includes, refers to), property relations (e.g., owns, comprises), explanatory relations (e.g., is used for), procedure or method relations (e.g., is achieved by), contingencies and dependencies (e.g., requires), probabilistic relations (e.g., is more likely than), event relations (e.g., occurs before), and uncertainty or frequency relations (e.g., is more common than).

Step 9: Review and Refine Concept Map

The concept map is a highly iterative approach and many revisions are normally required. The next step therefore involves reviewing and refining the map until the analysts and participant are happy with it. Refining the map might include adding concepts, subtracting concepts, adding further subordinate concepts and links, and/or changing the links. One important factor is to check that all node-link-node triples express propositions (Crandall, Klein, and Hoffman, 2006).

ADVANTAGES

1. The concept maps procedure can be used to elicit information regarding the knowledge used during task performance.
2. The method is relatively easy to learn and quick to apply, at least for simple concepts and tasks.
3. The approach is particularly suited to comparing the knowledge used by different performers and could be applied in a sporting context for this purpose; for example, concept maps could be used for identifying the knowledge concepts used by elite performers, which in turn could be used for developing coaching and training interventions.

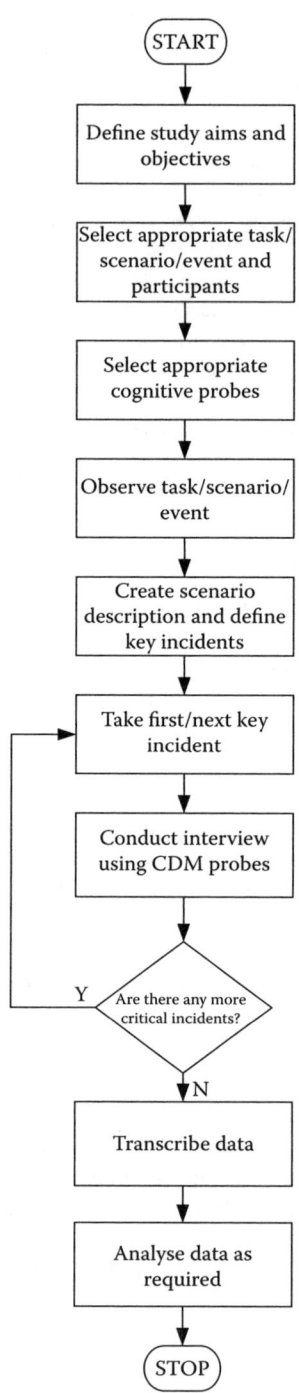

FLOWCHART 4.1 Critical decision method flowchart.

Cognitive Task Analysis

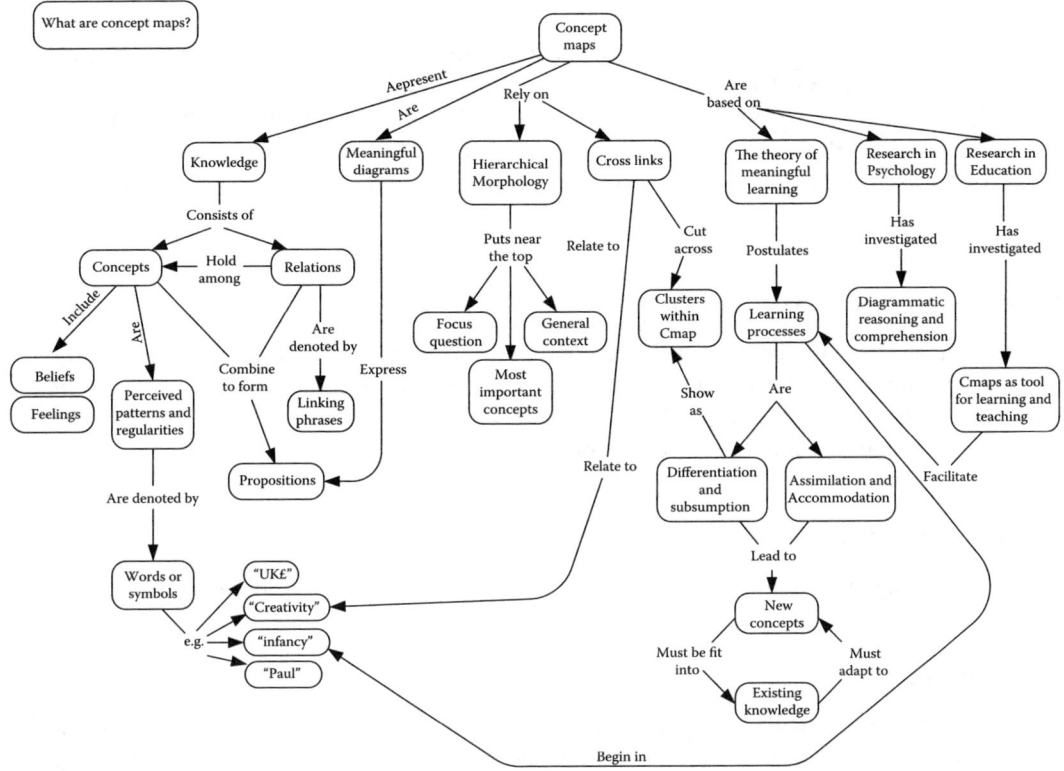

FIGURE 4.8 Concept map about concept maps. Adapted from Crandall et al. 2006.

4. The method is a popular one and has been applied in a number of different domains.
5. The flexibility of the approach allows all manner of Human Factors concepts to be studied, including decision making, situation awareness, human error, workload, and distraction.
6. The output can be used for a range of purposes, including performance evaluation, training and teaching materials development, and concept analysis.
7. The approach has sound underpinning theory.
8. The concept map output provides a neat representation of participant knowledge.

Disadvantages

1. The output offers little direct input into design.
2. For complex concepts the process may be difficult and time consuming and the output may become complex and unwieldy.
3. Many revisions and iterations are normally required before the concept map is complete, even for simplistic analyses.
4. The data obtained are highly dependent upon the skill of the interviewer and the quality and willingness to participate of the interviewee. A high level of skill and expertise is required in order to use the concept map method to its maximum effect.
5. It is often difficult to gain the levels of access to SMEs that are required for the concept map method.

Related Methods

The concept map method is a knowledge elicitation approach that is often used in CTA efforts. Other network-based knowledge representation methods exist, including the propositional network approach (Salmon et al., 2009) and semantic networks (Eysenck and Keane, 1990). Concept maps use interviews and walkthrough type analyses as the primary form of data collection. Concept maps might also be used to identify the situation awareness requirements associated with a particular task or concept.

Approximate Training and Application Times

The training time for the concept map method is low. The application time is also typically low, although for more complex concepts or tasks this may increase significantly. Crandall, Klein, and Hoffman (2006) suggest that typical concept map knowledge elicitation sessions take around 1 hour, which normally produces two semi-refined concept maps.

Reliability and Validity

No data regarding the reliability and validity of the method are presented in the literature.

Tools Needed

Primarily a representational method, concept maps can be constructed simply using pen and paper, whiteboard, or flipcharts; however, for the purposes of reports and presentations it is normally useful to construct them using a drawing software package such as Microsoft Visio. Crandall, Klein, and Hoffman (2006) point to a number of software packages that have been developed to support concept map analyses (e.g., Chung, Baker, and Cheak, 2002; Hoeft et al., 2002; both cited in Crandall, Klein, and Hoffman, 2006) and also CmapTools, which is a free downloadable tool that can be found at www.ihmc.us.

Example

An example concept map for identifying the factors that a golfer considers when determining an appropriate strategy for an approach shot is presented in Figure 4.9. The concept map was developed based on discussions and walkthrough analyses held with one golf SME. The aim of constructing the concept map was to identify the knowledge, and underpinning information, that an elite golfer uses when making an approach shot, including the process of shot and club selection, shot visualization, and shot completion.

Flowchart

(See Flowchart 4.2.)

Recommended Text

Crandall, B., Klein, G., and Hoffman, R. (2006). *Working minds: A practitioner's guide to cognitive task analysis.* Cambridge, MA: MIT Press.

Cognitive Task Analysis

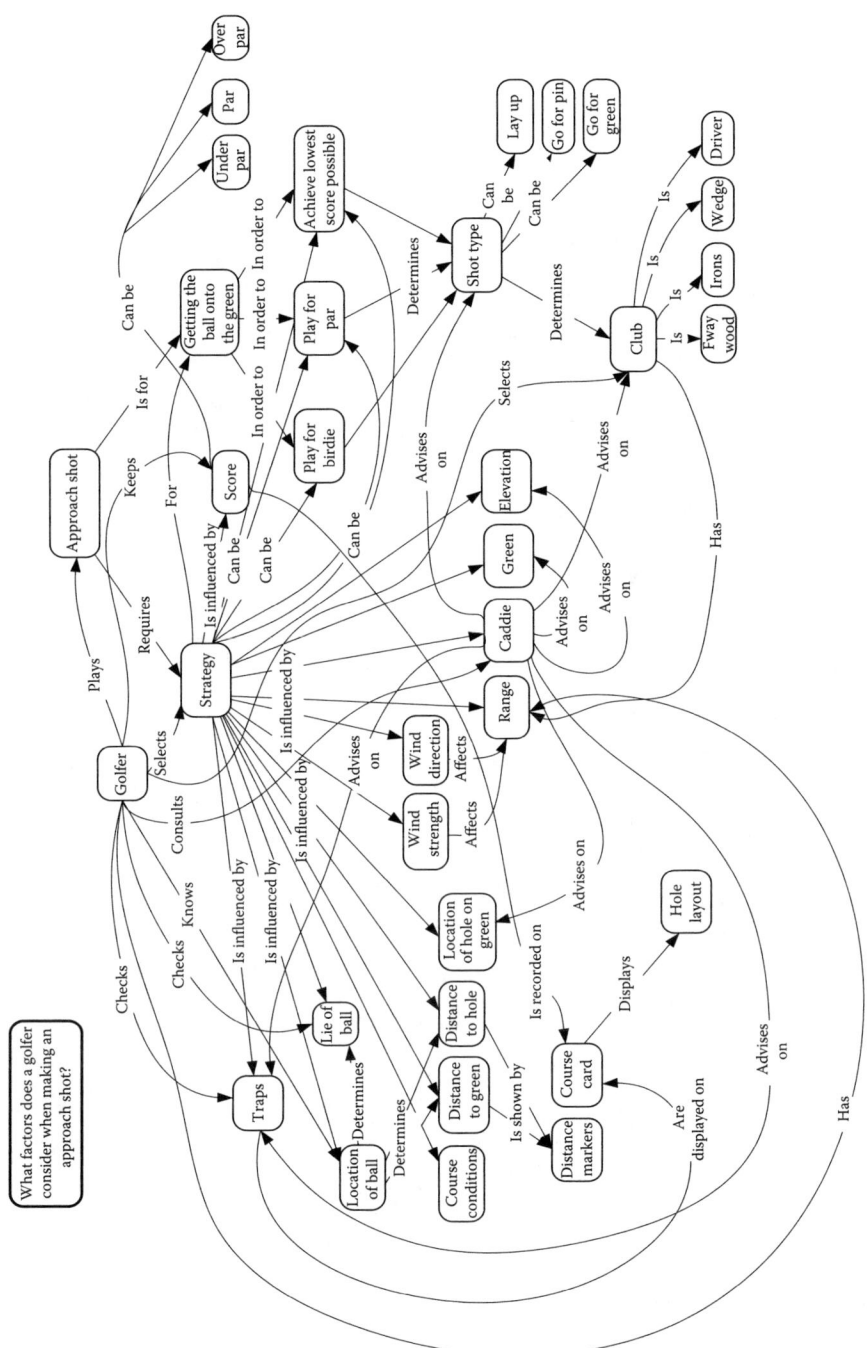

FIGURE 4.9 Example concept map for identifying the factors considered by golfers when determining approach shot strategy.

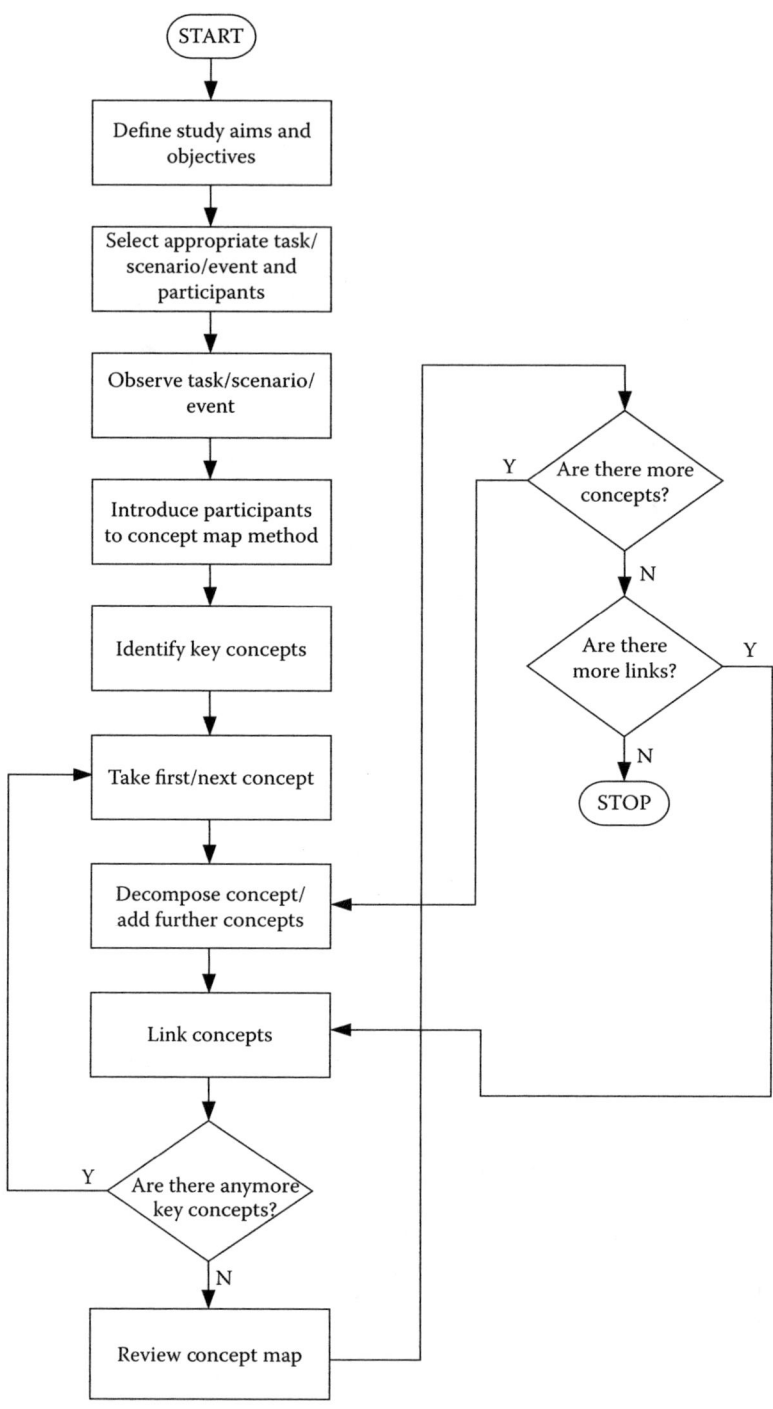

FLOWCHART 4.2 Concept map flowchart.

APPLIED COGNITIVE TASK ANALYSIS

BACKGROUND AND APPLICATIONS

Applied Cognitive Task Analysis (ACTA; Militello and Hutton, 2000) offers a toolkit of complementary interview methods that can be used to identify the cognitive demands associated with tasks. The ACTA framework was originally developed as a solution to the inaccessibility and difficulty of existing CTA methods and was designed with the aim that analysts with little or no training in cognitive psychology could use it (Militello and Hutton, 2000). ACTA comprises the following three interview methods designed to allow the analyst to elicit pertinent information surrounding operator decision making:

1. *Task diagram interview* The task diagram interview is used to provide the analyst with an in-depth overview of the task under analysis. In particular, those elements of the task that are cognitively challenging are identified.
2. *Knowledge audit interview* The knowledge audit interview is used to highlight those parts of the task under analysis where expertise is required. Once examples of expertise are highlighted, the SME is probed for specific examples within the context of the task.
3. *Simulation interview* The simulation interview is used to identify and investigate the cognitive processes used by the SME during the task under analysis.

Upon completion of the three interviews, a cognitive demands table is used to integrate and present the data obtained from the ACTA interviews.

DOMAIN OF APPLICATION

ACTA was developed as part of a Naval research programme; however, the approach is generic and can be applied in any domain. Militello and Hutton (2000), for example, describe an application focussing on fire-fighters.

APPLICATION IN SPORT

There is great potential for applying ACTA in a sporting context; the approach offers a means of gathering data regarding the cognitive processes used during task performance and is a generic approach so is suited to the analysis of sports performance. In particular, the method appears to be suited to identifying and analysing the differences between the processes adopted by sports performers of differing abilities and expertise levels (e.g., elite versus amateur sports performers). Further, it may also be useful for identifying training, decision making, and situation awareness requirements for different sports tasks.

PROCEDURE AND ADVICE

Step 1: Define the Task under Analysis
The first part of an ACTA analysis is to select and define the task or scenario under analysis. This is dependent upon the nature and focus of the analysis. Exactly what the aims of the analysis to be undertaken are should be clearly defined so that appropriate scenarios and participants can be focussed on.

Step 2: Select Appropriate Participant(s)

Once the scenario under analysis is defined, the analyst(s) should proceed to identify an appropriate SME or set of SMEs. Typically, operators of the system under analysis or SMEs for the task under analysis are used.

Step 3: Observe Task or Scenario under Analysis

If the analysis is based on direct observation, then this step involves the analyst(s) observing the task or scenario. An observational transcript should be completed for the scenario, including a description of the scenario, the actors and activities involved, the technology used, the interactions between actors and between actors and technology, and the outcome of the scenario. Otherwise, the task can be described retrospectively from memory by an appropriate SME, or a walkthrough of the task may suffice. Following step 3, the analyst should fully understand the task and the participant's role during task performance.

Step 4: Conduct Task Diagram Interview

The purpose of the task diagram interview is to elicit a broad overview of the task under analysis in order to focus the knowledge audit and simulation interview parts of the analysis. Once the task diagram interview is complete, the analyst should have created a diagram representing the component task steps involved and those task steps that require the most cognitive skill. According to Militello and Hutton (2000), the SME should first be asked to decompose the task into relevant task steps. The analyst should use questions like, "Think about what you do when you (perform the task under analysis)," and "Can you break this task down into less than six, but more than three steps?" (Militello and Hutton, 2000). Once the task is broken down, the SME should then be asked to identify which of the task steps require cognitive skills. Militello and Hutton (2000) define cognitive skills as judgements, assessments, and problem-solving and thinking skills.

Step 5: Conduct Knowledge Audit Interview

Next, the analyst should proceed with the knowledge audit interview. This allows the analyst to identify instances during the task under analysis where expertise is required, and also what sort of expertise is used. The knowledge audit interview uses a series of predefined probes based upon the following knowledge categories that characterise expertise (Militello and Hutton, 2000):

- Diagnosing and predicting
- Situation awareness
- Perceptual skills
- Developing and knowing when to apply tricks of the trade
- Improvising
- Meta-cognition
- Recognising anomalies
- Compensating for equipment limitations

Once a probe has been administered, the analyst should then query the SME for specific examples of critical cues and decision-making strategies. Potential errors should also be discussed. The list of knowledge audit interview probes typically used is presented below (Militello and Hutton, 2000).

Basic Probes
1. *Past and Future*   Is there a time when you walked into the middle of a situation and knew exactly how things got there and where they were headed?
2. *Big Picture*   Can you give me an example of what is important about the big picture for this task? What are the major elements you have to know and keep track of?

3. *Noticing*   Have you had experiences where part of a situation just "popped" out at you, where you noticed things going on that others didn't catch? What is an example?
4. *Job Smarts*   When you do this task, are there ways of working smart or accomplishing more with less that you have found especially useful?
5. *Opportunities/Improvising*   Can you think of an example when you have improvised in this task or noticed an opportunity to do something better?
6. *Self-Monitoring*   Can you think of a time when you realised that you would need to change the way you were performing in order to get the job done?

Optional Probes
1. *Anomalies*   Can you describe an instance when you spotted a deviation from the norm, or knew something was amiss?
2. *Equipment Difficulties*   Have there been times when the equipment pointed in one direction but your own judgement told you to do something else? Or when you had to rely on experience to avoid being led astray by the equipment?

Step 6: Conduct Simulation Interview

The simulation interview allows the analyst to determine the cognitive processes involved during the task under analysis. The SME is presented with a typical scenario. Once the scenario is completed, the analyst should prompt the SME to recall any major events, including decisions and judgements that occurred during the scenario. Each event or task step in the scenario should be probed for situation awareness, actions, critical cues, potential errors, and surrounding events. Militello and Hutton (2000) present the following set of simulation interview probes:

For each major event, elicit the following information:
- As the (job you are investigating) in this scenario, what actions, if any, would you take at this point in time?
- What do you think is going on here? What is your assessment of the situation at this point in time?
- What pieces of information led you to this situation assessment and these actions?
- What errors would an inexperienced person be likely to make in this situation?

Any information elicited here should be recorded in a simulation interview table. An example simulation interview table is shown in Table 4.6.

Step 7: Construct Cognitive Demands Table

Once the knowledge audit and simulation interview are completed, it is recommended that a cognitive demands table be used to integrate the data collected (Militello and Hutton, 2000). This table is used to help the analyst focus on the most important aspects of the data obtained. The analyst should prepare the cognitive demands table based upon the goals of the particular project involved. An example of a cognitive demands table is presented in Table 4.7 (Militello and Hutton, 2000).

Advantages

1. ACTA offers a structured approach to CTA and aids the analyst through the provision of pre-defined probes.
2. The ACTA probes are quite comprehensive, covering various aspects of the cognitive elements of task performance.
3. The ACTA approach is generic and seems suited to analysing the cognitive requirements associated with sporting tasks.

TABLE 4.6
Example Simulation Interview Table

Events	Actions	Assessment	Critical Cues	Potential Errors
On scene arrival	Account for people (names) Ask neighbours Must knock on or knock down to make sure people aren't there	It's a cold night, need to find place for people who have been evacuated	Night time Cold >15° Dead space Add on floor Poor materials, metal girders Common attic in whole building	Not keeping track of people (could be looking for people who are not there)
Initial attack	Watch for signs of building collapse If signs of building collapse, evacuate and throw water on it from outside	Faulty construction, building may collapse	Signs of building collapse include: What walls are doing—cracking What floors are doing—groaning What metal girders are doing—clicking, popping Cable in old buildings hold walls together	Ventilating the attic, this draws the fire up and spreads it through the pipes and electrical system

Source: Militello, L. G., and Hutton, J. B. (2000). Applied cognitive task analysis (ACTA): A practitioner's toolkit for understanding cognitive task demands. In *Task analysis*, eds. J. Annett and N. S. Stanton, 90–113. London: Taylor & Francis.

4. ACTA was originally conceived for use by analysts with no training in cognitive psychology and so is relatively easy to learn and apply.
5. The training time for such a comprehensive method is minimal. In a validation study, Militello and Hutton (2000) gave participants an initial 2-hour workshop introducing CTA followed by a 6-hour workshop introducing the ACTA method.
6. The use of three different interviews ensures the method's comprehensiveness.
7. Militello and Hutton (2000) report positive ratings for the approach in a usability study. The findings indicate that the method was flexible and easy to use, and that the outputs generated were both clear and useful.
8. The provision of probes and questions facilitates relevant data extraction.

DISADVANTAGES

1. Compared to other CTA methods, such as CWA and CDM, there are only limited applications of the method reported in the academic literature.
2. The use of three interviews ensures that ACTA is time consuming in its application.
3. The process may be laborious, both for analysts and interviewees, since there appears to be some repetition between the three interviews used. The data analysis component may also be highly laborious and time consuming.
4. The output offers little direct input into design.
5. The data obtained are highly dependent upon the skill of the interviewer and the quality and willingness to participate of the interviewee.
6. Designing, conducting, transcribing, and analysing interviews is an extremely time-consuming process, and the large amount of time required may limit the number of participants that can be used.

TABLE 4.7
Example Cognitive Demands Table

Difficult Cognitive Element	Why Difficult?	Common Errors	Cues and Strategies Used
Knowing where to search after an explosion	Novices may not be trained in dealing with explosions. Other training suggests you should start at the source and work outward	Novice would be likely to start at the source of the explosion. Starting at the source is a rule of thumb for most other kinds of incidents	Start where you are most likely to find victims, keeping in mind safety considerations. Refer to material data sheets to determine where dangerous chemicals are likely to be. Consider the type of structure and where victims are likely to be. Consider the likelihood of further explosions. Keep in mind the safety of your crew
Finding victims in a burning building	There are lots of distracting noises. If you are nervous or tired, your own breathing makes it hard to hear anything else	Novices sometimes don't recognise their own breathing sounds; they mistakenly think they hear a victim breathing	Both you and your partner stop, hold your breath, and listen. Listen for crying, victims talking to themselves, victims knocking things over, etc.

Source: Militello, L. G., and Hutton, J. B. (2000). Applied cognitive task analysis (ACTA): A practitioner's toolkit for understanding cognitive task demands. In *Task analysis*, eds. J. Annett and N. S. Stanton, 90–113. London: Taylor & Francis.

7. The reliability and validity of interview methods is difficult to address. Klein and Armstrong (2004), for example, point out that methods that analyse retrospective incidents may have limited reliability due to factors such as memory degradation.
8. Since it involves the conduct of interviews, the ACTA approach is susceptible to a range of interviewer and interviewee biases.
9. When analysing real-world tasks, participants often do not have sufficient free time to fully engage in interviews. Often interviewers are given only a short period of time to collect their data.
10. It is often difficult to gain the levels of access to SMEs that are required for successful completion of ACTA interviews.

RELATED METHODS

ACTA is an interview-based CTA method and so is similar to the CDM approach. ACTA can also utilise other data collection methods such as direct observation and walkthrough methods.

APPROXIMATE TRAINING AND APPLICATION TIMES

In a validation study (Militello and Hutton, 2000), participants were given 8 hours of training, comprising a 2-hour introduction to CTA and a 6-hour workshop on the ACTA method. In the same study the application time was 7 hours, consisting of 3 hours for the conduct of the interviews and 4 hours for data analysis.

Reliability and Validity

Militello and Hutton (2000) point out that there are no well-established metrics for testing the reliability and validity of CTA methods. In an attempt to establish the reliability and validity of the ACTA method they addressed the following questions:

1. Does the information gathered address cognitive issues?
2. Does the information gathered deal with experience-based knowledge as opposed to classroom-based knowledge?
3. Do the instructional materials generated contain accurate information that is important for novices to learn?

Each item in the cognitive demand tables was examined for its cognitive content. The analysis found that 93% of the items were related to cognitive issues. To establish the level of experience-based knowledge elicited, participants were asked to subjectively rate the proportion of information that only highly experienced SMEs would know. In one study, the average proportion was 95%, and in another it was 90%. Finally, the importance of the instructional materials generated was validated via domain experts rating their importance and accuracy. The findings indicated that the majority (70% in one study, 90% in another) of the materials generated contained important information for novices. Reliability was assessed by determining whether the analysts involved generated similar information. Militello and Hutton (2000) concluded that analysts were able to consistently elicit relevant cognitive information using ACTA.

Tools Needed

ACTA can be applied using pen and paper only, providing the analyst has access to the ACTA probes required during the knowledge audit and simulation interviews. It is recommended that an audio recording device be used to record the interviews conducted. A word processing software programme such as Microsoft Word is required for creating interview transcripts and output tables.

Flowchart

(See Flowchart 4.3.)

Recommended Text

Militello, L. G., and Hutton, J. B. (2000). Applied cognitive task analysis (ACTA): A practitioner's toolkit for understanding cognitive task demands. In *Task Analysis*, eds. J. Annett, and N. A. Stanton, 90–113. London: Taylor & Francis.

Cognitive Task Analysis

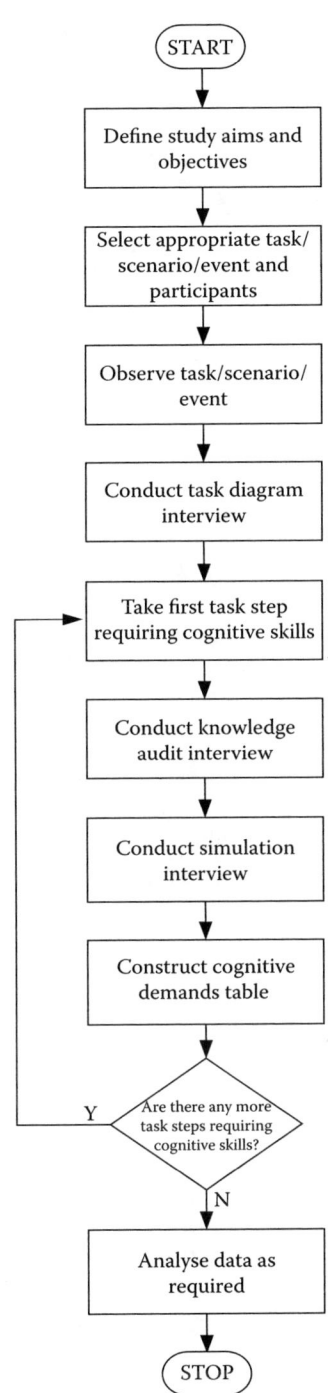

FLOWCHART 4.3 ACTA flowchart.

5 Human Error Identification and Analysis Methods

INTRODUCTION

Human error is probably the most well known and widely publicised of all Human Factors concepts. High profile incidents such as the Three Mile Island and Chernobyl nuclear power disasters, the Tenerife and Kegworth air disasters, the Herald of Free Enterprise ferry disaster, and the Ladbroke Grove rail disaster have ensured that human error has received considerable attention, not only from the relevant safety and research communities, but also from the public. As a corollary, human error has been investigated across a wide range of safety critical domains; however, it continues to be identified as a contributory factor in a high proportion of accidents and incidents. For example, typical estimates suggest human error plays a role in around 70% of all commercial aviation accidents and incidents (e.g., BASE, 1997; cited in McFadden and Towell, 1999), 95% of all road traffic incidents (Rumar, 1995), and over half of all UK rail network collisions (Lawton and Ward, 2006). Further, within the health care domain, the U.S. Institute of Medicine estimates that between 44,000 and 88,000 people die each year as a result of human errors (Helmreich, 2000).

Although the costs are not typically as severe, the concept is just as captivating within the sporting domains, and the history books in most sports are littered with accounts of players, coaches, and game officiators making errors that have ultimately decided the outcome of games and cost titles, championships, cups, careers, and in the worst cases even human life. Recent high profile examples include golfer Ian Woosnam, leading in the final round of the British Open tournament in 2001, incurring a two-shot penalty for carrying 15 clubs instead of the regulation 14 (the error was blamed on an oversight by Woosnam's caddie). In soccer, the England international team have missed numerous penalties during recent World Cup and European championship knockout stages (e.g., in the semifinals of World Cup 1990 and Euro 1996, and in the quarterfinals of World Cup 1998, Euro 2004, and World Cup 2006). In golf's Ryder Cup, Bernhard Langer missed a 3-foot putt on the final green, costing the European team victory at Kiawah Island in 1991. In one of the biggest sporting collapses of recent times, Jean Van de Velde, leading by three shots in the 1999 British Open, inexplicably triple bogeyed the final hole and eventually lost golf's most prestigious title in a three-way playoff.

Errors are not confined to sporting performance alone; errors in tactical decisions, team selection, training, equipment selection, and preparation can all be catastrophic in terms of the resulting performance. A recent example was the England international soccer team manager Steve McLaren's selection of rookie goalkeeper Steve Carson for England's final Euro 2008 qualification game (Carson went on to make a catastrophic error which led to Croatia's first goal and a subsequent 3-2 defeat). In addition, errors made by match officials (e.g., referees) are currently controversial in light of the continuing debate over whether technology should be used to aid officiator decisions. Recent high profile examples of match official errors in soccer include referee Mark Clattenburg's (and his assistant's) decision not to award a goal for Pedro Mendes' strike for Tottenham Hotspur against Manchester United, even though the ball was clearly over the line and referee Graham Poll's booking of the same player three times in the World Cup 2006 game between Australia and Croatia (a second booking should result in the player being sent off the field of play). More recently, in what has been labelled as a "blunder of monumental ineptness" (Montgomery, 2008), referee Stuart Attwell awarded a goal, on the advice of his assistant, in a Coca Cola Championship match between Watford and Reading, when the ball had actually gone out of play 4 yards wide of the goal.

It is notable that, within tennis and rugby union, technology is now used as a backup for umpiring (e.g., Hawk-Eye) or refereeing and touch judge decisions; however, despite repeated calls for similar technology to be introduced in soccer, at the time of writing this has not yet happened.

Despite its prolificacy, there has been little investigation into the concept of human error within the sporting arena. This is despite the fact that sporting errors are often the subject of great debate amongst sports fans and professionals; how, they ask, can highly trained elite professionals make such basic errors? Why do these errors continue to occur from game to game? How can they be eradicated? As a result of over three decades of research into the concept, Human Factors researchers now have a considerable understanding of human error, its causes and consequences, and how it can be eradicated or managed. Further, various approaches have been developed that allow us to comprehensively analyse the errors made and even predict the errors that are likely to occur in particular situations. Most of the research into human error, and the methods developed to investigate it, are applicable across domains and there is great scope to apply these approaches within the sporting domain. This chapter presents a brief summary of the concept of human error along with a description of the different approaches that Human Factors researchers use to understand what errors are made or are likely to be made, why they are made, and how they can be prevented or their consequences mitigated.

DEFINING HUMAN ERROR

At its simplest, human error can be defined as the performance of an incorrect or inappropriate action, or a failure to perform a particular action, that leads to an unwanted outcome. Hollnagel (1993; cited in Strauch, 2002) define errors as "an action which fails to produce the expected result and which therefore leads to an unwanted consequence." Senders and Moray (1991) suggest that an error is something that has been done which was:

- Not intended by the actor;
- Not desired by a set of rules or an external observer; or
- That led the task or system outside of its acceptable limits.

Probably the most widely recognised definition of human error is offered by Reason (1990), who formally defines human error as "a generic term to encompass all those occasions in which a planned sequence of mental or physical activities fails to achieve its intended outcome, and when these failures cannot be attributed to the intervention of some chance agency."

ERROR CLASSIFICATIONS

At the most basic level of error classification, a distinction between errors of omission and errors of commission is proposed. Errors of omission are those instances where an actor fails to act at all, whereas errors of commission are those instances where an actor performs an action incorrectly or at the wrong time. Payne and Altman (1962; cited in Isaac, Shorrock, Kennedy, Kirwan, Anderson, and Bove, 2002) proposed a simplistic information processing theory–based error classification scheme containing the following error categories:

1. *Input errors*—those errors that occur during the input sensory and perceptual processes, e.g., visual perception and auditory errors
2. *Mediation errors*—those errors that occur or are associated with the cognitive processes employed between the perception and action stages

3. *Output errors*—those errors that occur during the selection and execution of physical responses

The most commonly referred to error classification within the literature, however, is the slips and lapses, mistakes and violations classification proposed by Reason (1990), an overview of which is presented below.

SLIPS AND LAPSES

The most common form of human error is slip-based errors. Slips are categorised as those errors in which the intention or plan was correct but the execution of the required action was incorrect. In a sporting context, examples of slip-based errors would be when a soccer player intending to kick the ball inadvertently misses the ball completely or strikes it incorrectly, or when a golfer intending to drive onto the fairway inadvertently drives the ball out of bounds. In both cases, the intention (i.e., to kick the ball or drive the ball onto the fairway) was correct, but the physical execution of the required action was incorrect (i.e., failing to make contact with the ball or striking the golf ball erroneously). Slips are therefore categorised as actions with the appropriate intention followed by the incorrect execution, and are also labelled action execution failures (Reason, 1990).

Lapse-based errors refer to more covert error forms that involve a failure of memory that may not manifest itself in actual behaviour (Reason, 1990). Lapses typically involve a failure to perform an intended action or forgetting the next action required in a particular sequence. Examples of lapses within the sporting context include a defender failing to mark his allotted opposition striker on a set piece, or a caddie failing to check the golfer's clubs prior to commencing the round. Whereas slips occur at the action execution stage, lapses occur at the storage stage, whereby intended actions are formulated prior to the execution stage of performing them.

MISTAKES

While slips reside in the observable actions made by operators, mistakes reside in the unobservable plans and intentions that they form, and are categorised as an inappropriate intention or wrong decision followed by the correct execution of the required action. Mistakes occur when actors intentionally perform a wrong action and therefore originate at the planning level, rather than the execution level (Reason, 1990). Mistakes are rife in the sporting arena and examples include choosing to play the wrong pass, selecting the wrong shot to play, selecting the wrong tactics, and choosing to make an incorrect manoeuvre. According to Reason (1990), mistakes involve a mismatch between the prior intention and the intended consequences and are likely to be more subtle, more complex, less well understood, and harder to detect than slips. Reason (1990) defines mistakes as "deficiencies or failures in the judgmental and/or inferential processes involved in the selection of an objective or in the specification of the means to achieve it, irrespective of whether or not the actions directed by this decision-scheme run according to plan."

VIOLATIONS

Another, altogether more complex category of error is violations. Violations are categorised as any behaviour that deviates from accepted procedures, standards, and rules. Violations can be either deliberate or unintentional (Reason, 1997). Deliberate violations occur when an actor deliberately deviates from a set of rules or procedures. This form of violation is rife in sport, and typically involves players knowingly breaking the rules, for example, tackling from behind in soccer or making a high tackle in rugby. Erroneous or unintentional violations, whereby players unintentionally break the

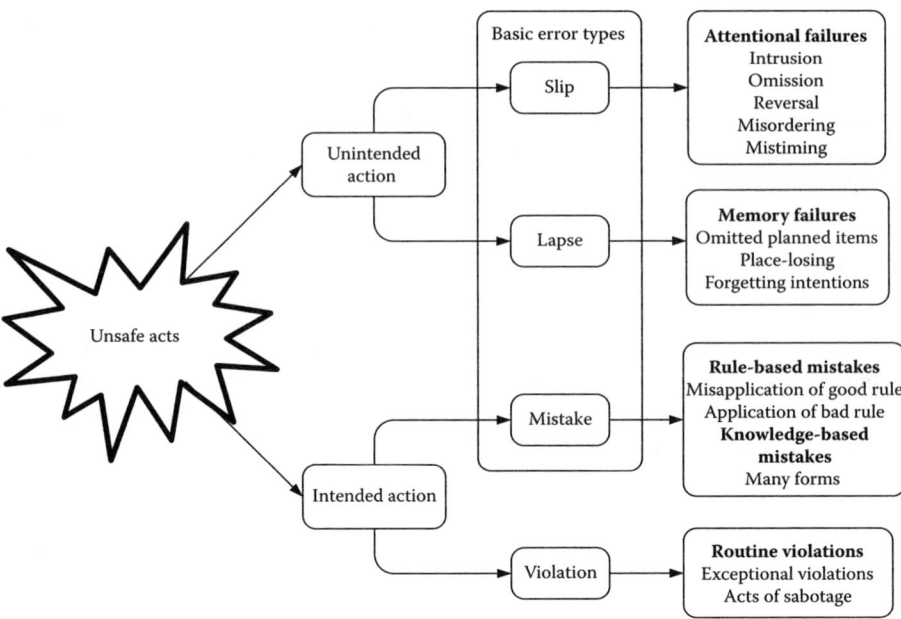

FIGURE 5.1 Unsafe acts taxonomy. *Source*: Reason, J. (1990). *Human error*. New York: Cambridge University Press.

rules, are also common; one example of an unintentional violation is when players attempting to make a legitimate tackle instead commit a foul of some sort.

In addition to the simplistic slips and lapses, mistakes and violations classification described above, further error types have been specified within each category. For example, Reason (1990) proposed a taxonomy of unsafe acts, which prescribes a number of different error types within each of the four error categories. The taxonomy of unsafe acts is presented in Figure 5.1.

THEORETICAL PERSPECTIVES ON HUMAN ERROR

There are two distinct schools of thought concerning human error, namely the *person* and *system perspective* approaches (Reason, 2000). Early research into human error focused upon the tendency that operators had for making errors at the so-called "sharp end" of system operation; it looked at the problem from an individual operator or *person* perspective. Johnson (1999) describes how public attention was focused upon the human contribution to system failure during the 1970s and 1980s due to a number of high profile catastrophes, including the Flixborough, Seveso, Three Mile Island, Bhopal, and Chernobyl disasters. In recent years, however (and in line with most other Human Factors constructs), the focus on human error in complex systems has shifted from the individual operator onto the system as a whole, to consider the complex interaction between error-causing conditions residing within systems and the consequential errors made by operators performing activity within them. No longer seen to be at fault for errors made, individual operators are now seen as victims of poorly designed systems and procedures, and errors are treated as the end product of a chain of systemic failures. This so-called *systems* view on human error and accident causation first began to gain recognition in the late 1980s due to a number of high profile accident investigations that highlighted the contribution of so-called latent failures (i.e., error-causing conditions). For example, Johnson (1999) points out how investigators focussed upon managerial factors in the wake of the Challenger, Piper Alpha, Hillsborough, and Narita catastrophes.

Making the same distinction but with different labels, Dekker (S. W. A., 2002) describes the *old view* and *new view* on human error. In the old view, human error is treated as the cause of most accidents; the systems in which people work are safe; the main threat to system safety is human unreliability; and safety progress is achieved by protecting systems from human unreliability through automation, training, discipline, selection, and proceduralisation. In the new view, however, human error is treated as a symptom of problems within the system; safety is not inherent within systems; and human error is linked to the tools used, tasks performed, and operating environment. A brief overview of both perspectives is given below.

THE PERSON APPROACH

The person approach focuses upon the errors that operators make at the "sharp end" of system operation (Reason, 2000) and views error occurrence as the result of psychological factors within an individual. According to the person approach, errors arise from aberrant mental processes such as forgetfulness, inattention, poor motivation, carelessness, negligence, and recklessness (Reason, 2000). Person approach–related research typically attempts to identify the nature and frequency of the errors made by operators within complex systems, the ultimate aim being to propose strategies, remedial measures, and countermeasures designed to prevent future error occurrence. While person-based research is worthwhile for these reasons, it is often criticised for its contribution to individualistic blame cultures within organisations (e.g., operator X made these errors so the incident was operator X's fault). Also, person approach–based error countermeasures, such as poster campaigns, additional procedures, disciplinary measures, threat of litigation, retraining, and naming, blaming, and shaming (Reason, 2000), are ultimately focussed upon reducing the variability in human behaviour rather than flaws in the system itself. Despite these failings, however, it still remains the dominant approach in a number of domains, including healthcare (Reason, 2000) and road transport (Salmon, Regan, and Johnston, 2007). Person-based approach models of human error include Rasmussen's SRK framework (Rasmussen, 1983; cited in Vicente, 1999), Reason's Generic Error Modelling System (GEMS; Reason, 1990), and Rasmussen's model of human malfunction (Rasmussen, 1982).

THE SYSTEMS PERSPECTIVE APPROACH

The systems perspective approach treats error as a systems failure, rather than an individual operator's failure. Systems approaches consider the presence of latent or error-causing conditions within systems and their role in the errors made at the sharp end by operators, purporting that the errors are a consequence of the error-causing conditions present within the system. Unlike the person approach, human error is no longer seen as the primary cause of accidents; rather, it is treated as a consequence of the latent failures residing within the system. The notion that human error is a consequence of latent failures rather than a cause of catastrophes was first entertained by Chapanis in the 1940s (1949; cited in Stanton and Baber, 2002) who, in conclusion to an analysis of a series of crashes in which the landing gear was not lowered, found that pilots were erroneously adjusting the flaps. Chapanis suggested that in this case "pilot error" was really "designer error," since the landing gear and flap controls were identical and were located adjacent to one another.

The systems perspective model of human error and accident causation (Reason, 1990) is the most influential and widely recognised systems approach model. Reason's (1990) "Swiss cheese" model, as it is more commonly known, focuses on the interaction between latent conditions and errors and their contribution to organisational accidents. According to the model, complex systems consist of various levels, each of which contributes to production (e.g., decision makers, line management, productive activities, and defences). Each layer has defences, such as protective equipment, rules and regulations, training, checklists, and engineered safety features, which are designed to prevent the occurrence of occupational accidents. Weaknesses in these defences, created by latent conditions and unsafe acts, create "windows of opportunity" for accident trajectories to breach the defences and cause an accident. Accidents and incidents occur when the holes in the systems defences line up

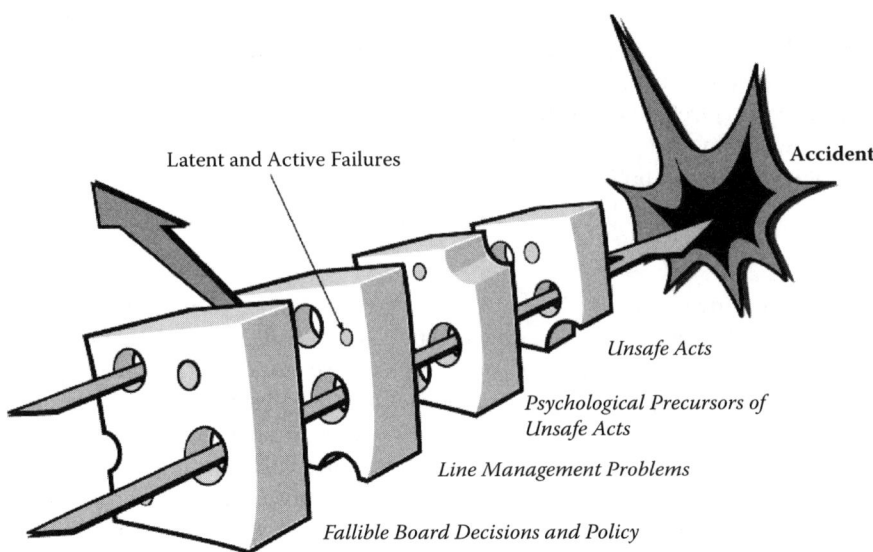

FIGURE 5.2 Reason's Swiss cheese systems perspective on error and accident causation. *Source*: Adapted from Reason, J. (2000) Human error: Models and management. *British Medical Journal* 320:768–70.

in a way that allows the accident trajectory to breach each of the different layers. Reason's systems perspective model is presented in Figure 5.2.

Within Reason's model, latent conditions and unsafe acts combine to produce organisational accidents and incidents. Latent conditions are inadequate conditions or failures residing throughout a system and include poor designs, inadequate supervision, manufacturing defects, maintenance failures, inadequate training, clumsy automation, inappropriate or ill-defined procedures, inadequate equipment, and procedural short cuts, to name only a few (Reason, 1997). Unsafe acts, on the other hand, represent those errors that are made by human operators that have an immediate impact upon system safety.

At the fallible board decisions and policy level, fallible decisions made by system designers and higher-level management create latent conditions within the system. Examples of fallible board decisions and policy include the selection of inadequately designed equipment, the vetoing of system improvement measures, and the use of policies that incur time pressure on actors within the system. The next level of Reason's model is the line management level. According to Reason (1990), line management problems arise from incompetent management and also the fallible board decisions and policy from the preceding level in the model. Line management problems represent those instances where management is either inadequate and/or inappropriate. Examples of line management problems include inadequate management or supervision and the use of inadequate or inappropriate training and procedures. The psychological precursors of unsafe acts failure level refers to latent states that create the potential for unsafe acts. According to Reason (1990), the precise nature of unsafe acts is defined through the complex combination of a number of factors, including the task being performed, the environmental conditions, and the presence of hazards. Each precursor can contribute to a great number of unsafe acts, depending upon the associated conditions. Examples of these precursors include poor motivation, negative attitudes, and a failure to perceive hazards.

Reason's model can be easily applied to the sporting arena. Just like complex safety-critical systems, sporting systems too have error-causing latent conditions residing within them. Inadequate management, training, preparation, tactics, and equipment, and poor player motivation and condition are all examples of latent conditions that can cause errors in different sporting domains. For example, within the UK soccer domain, the fallible board decisions and policy level could represent the level of government and sporting bodies, such as the Football Association (FA) and the

Professional Footballers Association (PFA). Examples of poor decisions and policies here might include a lack of investment in the development of coaching and in grassroots coaching. The line management level could represent the board and management of a particular sporting team or organization, such as the chairman, the board, manager, and coaches. Examples of inadequate management at this level include inadequate coaching and training techniques, lack of investment, and poor tactics selection on behalf of the management. The psychological precursors level relates to the psychological conditions of the managers and players and includes factors such as lack of motivation, negative attitude, and mental fatigue. Finally, the unsafe acts level represents those errors made by sports players as a result of the preceding latent conditions.

HUMAN ERROR METHODS

Such has been the level of investigation into the concept over the past three decades, Human Factors researchers now have a variety of approaches designed to investigate, eradicate, remove, or mitigate human error. These approaches include error management systems, human error analysis, and human error identification methods. This chapter focuses on human error identification and human error analysis methods.

HUMAN ERROR ANALYSIS METHODS

One of the most obvious means of collecting and analysing error data is the retrospective investigation and analysis of accidents and incidents involving human error. Accident investigation is used to reconstruct accidents and identify the human and system contribution to a particular accident or incident and allows researchers to identify exactly what happened and why, and then use the findings to ensure similar accidents do not occur again. There are various accident investigation and analysis methods available (our review identified over 30 accident analysis–related methods), although most are domain specific. For the purposes of this chapter, we focus on two generic approaches: fault tree analysis (Kirwan and Ainsworth, 1992) and Accimaps (Svedung and Rasmussen, 2002). Fault tree analysis is used to depict accident or failure scenarios. Most commonly used for probabilistic safety assessment purposes in the nuclear power domain, fault trees are tree-like diagrams that define failure events and display the possible causes of an accident in terms of hardware failure or human error (Kirwan and Ainsworth, 1992). Accimap (Svedung and Rasmussen, 2002) is an accident analysis method that is used to graphically represent the systemic causal factors involved in a particular accident. An Accimap identifies and represents the failures involved in an incident at the following six main levels: government policy and budgeting; regulatory bodies and associations; local area government planning and budgeting (including company management); technical and operational management; physical processes and actor activities; and equipment and surroundings.

HUMAN ERROR IDENTIFICATION METHODS

Human Error Identification (HEI) or error prediction offers a proactive strategy for investigating human error in complex sociotechnical systems. The prediction of human error is used within risk assessments in order to identify potential error occurrence and determine the causal factors, consequences, and recovery strategies associated with the errors identified. The information produced is then typically used to highlight system design flaws, propose remedial design measures, identify procedural deficiencies, and quantify error incidence probabilities. HEI works on the premise that an understanding of a work task and the characteristics of the technology being used allows us to indicate potential errors that may arise from the resulting interaction (Stanton and Baber, 1996). First attempted in response to a number of high profile catastrophes attributed to human error in the nuclear and chemical processing domains (e.g., the Three Mile Island, Bhopal, and Chernobyl disasters), the use of HEI methods is now widespread, with applications in a wide range of domains, including nuclear power and petrochemical processing (Kirwan, 1996), air traffic control (Shorrock and Kirwan, 2002), aviation (Stanton et al., 2006), space operations (Nelson, Haney, Ostrom, and Richards, 1998), healthcare (Lane et al., 2007), and public technology (Baber and Stanton, 1996a).

HEI methods can first be broadly classified into two groups: qualitative and quantitative. Qualitative HEI methods are used to predict the different types of errors that may occur, while quantitative HEI methods are used to predict the numerical probability of the different errors occurring. There are a range of HEI methods available, from simple error taxonomy-based methods, which offer error modes linked to operator behaviours, to more sophisticated error quantification and computational error modelling methods. Our literature review identified over 50 HEI methods, which were subsequently broadly classified into the following categories:

- Taxonomy-based methods
- Error identifier prompt methods
- Error quantification methods
- Cognitive modelling methods
- Cognitive simulation methods

For the purposes of this chapter, only taxonomy-based methods and error identifier prompt methods are considered. A brief overview of the different approaches considered is presented below.

Taxonomy-Based Methods

Taxonomy-based HEI methods use External Error Mode (EEM) taxonomies to identify potential errors. Typically, EEMs are considered for each step in a particular task or scenario in order to identify credible errors that may arise. Taxonomic approaches are typically the most successful in terms of sensitivity and are the cheapest, quickest, and easiest to use. However, these methods depend greatly on the judgement of the analyst, and their reliability and validity may at times be questionable. For example, different analysts with different experience may make different error predictions for the same task (inter-analyst reliability). Similarly, the same analyst may make different judgements on different occasions (intra-analyst reliability).

This chapter focuses on the following taxonomy-based HEI methods: SHERPA (Embrey, 1986) and the Human Error Template (HET; Stanton et al., 2006). The SHERPA method (Embrey, 1986) was originally developed for use in the nuclear reprocessing domain and uses an EEM taxonomy linked to a behavioural taxonomy in order to identify the errors that are likely to emerge from man-machine interactions. The HET approach (Stanton et al., 2006) was developed for the civil aviation domain and uses a generic EEM taxonomy in order to identify the design-induced errors that are likely to emerge during task performance with a particular device or interface.

Error Identifier Methods

Error identifier methods use predefined prompts or questions in order to aid the analyst in identifying potential errors. Examples of error identifier prompts include, "Could the operator fail to carry out the act in time?" "Could the operator carry out the task too early?" and "Could the operator carry out the task inadequately?" (Kirwan, 1994). Prompts are typically linked to a set of error modes and associated error reduction strategies. While these methods attempt to remove the reliability problems associated with taxonomy-based approaches, they add considerable time to the analysis because each prompt must be considered. For the purposes of this chapter we focus on the Technique for Human Error Assessment (THEA; Pocock, Harrison, Wright, and Johnson, 2001). In addition, we also describe the Task Analysis for Error Identification (TAFEI; Baber and Stanton, 2002) method, which does not fall in either category. THEA (Pocock et al., 2001) is a highly structured error identifier–based approach that employs cognitive error analysis based upon Norman's (1988) model of action execution. THEA uses a scenario analysis to consider context and then employs a series of questions in a checklist-style approach based upon goals, plans, performing actions, and perception/evaluation/interpretation. Using a slightly different approach, TAFEI (Baber and Stanton, 1996a) is used to identify errors that occur when using technological devices, and combines an HTA of the task under analysis with State Space Diagrams (SSDs) for the device in question, to identify illegal interactions between the human and the device. TAFEI is suited only to those sports where performers use a technological device of some sort.

A summary of the human error methods described is presented in Table 5.1.

ACCIMAPS

BACKGROUND AND APPLICATIONS

Accimap (Rasmussen, 1997; Svedung and Rasmussen, 2002) is an accident analysis method that is used to graphically represent the systemic failures involved in accidents and incidents. The Accimap method differs from typical accident analysis approaches in that, rather than identifying and apportioning blame at the sharp end, it is used to identify and represent the causal flow of events upstream from the accident and looks at the planning, management, and regulatory bodies that may have contributed to the accident (Svedung and Rasmussen, 2002). Although the number of organisational levels can vary according to domain, a typical Accimap uses the following six main levels: government policy and budgeting; regulatory bodies and associations; local area government planning and budgeting (including company management); technical and operational management; physical processes and actor activities; and equipment and surroundings. Failures at each of the levels are identified and linked between and across levels based on cause-effect relations. Starting from the bottom of the list, the equipment and surroundings level provides a description of the accident scene in terms of the configuration and physical characteristics of the landscape, buildings, equipment, tools, and vehicles involved. The physical processes and actor activities level provides a description of the failures involved at the "sharp end." The remaining levels above the physical processes level represent all of the failures by decision makers that, in the course of the decision making involved in their normal work context, did or could have influenced the accident flow during the first two levels.

DOMAIN OF APPLICATION

Accimap is a generic approach that was designed for use in any complex safety-critical system. The method has been applied to a range of accidents and incidents, including gas plant explosions (Hopkins, 2000), police firearm mishaps (Jenkins et al., 2009), loss of space vehicles (Johnson and de Almeida, 2008), aviation accidents (Royal Australian Air Force, 2001), public health incidents (Woo and Vicente, 2003; Vicente and Christoffersen, 2006), and road and rail accidents (Svendung and Rasmussen, 2002; Hopkins, 2005).

APPLICATION IN SPORT

Accimap is suited to the analysis of failures in sports performance at the individual, team, and organisation levels; for example, at an organisational level, Accimaps might be used to investigate the failure of a sporting body or organisation to produce elite players in a particular sport, or a national team's poor performance in a major tournament. At the individual level, sporting failures can be evaluated in order to identify the wider systemic failures, such as inadequate training or poorly designed equipment.

TABLE 5.1
Human Error Assessment Methods Summary Table

Name	Domain	Application in Sport	Training Time	App. Time	Input Methods	Tools Needed	Main Advantages	Main Disadvantages	Outputs
Accimaps (Svedung and Rasmussen, 2002)	Generic	Performance evaluation Accident/incident analysis	Low	High	Interviews Observational study	Pen and paper Flipchart Microsoft Visio	1. Considers both the errors at the sharp end and also the system-wide contributory factors involved 2. The output is visual and easily interpreted 3. Considers contributory factors at six different levels	1. Time consuming due to its comprehensiveness 2. Suffers from problems with analyst hindsight 3. The quality of the analysis produced is entirely dependent upon the quality of the input data used	Graphical representation of the incident in question including errors at the sharp end and the causal factors at six organisational levels
Fault tree analysis (Kirwan and Ainsworth, 1992)	Generic	Performance evaluation Accident/incident analysis	Low	Med	Interviews Observational study	Pen and paper Flipchart Microsoft Visio	1. Defines possible failure events and associated causes; this is especially useful when looking at failure events with multiple causes 2. Quick to learn and use in most cases 3. When completed correctly they are potentially very comprehensive	1. For complex scenarios the method can be complex, difficult, and time consuming to construct, and the output may become unwieldy 2. Offers no remedial measures or countermeasures 3. Little evidence of their application outside of the nuclear power domain	Graphical representation of the incident in question
Systematic Human Error Reduction and Prediction Approach (SHERPA; Embrey, 1986)	Generic	Human error identification Sports product design and evaluation	Low	Med	Hierarchical Task Analysis Observational study Walkthrough Interviews	Pen and paper Microsoft Word	1. Offers a structured and comprehensive approach to the prediction of human error 2. The SHERPA behaviour and error mode taxonomy lend themselves to the sporting domain 3. According to the Human Factors literature, SHERPA is the most promising HEI method available	1. Can be tedious and time consuming for large, complex tasks 2. The initial HTA adds additional time to the analysis 3. SHERPA only considers errors at the sharp end of system operation and does not consider system or organisational errors	Description of errors likely to occur during task performance, including the error, its consequences, probability and criticality, recovery steps, and any remedial measures

Method	Type	Domain	Training time	Application time	Related methods	Tools needed	Pros	Cons	Output
Human Error Template (Stanton et al., 2006)	Generic	Human error identification Sports product design and evaluation	Low	Med	Hierarchical Task Analysis Observational study Walkthrough Interviews	Pen and paper Microsoft Word	1. Quick, simple to learn and use, and requires very little training 2. The HET taxonomy prompts the analyst for potential errors 3. Encouraging reliability and validity data	1. For large, complex tasks HET analyses may become overly time consuming and tedious to perform 2. HET analyses do not provide remedial measures or countermeasures 3. Extra work is required if the HTA is not already available	Description of errors likely to occur during task performance, including the error, its consequences, probability and criticality
Task Analysis for Error Identification (TAFEI; Baber and Stanton, 2002)	Generic	Human error identification Sports product design and evaluation	High	High	Hierarchical Task Analysis State Space Diagrams Observational study Walkthrough Interviews	Pen and paper Microsoft Word Microsoft Visio	1. Offers a structured and thorough procedure for predicting errors that are likely to arise during human-device interaction 2. Has a sound theoretical underpinning 3. TAFEI is a flexible and generic methodology that can be applied in any domain in which humans interact with technology	1. Time consuming compared to other error prediction methods 2. For complex systems, the analysis is likely to be complex and the outputs large and unwieldy 3. The number of states that a complex device can potentially be in may overwhelm the analyst	TAFEI matrix depicting illegal (i.e., errors) interactions between man and machine
Technique for Human Error Assessment (THEA; Pocock, Harrison, Wright and Johnson, 2001)	Generic	Human error identification Sports product design and evaluation	Low	High	Hierarchical Task Analysis Observational study Walkthrough Interviews	Pen and paper Microsoft Word	1. HEA offers a structured approach to HEI 2. Easy to learn and use 3. Error identifier prompts the analyst for errors	1. THEA does not use error modes and so the analyst may be unclear on the types of errors that may occur 2. Highly time consuming and may be difficult to grasp for analysts with little experience of human error theory 3. Little evidence of validation or uptake of the method in the literature	Description of errors likely to occur during task performance, their consequences, and any associated design issues

Procedure and Advice

Step 1: Data Collection

Being a retrospective approach, the Accimap approach is dependent upon accurate data regarding the incident under analysis. The first step therefore involves collecting data regarding the incident in question. Data collection for Accimaps can involve a range of activities, including interviews with those involved in the incident or SMEs for the domain in question, analysing reports or inquiries into the incident, and observing recordings of the incident. Since the Accimap method is so comprehensive, the data collection phase is typically time consuming and involves analysing numerous data sources.

Step 2: Identify Physical Process/Actor Activities Failures

Essentially Accimap analyses involve identifying the failures and errors involved in an incident and then identifying and linking the causes of the errors at each of the different organisational levels; the process therefore begins with the identification of the failures at the "sharp end" of system operation. These failures should be placed at the physical processes and actor activities level.

Step 3: Identify Causal Factors

For each error or failure identified during step 2, the analyst should then use the data collected during step 1 to identify the contributory factors at each of the following levels: government policy and budgeting; regulatory bodies and associations; local area government planning and budgeting; physical processes and actor activities; and equipment and surroundings. This involves taking each failure at the physical processes and actor activities level and identifying related failures at the other five levels.

Step 4: Identify Failures at Other Levels

The process of identifying failures at the physical processes and actor activities level and the causal factors from the other levels is normally sufficient; however, it is often useful to step through the other levels to see if any failures have been missed. If any failures are identified at the other levels then the consequences and causal factors should be identified.

Step 5: Finalise and Review Accimap Diagram

It is normally best to construct the Accimap diagram as one proceeds through the analysis. The final step involves reviewing the Accimap diagram and making sure the links between the causal factors and errors identified are appropriate. It is important to use SMEs during this process to ensure the validity of the analysis. Normally an Accimap requires multiple revisions before it is acceptable.

Advantages

1. Accimap offers an approach to analysing both the errors at the sharp end and also the system-wide contributory factors involved. When undertaken properly the entire sequence of events is exposed.
2. It is simple to learn and use.
3. Accimap permits the identification of system failures or inadequacies, such as poor preparation, inappropriate or inadequate government policy, inadequate management, bad design, inadequate training, and inadequate equipment.
4. Accimap considers contributory conditions at six different levels.
5. Accimap outputs offer an exhaustive analysis of accidents and incidents.
6. The output is visual and easily interpreted.
7. Accimap is a generic approach and can be applied in any domain.
8. It has been used in a variety of different domains to analyse accidents and incidents.

9. The different levels analysed allow causal factors to be traced back over months and even years.
10. It removes the apportioning of blame to individuals and promotes the development of systematic (as opposed to individual-based) countermeasures.

Disadvantages

1. Accimap can be time consuming due to its comprehensiveness.
2. Accimap suffers from problems with analyst hindsight; for example, Dekker (S. W. A., 2002) suggests that hindsight can potentially lead to oversimplified causality and counterfactual reasoning.
3. The quality of the analysis produced is entirely dependent upon the quality of the input data used. Accurate and comprehensive data are not always available, so much of the investigation may be based on assumptions, domain knowledge, and expertise.
4. The output does not generate remedial measures or countermeasures; these are based entirely on analyst judgement.
5. The errors involved are not classified into error types or modes.
6. The approach can only be used retrospectively.

Example

The Accimap approach can be used to analyse sporting failures, accidents, and incidents. The example presented is an Accimap, completed for representational purposes, of the Hillsborough Stadium disaster. On April 15, 1989, the Liverpool Football Club was scheduled to play Nottingham Forest in the FA Cup semi-final at Hillsborough Stadium in Sheffield. As a result of severe overcrowding just prior to kick-off an attempt was made to facilitate the entry of the fans into the stadium. As a result, a major crush developed, and 96 fans lost their lives, mainly due to asphyxiation, and over 400 required hospital treatment (Riley and Meadows, 1995). The disaster remains the UK's worst football tragedy.

For the purposes of this example, an Accimap was produced for the Hillsborough soccer disaster. The Accimap was developed based on the data contained in Lord Justice Taylor's Inquiry Report (Taylor, Lord Justice, 1990) and is presented in Figure 5.3 The analysis reveals a number of failures at the physical actor and processes levels, including communications failures, a failure (despite requests) to call off the game prior to kick-off, inadequate leadership and command, and a delay in initiating a major disaster plan. Various systemic failures that contributed to the failures on the day were also identified, including failures in the Police Force's planning for the game (e.g., a failure to review a previous operations order and the production of an inadequate operations order), a change of command part way through the planning process, and a lack of experience of handling similar events on behalf of the new Commanding Officer. At the local area and government level and regulatory bodies level a number of failures also allowed the continued use of an inadequate stadium design (e.g., the division of the terraces into pens).

Related Methods

The Accimap method adopts a systems perspective on human error as advocated by Rasmussen (1997) and Reason's (1990) Swiss cheese model. Using Accimaps also involves considerable data collection activities and might involve the use of data collection methods such as interviews, questionnaires, and/or observational study. Another Swiss cheese–inspired analysis method is the Human Factors Analysis and Classification System (HFACS; Wiegmann and Shappell, 2003).

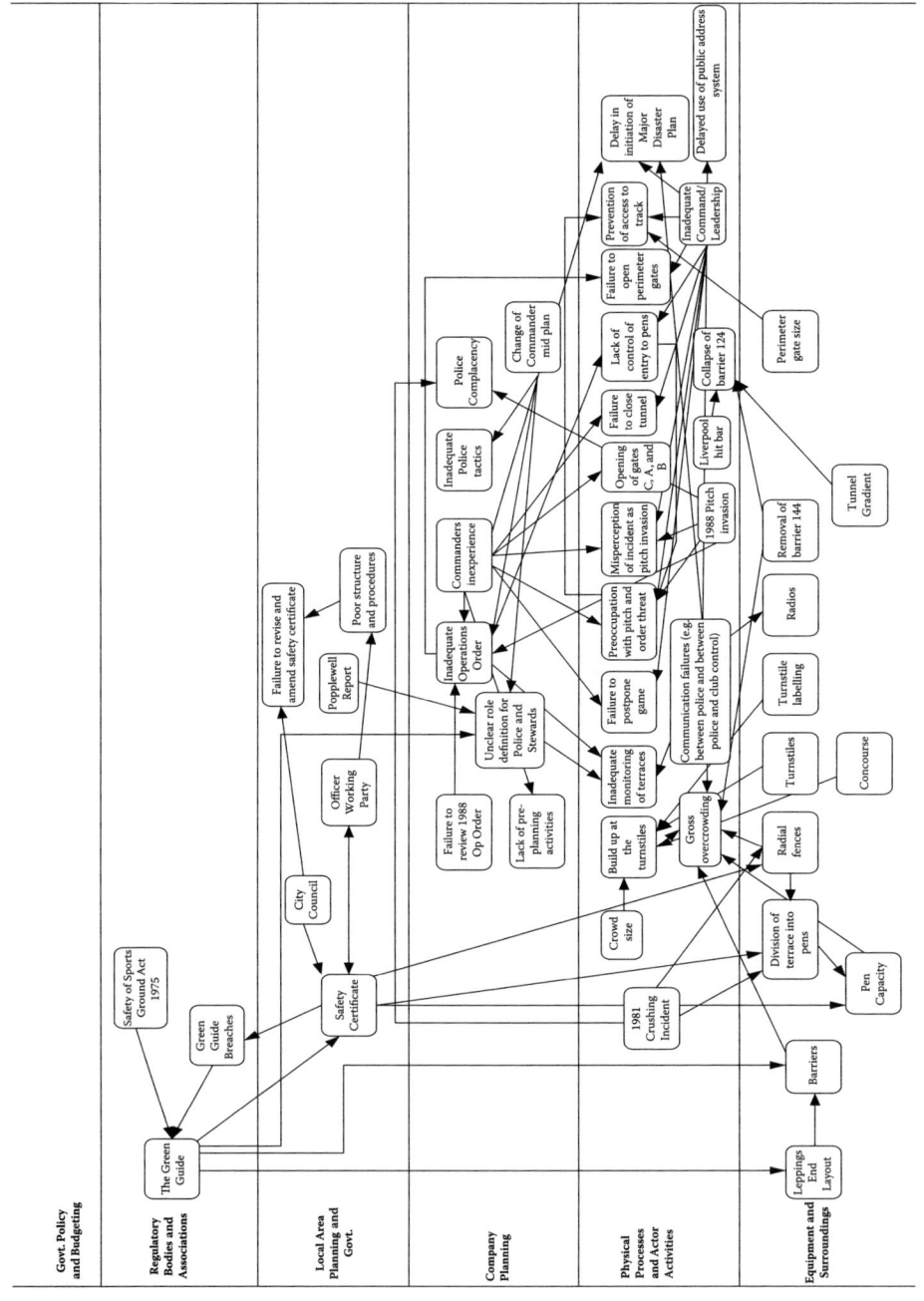

FIGURE 5.3 Hillsborough disaster Accimap.

Approximate Training and Application Times

Accimaps are relatively easy to learn; however, depending upon the incident under analysis they can be highly time consuming to apply, with both the data collection procedure and the analysis itself requiring substantial effort on the behalf of the analyst.

Reliability and Validity

No reliability and validity data are presented in the literature. Since the analysis is only guided by the different failure levels, it may be that the reliability of the approach is low, as different analysts may classify events differently and also may miss contributory factors.

Tools Needed

Accimaps can be conducted using pen and paper. Typically, a rough Accimap is produced using pen and paper or a flipchart and then drawing software tools such as Microsoft Visio or Adobe Illustrator are used for constructing the final Accimap.

Flowchart

(See Flowchart 5.1.)

Recommended Texts

Rasmussen, J. (1997). Risk management in a dynamic society: A modelling problem. *Safety Science* 27(2/3):183–213.

Svedung, J., and Rasmussen, J. (2002). Graphic representation of accident scenarios: Mapping system structure and the causation of accidents. *Safety Science* 40:397–417.

FAULT TREE ANALYSIS

Background and Application

Fault trees are used to graphically represent failures and their causes. They use tree-like diagrams to define failure events and possible causes in terms of hardware failures and/or human errors (Kirwan and Ainsworth, 1992). The fault tree approach was originally developed for the analysis of complex systems in the aerospace and defence industries (Kirwan and Ainsworth, 1992) and is now most commonly used in probabilistic safety assessment. Fault trees begin with the failure or top event, which is placed at the top of the fault tree, and the contributing events are placed below (Kirwan and Ainsworth, 1992). The fault tree is held together by AND and OR gates, which link contributory events together. An AND gate is used when more than one event causes a failure. The events placed directly underneath an AND gate must occur together for the failure event above to occur. An OR gate is used when the failure event could be caused by more than one contributory event in isolation, but not together. The event above the OR gate may occur if any one of the events below the OR gate occurs. Fault tree analysis can be used for the retrospective analysis of incidents or for the prediction of failure in a particular scenario.

Domain of Application

Fault tree analysis was originally applied in the nuclear power and chemical processing domains. However, the method is generic and could potentially be applied in any domain.

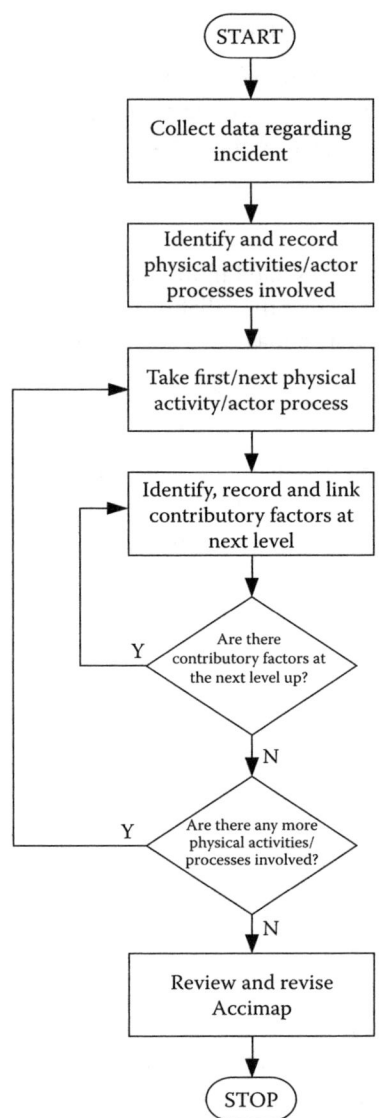

FLOWCHART 5.1 Accimap flowchart.

Application in Sport

Fault trees provide a powerful way of representing system or human failures, and could be used in a sporting context to depict why a particular failure occurred. Fault trees could also be used to predict potential failures and their causes.

Procedure and Advice

Step 1: Define Failure Event

The failure or event under analysis should be defined first. This may be either an actual event that has occurred (retrospective incident analysis) or a projected failure event (predictive analysis). This event then becomes the top event in the fault tree.

Step 2: Collect Data Regarding Failure Event

Fault tree analysis is dependent upon accurate data regarding the incident under analysis. The first step therefore involves collecting data regarding the incident in question. This might involve a range of activities, including conducting interviews with those involved in the incident or SMEs, analysing reports or inquiries into the incident, and observing recordings of the incident.

Step 3: Determine Causes of Failure Event

Once the failure event has been defined, the contributory causes associated with the event should be defined. The nature of the causes analysed is dependent upon the focus of the analysis. Typically, human error and hardware failures are considered (Kirwan and Ainsworth, 1992). It is useful during this phase to use various supporting materials, such as documentation regarding the incident, task analyses outputs, and interviews with SMEs or those involved in the incident.

Step 4: AND/OR Classification

Once the cause(s) of the failure event is defined, the analysis proceeds with the AND/OR causal classification phase. Each causal factor identified during step 3 of the analysis should be classified as either an AND or an OR event. If two or more contributory events contribute to the failure event, then they are classified as AND events. If two or more contributory events can cause the failure even when they occur separately, then they are classified as OR events. Again, it is useful to use SMEs or the people involved in the incident under analysis during this phase.

Steps 3 and 4 should be repeated until each of the initial causal events and associated causes are investigated and described fully.

Step 5: Construct Fault Tree Diagram

Once all events and their causes have been defined fully, they should be put into the fault tree diagram. The fault tree should begin with the main failure or top event at the top of the diagram with its associated causes linked underneath as AND/OR events. Then, the causes of these events should be linked underneath as AND/OR events. The diagram should continue until all events and causes are exhausted fully or until the diagram satisfies its purpose.

Step 6: Review and Refine Fault Tree Diagram

Constructing fault trees is a highly iterative process. Once the fault tree diagram is complete, it should be reviewed and refined, preferably using SMEs or the people involved in the incident.

ADVANTAGES

1. Fault trees are useful in that they define possible failure events and associated causes. This is especially useful when looking at failure events with multiple causes.
2. A simple approach, fault trees are quick and easy to learn and use.
3. When completed correctly they are potentially very comprehensive.
4. Fault trees could potentially be used both predictively and retrospectively.
5. Although most commonly used in the analysis of nuclear power plant events, the method is generic and can be applied in any domain.
6. Fault trees can be used to highlight potential weak points in a system design concept (Kirwan and Ainsworth, 1992).
7. The method could be particularly useful in modelling team-based errors, where a failure event is caused by multiple events distributed across a team of personnel.

Disadvantages

1. When used in the analysis of large, complex systems, fault trees can be complex, difficult, and time consuming to construct, and the output may become unwieldy.
2. The method offers no remedial measures or countermeasures. These are based entirely on the judgement of the analysts involved.
3. To utilise the method quantitatively, a high level of training may be required (Kirwan and Ainsworth, 1992).
4. The use of fault trees as a predictive tool remains largely unexplored.
5. There is little evidence of their application outside of the nuclear power domain.

Related Methods

The fault tree method is often used with event tree analysis (Kirwan and Ainsworth, 1992). Fault trees are similar to many other charting methods, including cause-consequence charts, decision action diagrams, and event trees. Data collection methods, such as interviews and observational study, are also typically used during the construction of fault tree diagrams.

Approximate Training and Application Times

A simplistic method, the training time required for the fault tree method is minimal. The application time is dependent upon the incident under analysis. For complex failure scenarios, the application time is high; however, for simpler failure events, the application time is often very low.

Reliability and Validity

No data regarding the reliability and validity of the fault tree approach are presented in the literature.

Tools Needed

Fault tree analysis can be conducted using pen and paper. If the analysis were based upon an existing system, an observational study of the failure event under analysis would be useful. This would require video and audio recording equipment. It is also recommended that when constructing fault tree diagrams, a drawing package such as Microsoft Visio or Adobe Illustrator is used to produce the final fault tree diagram.

Example

One of golf's most famous failures in recent times is Frenchman Jean Van de Velde's catastrophic collapse on the final hole of the British Open at Carnoustie in 1999. Leading the 72-hole tournament by three shots, Van de Velde needed only a double-bogey 6 on the par 4 final hole to win the tournament. Unfortunately, Van de Velde triple-bogeyed the hole and eventually lost the title in the resulting three-way playoff. Using an observation of a video recording of the incident and descriptions of the incident from both eyewitnesses and Van de Velde himself, a simple fault tree diagram analysis of the failure is presented in Figure 5.4.

The fault tree diagram shows that Van de Velde's failure was ostensibly down to three main causal events: his pushed second shot, which landed in deep rough; his third shot into Carnoustie's famous Barry Burn water trap; and his fourth shot into a greenside bunker. Each of these failures led to the overall failure event, which was the triple-bogey 7 that he carded. Each causal event can be further decomposed using fault tree AND/OR logic. For example, his second shot into deep rough was caused by five events: his decision to drive off the tee to gain maximum distance rather than play a mid-iron onto the fairway (with such a big lead he could have played safe with a mid-iron onto

Human Error Identification and Analysis Methods

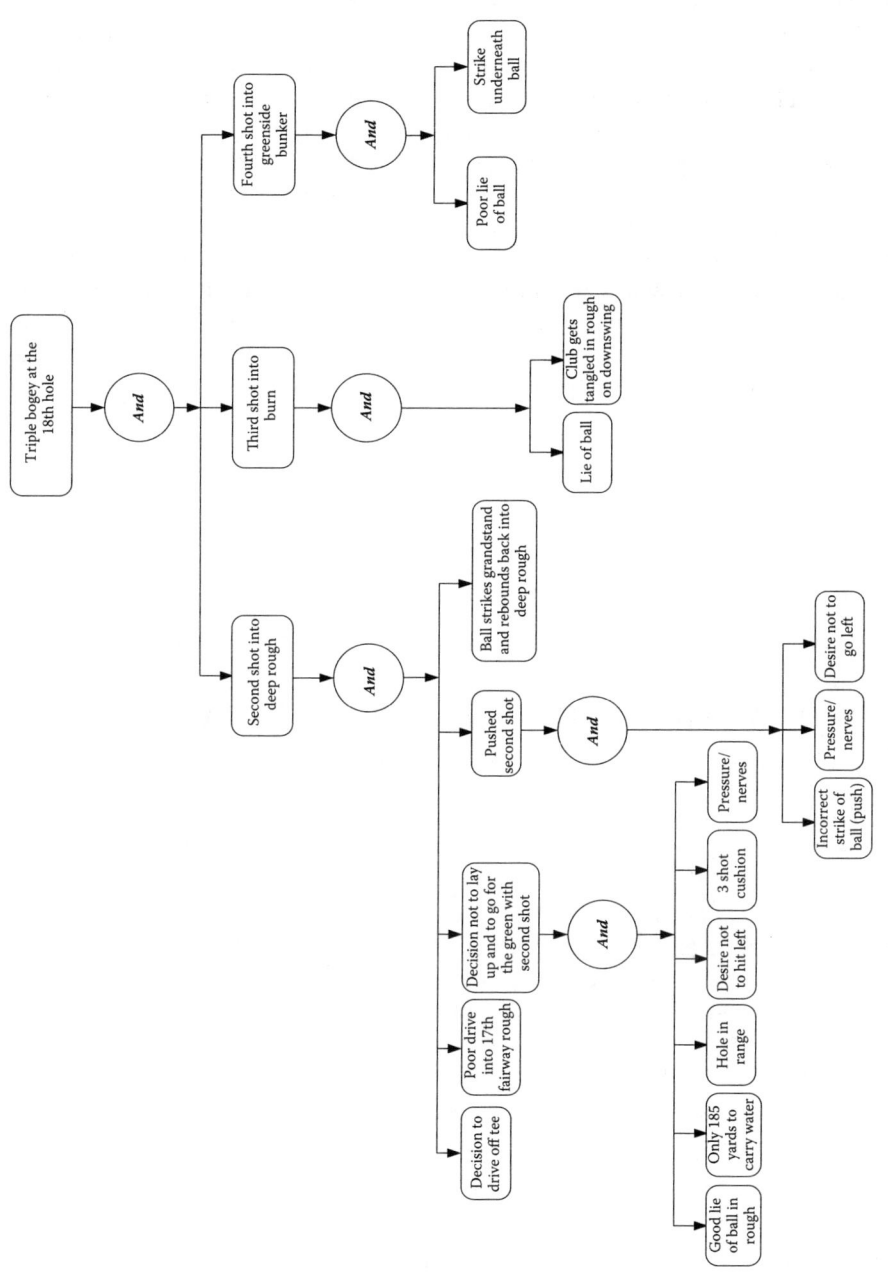

FIGURE 5.4 Fault tree diagram of Jean Van de Velde's final-hole triple bogey in the 1999 British Open.

the fairway); the ensuing poorly executed drive which landed in the 17th hole fairway rough; his decision to go for the green with his second shot rather than lay up before the green (which is what most players would have done in the circumstances); his poor execution of the second shot (the shot was "pushed"); and finally the ball striking the grandstand and rebounding backwards into the deep rough. These causal events too are decomposed. For example, his decision to go for the green with his second shot instead of laying up was a function of six factors: despite the ball landing in the 17th fairway rough it had a good lie; he only needed to hit his second shot 185 yards in order to clear the water trap (which should have been easily achieved given the lie of the ball); the green was easily in range; Van de Velde had a desire not to hit left, which he would have done had he chosen to lay up; he had a three-shot lead, which meant he could afford to play risky; and finally, the pressure of the situation also undoubtedly had some bearing—this was the final hole in the British Open, golf's most prestigious major tournament, and Van de Velde was inexperienced in this situation, never before having led a major tournament at the final hole. The next primary failure event was his third shot, which ended up in Barry Burn. This was caused by the poor lie of the ball in the deep rough and also the fact that, on his downswing, Van de Velde's club got tangled in the rough and reduced the power of his shot. Finally, the fourth shot into the bunker was caused by the poor lie of the ball and poor execution of the shot, with Van de Velde getting underneath the ball, which caused it to drop short of the green and land in the greenside bunker.

FLOWCHART

(See Flowchart 5.2.)

RECOMMENDED TEXT

Kirwan, B., and Ainsworth, L. K. (1992). *A guide to task analysis*. London: Taylor & Francis.

SYSTEMATIC HUMAN ERROR REDUCTION AND PREDICTION APPROACH

BACKGROUND AND APPLICATIONS

The most popular of all HEI approaches is the Systematic Human Error Reduction and Prediction Approach (SHERPA; Embrey, 1986). Originally developed for use in the nuclear reprocessing industry, to date SHERPA has had further application in a number of domains, including the military (Salmon, Jenkins, Stanton, and Walker, in press), aviation (Stanton et al., 2009), healthcare (Lane et al., 2007), public technology (Baber and Stanton, 1996a; Stanton and Stevenage, 1998), and in-car entertainment (Stanton and Young, 1999). SHERPA uses an EEM taxonomy linked to a behavioural taxonomy and is applied to an HTA of the task under analysis in order to predict potential human- or design-induced error. In addition to being the most commonly used of the various HEI methods available, according to the literature SHERPA is also the most successful in terms of accuracy of error predictions (Kirwan, 1992, 1998a; Baber and Stanton, 1996a, 2001; Stanton et al., 2009; Stanton and Stevenage, 1998).

DOMAIN OF APPLICATION

Despite being developed originally for use in the nuclear process industries, the SHERPA behaviour and error taxonomies are generic and can be applied in any domain. To date SHERPA has been applied in a range of domains for error prediction purposes, including process control, the military, aviation, healthcare, public technology, and in-vehicle entertainment design.

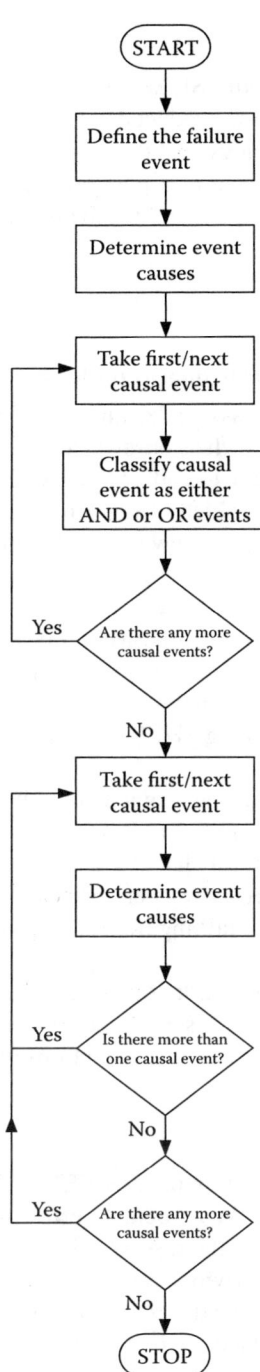

FLOWCHART 5.2 Fault tree flowchart.

Application in Sport

There is great potential for applying the SHERPA method in a sporting context. For example, SHERPA could be used during the sports product design concept phases to identify potential errors that could be made with sports technology and devices, or it could be applied to identify, a priori, errors that performers might make during task performance, the findings of which could be used to develop training interventions designed to reduce the probability that the errors are made during future performances.

Procedure and Advice

Step 1: Conduct HTA for the Task or Scenario under Analysis

The first step involves describing the task under analysis. For this purpose, an HTA is normally constructed. Data collection methods, such as observational study, walkthrough analysis, and interviews may be employed during this phase to collect the data required for constructing the HTA. The SHERPA method works by indicating which of the errors from the SHERPA error taxonomy are credible during performance of each bottom-level task step in the HTA.

Step 2: Task Classification

SHERPA is best conducted in conjunction with SMEs; however, if they are not available to undertake the analysis initially then it is normal practice to involve them in the review and validation steps. Once the HTA is constructed, the analyst should take each bottom-level task step from the HTA and classify it as one of the following SHERPA behaviours:

- Action (e.g., pressing a button, pulling a switch, opening a door)
- Retrieval (e.g., getting information from a screen or manual)
- Check (e.g., conducting a procedural check)
- Selection (e.g., choosing one alternative over another)
- Information communication (e.g., talking to another party)

For example, for a golf HEI analysis the task step "check distance from hole" would be classified as a "Check" behaviour, whereas the task steps "set up swing" and "make shot" would both be "Action" type behaviours, and the task step "discuss options with caddie" would be an "Information communication" behaviour.

Step 3: Identify Likely Errors

Each SHERPA behaviour has a series of associated EEMs. The SHERPA EEM taxonomy is presented in Figure 5.5. The analyst uses the associated error mode taxonomy and domain expertise to identify any credible error modes for the task step in question. For example, if the task step in question is classified as an "Action" behaviour during step 2, then the analyst should consider each of the 10 "Action" EEMs and consider whether they could potentially occur or not. A performance-shaping factor taxonomy, which introduces error-causing conditions such as time pressure, fatigue, and inadequate equipment, might also be used to aid this process. For each error identified, the analyst should give a description of the form that the error would take, such as, "golfer miscalculates distance to hole," "golfer selects the wrong club for shot," or "golfer over hits shot."

Step 4: Determine and Record Error Consequences

The next step involves determining and describing the consequences associated with the errors identified in step 3. The analyst should determine the consequences associated with each error identified and provide clear descriptions in relation to the task under analysis. For example, for the error "golfer miscalculates distance to hole" would have the consequences of "golfer selects wrong club" and

Human Error Identification and Analysis Methods

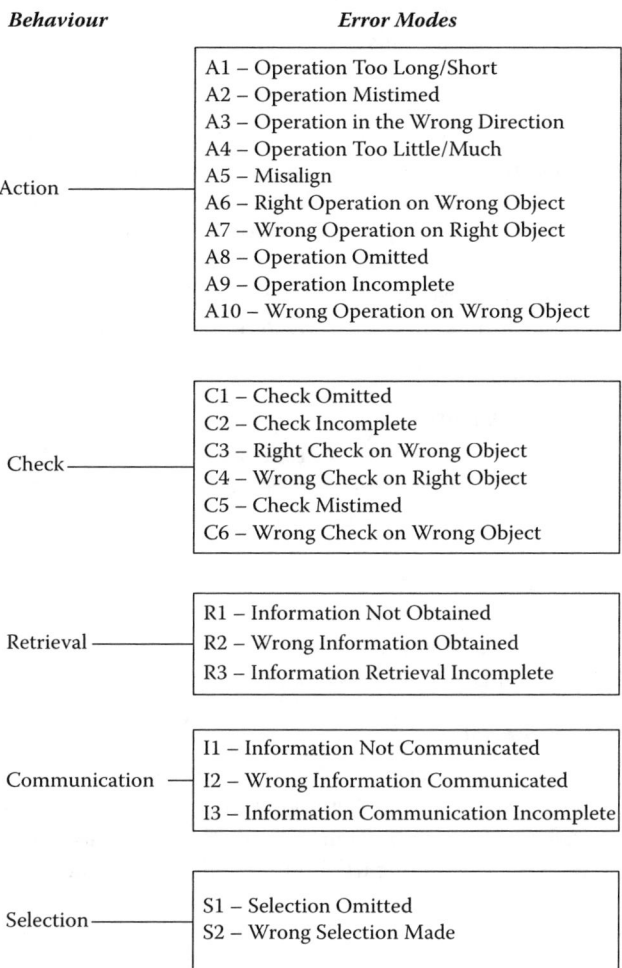

FIGURE 5.5 SHERPA external error mode taxonomy.

"golfer leaves shot short or overshoots the green," and for the error "golfer over hits shot" the consequence would be "ball travels through the back of the green and into rough/greenside bunker."

Step 5: Recovery Analysis

Next, the analyst should determine the recovery potential of the identified error. If there is a later task step in the HTA at which the error could be recovered, it is entered here; if there is no recovery step then "None" is entered; finally, if the error is recognised and recovered immediately, the analyst enters "immediate." For example, there would be no recovery for the error of over hitting the shot; however, the recovery for the error of "selecting the wrong club for shot" may be recovered either when the golfer sets up his swing or when the golfer takes a practice swing. The corresponding task steps should be entered in the SHERPA output table.

Step 6: Ordinal Probability Analysis

Once the consequences and recovery potential of the error have been identified, the analyst should rate the probability of the error occurring. An ordinal probability scale of low, medium, or high is typically used. If the error has not occurred previously then a low (L) probability is assigned. If the error has occurred on previous occasions then a medium (M) probability is assigned. Finally, if the error has occurred on frequent occasions, a high (H) probability is assigned. SME judgements are particularly useful for this part of the analysis.

Step 7: Criticality Analysis

Next, the analyst rates the criticality of the error in question. A scale of low, medium, and high is also used to rate error criticality. Normally, if the error would lead to a critical incident (in relation to the task in question), or if the error results in the task not being successfully performed, then it is rated as a highly critical error. For example, if the error resulted in the golfer hitting the ball out of bounds or into a water trap, then the error criticality would be rated as "High," since it would incur a two shot penalty and invariably lead to a bogey score for the hole. On the contrary, if the error led to a slightly pushed shot then criticality would be rated as "Low." SME judgements are particularly useful for this part of the analysis.

Step 8: Propose Remedial Measures

The final stage in the process is to propose error reduction strategies. Normally, remedial measures comprise suggested changes to the design of the process or system. Remedial measures are normally proposed under the following four categories:

- Equipment (e.g., redesign or modification of existing equipment)
- Training (e.g., changes in training provided)
- Procedures (e.g., provision of new, or redesign of old, procedures)
- Organisational (e.g., changes in organisational policy or culture)

Step 9: Review and Refine Analysis

SHERPA is an iterative process and normally requires many passes before the analysis is completed to a satisfactory level. The analyst should therefore spend as much time as possible reviewing and refining the analysis, preferably in conjunction with SMEs for the task or scenario under analysis. The probability and criticality ratings in particular typically require numerous revisions.

ADVANTAGES

1. The SHERPA method offers a structured and comprehensive approach to the prediction of human error.
2. The SHERPA taxonomy prompts the analyst for potential errors.
3. The SHERPA behaviour and error mode taxonomy lend themselves to the sporting domain.
4. According to the Human Factors literature, SHERPA is the most promising HEI method available. SHERPA has been applied in a number of domains with considerable success. There is also a wealth of encouraging validity and reliability data available.
5. SHERPA is easy to learn and apply, requiring minimal training, and is also quick to apply compared to other HEI methods.
6. The method is exhaustive, offering error reduction strategies in addition to predicted errors, associated consequences, probability of occurrence, criticality, and potential recovery steps.
7. The SHERPA error taxonomy is generic, allowing the method to be used in any domain.
8. The outputs provided could be of great utility in a sporting context, potentially informing the design of sports technology, training, and tactical interventions.

DISADVANTAGES

1. SHERPA can be tedious and time consuming for large, complex tasks.
2. The initial HTA adds additional time to the analysis.

3. SHERPA only considers errors at the sharp end of system operation. The method does not consider system or organisational failures.
4. In its present usage SHERPA does not consider performance-shaping factors, although these can be applied by the analyst.
5. SHERPA does not model cognitive components of errors.
6. Some predicted errors and remedies are unlikely or lack credibility, thus posing a false economy.
7. The approach is becoming dated and the behaviour and error mode taxonomies may require updating.
8. While the approach handles physical tasks extremely well, it struggles when applied to purely cognitive tasks.

Related Methods

The initial data collection for SHERPA might involve a number of data collection methods, including interviews, observation, and walkthroughs. An HTA of the task or scenario under analysis is typically used as the input to a SHERPA analysis. The taxonomic approach to error prediction employed by the SHERPA method is similar to a number of other HEI approaches, such as HET (Harris, Stanton, Marshall, Young, Demagalski, and Salmon, 2005), Human Error HAZOP (Kirwan and Ainsworth, 1992), and the Technique for Retrospective Analysis of Cognitive Error (TRACEr; Shorrock and Kirwan, 2002).

Approximate Training and Application Times

The training and application time for SHERPA is typically low, although the application time can increase considerably for large, complex tasks. In order to evaluate the reliability, validity, and trainability of various methods, Stanton and Young (1999) compared SHERPA to 11 other Human Factors methods. Based on the application of the method to the operation of an in-car radio-cassette machine, Stanton and Young (1999) reported training times of around 3 hours (this is doubled if training in HTA is included). It took an average of 2 hours and 40 minutes for people to evaluate the radio-cassette machine using SHERPA. In a study comparing the performance of SHERPA, Human Error HAZOP, Human Error Identification in Systems Tool (HEIST), and HET when used to predict design induced pilot error, Salmon et al. (2002) reported that participants achieved acceptable performance with the SHERPA method after only 2 hours of training.

Reliability and Validity

SHERPA has enjoyed considerable success in HEI method validation studies. Kirwan (1992) reported that SHERPA was the most highly rated of five human error prediction methods by expert users. Baber and Stanton (1996a) reported a concurrent validity statistic of 0.8 and a reliability statistic of 0.9 in the application of SHERPA by two expert users to the prediction of errors on a ticket vending machine. It was concluded that SHERPA provided an acceptable level of validity based upon the data from two expert analysts. Stanton and Stevenage (1998) reported a concurrent validity statistic of 0.74 and a reliability statistic of 0.65 in the application of SHERPA by 25 novice users to prediction of errors on a confectionery vending machine. It was concluded that SHERPA provided a better means of predicting errors than a heuristics-type approach did. Harris et al. (2005) reported that SHERPA achieved acceptable performance in terms of reliability and validity when used by novice analysts to predict pilot error on a civil aviation flight scenario. Further, Harris et al. (2005) and Stanton et al. (2006) both reported that the SHERPA approach performed better than three other HEI approaches when used to predict pilot errors for an aviation approach and landing task.

Tools Needed

SHERPA can be conducted using pen and paper. A representation of the interface, device, or equipment under analysis is also required. This might be the device itself or can take the form of functional drawings or photographs. SHERPA output tables are normally constructed using word processing software such as Microsoft Word.

Example

A SHERPA analysis focussing on golf stroke play was conducted. Initially, a golf HTA was constructed in conjunction with an amateur golfer. Following this, a Human Factors analyst with considerable experience in the SHERPA approach applied SHERPA to the HTA to predict the errors likely to be made during a round of golf. An extract of the golf HTA is presented in Figure 5.6. An extract of the golf SHERPA output is presented in Table 5.2.

Flowchart

(See Flowchart 5.3.)

Recommended Texts

Embrey, D.E. (1986) SHERPA: A systematic human error reduction and prediction approach. Paper presented at the International Meeting on Advances in Nuclear Power Systems, Knoxville, Tennessee, USA.

Stanton, N. A., Salmon, P. M., Baber, C., and Walker, G. (2005). *Human Factors methods: A practical guide for engineering and design*. Aldershot, UK: Ashgate.

HUMAN ERROR TEMPLATE

Background and Applications

The Human Error Template (HET; Harris et al., 2005) was developed for use in the certification of civil flight deck technology in order to predict design-induced pilot error. The impetus for HET came from a US Federal Aviation Administration report (FAA, 1996), which, amongst other things, recommended that flight deck designs be evaluated for their susceptibility to design-induced flight crew errors and also to identify the likely consequences of those errors during the type certification process (Harris et al., 2005). The HET method is a simple checklist HEI approach that is applied to each bottom-level task step in an HTA of the task under analysis. Analysts use the HET EEMs and subjective judgement to identify credible errors for each task step. The HET EEM taxonomy consists of the following generic error modes:

- Fail to execute
- Task execution incomplete
- Task executed in the wrong direction
- Wrong task executed
- Task repeated
- Task executed on the wrong interface element
- Task executed too early
- Task executed too late
- Task executed too much
- Task executed too little
- Misread information
- Other

Human Error Identification and Analysis Methods

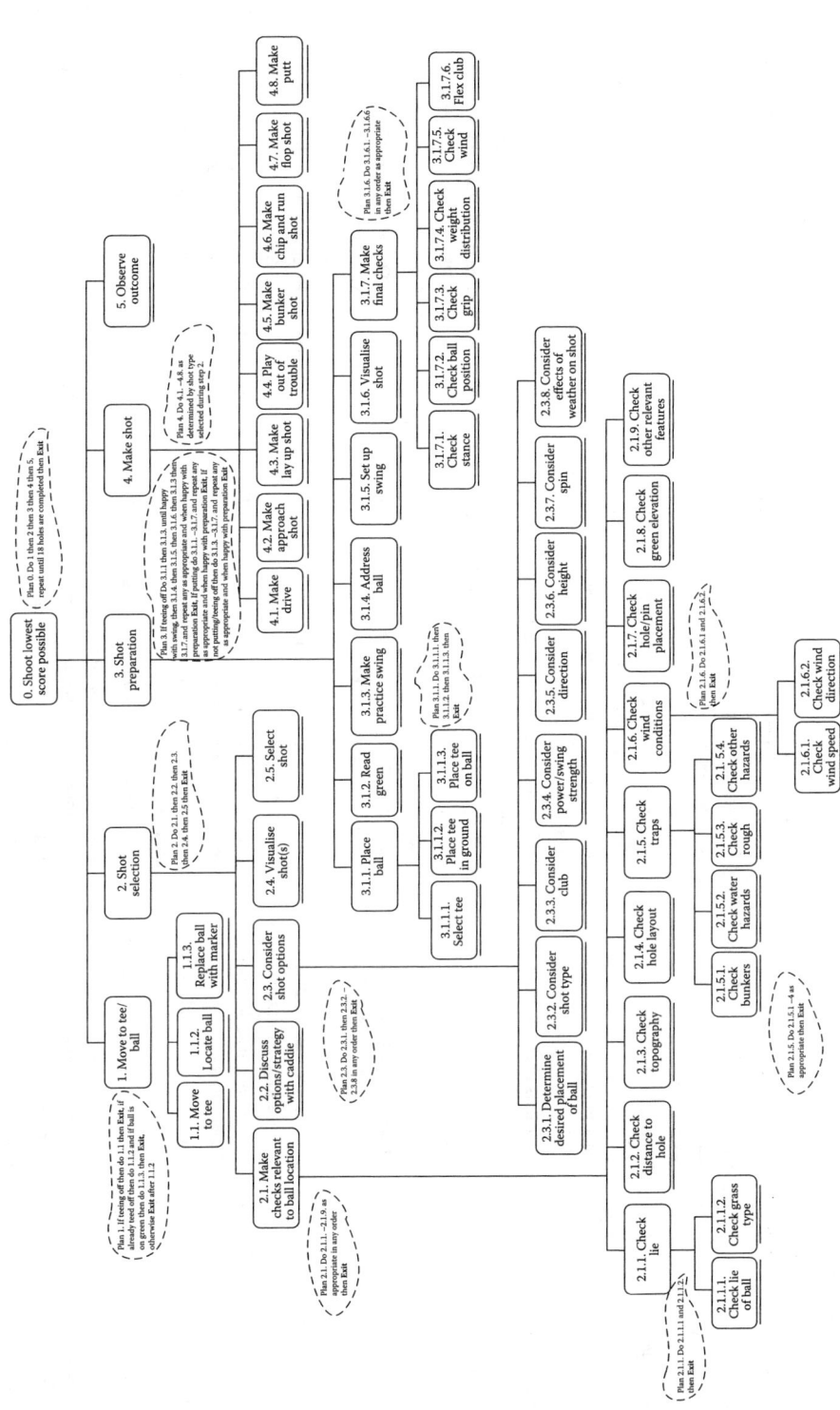

FIGURE 5.6 Golf HTA extract.

TABLE 5.2
Golf SHERPA Extract

Task Step	Error Mode	Description	Consequence	Recovery	P	C	Remedial Measures
1.2. Locate ball	A8	Golfer fails to locate ball (i.e., ball is in deep rough, trees, or water hazard)	Golfer has to take a drop and incurs a two shot penalty	None	L	M	N/A
2.1.1.1. Check lie of ball	R2	Golfer misreads lie of ball	Golfer misunderstands lie of ball and shot selection is affected	2.1.1.2.	L	M	Practise reading ball lies Assistance from caddie in reading ball lie
2.1.1.2. Check grass type	R2	Golfer misreads grass type	Golfer misunderstands grass type and shot selection is affected	2.1.1.2.	L	M	Practise reading ball lies Assistance from caddie in reading ball lie
2.1.1.2. Check grass type	C1	Golfer fails to check grass type	Golfer does not account for grass type and shot selection is inappropriate	2.2.	L	M	Practise reading ball lies Assistance from caddie in reading ball lie
2.1.2. Check distance to hole	R2	Golfer/caddie miscalculates distance to hole	Golfer misunderstands distance to hole and club and shot selection may be inappropriate	2.2.	M	M	Practise reading distances GPS device, e.g., "caddie aid"
2.1.5.1. Check bunkers	R2	Golfer/caddie miscalculates distance to bunker/bunker placement	Golfer misunderstands distance required to carry bunker	2.2.	M	M	Practise reading distances GPS device, e.g., "caddie aid"
2.1.5.2. Check water hazards	R2	Golfer/caddie miscalculates distance to water hazard	Golfer misunderstands distance required to carry water hazard	2.2.	M	M	Practise reading distances GPS device, e.g., "caddie aid"
2.1.5.3. Check rough	R2	Golfer/caddie miscalculates distance to rough	Golfer misunderstands distance required to carry rough	2.2.	M	M	Practise reading distances GPS device, e.g., "caddie aid"
2.1.6.1. Check wind speed	R2	Golfer/caddie misreads wind speed	Golfer misunderstands wind speed and shot selection is inappropriate	2.2.	M	M	Practise reading wind speeds
2.1.6.1. Check wind speed	C1	Golfer fails to check wind speed	Golfer does not account for wind speed and shot selection is inappropriate	2.2.	L	M	Practise reading wind speeds
2.1.6.2. Check wind direction	R2	Golfer/caddie misreads wind direction	Golfer misunderstands wind direction and shot selection is inappropriate	2.2.	M	M	Practise reading wind direction
2.1.6.2. Check wind direction	C1	Golfer fails to check wind direction	Golfer does not account for wind direction and shot selection is inappropriate	2.2.	L	M	Practise reading wind direction

Human Error Identification and Analysis Methods

Task	Code	Error Description	Consequence	Ref	P	C	Remedy
2.1.7. Check pin/hole placement	R2	Golfer misreads hole placement	Shot selection is inappropriate as golfer misunderstands hole placement	2.2.	L	M	Practise reading hole placements in context of green layout and green terrain
2.1.8. Check green elevation/terrain	R2	Golfer/caddie misreads green elevation/terrain	Shot selection is inappropriate as golfer misunderstands green elevation/terrain	2.2.	L	M	Practise reading greens (e.g., elevation and terrain)
2.2. Discuss options/strategy with caddie	I1	Caddie fails to inform golfer of critical information, e.g., green topography	Shot selection does not take into account critical information and is inappropriate for conditions	2.4.	L	M	Standard strategy discussion items, e.g., distance, wind, green topography, traps, etc.
2.2. Discuss options/strategy with caddie	I2	The information provided by the caddie is incorrect	Golfer is misinformed and shot selection is inappropriate	2.4.	M	M	N/A
2.2. Discuss options/strategy with caddie	I3	Not all critical information is given to the golfer by the caddie	Shot selection does not take into account critical information and is inappropriate for conditions	2.4.	L	M	Standard strategy discussion items, e.g., distance, wind, green topography, traps, etc. Golfer defines information required prior to round
2.2. Discuss options/strategy with caddie	R2	Golfer misunderstands some of the information given by the caddie	Golfer misunderstands critical information and shot selection is inappropriate	2.4.	L	M	Read back for confirmation of receipt of information
2.3.1. Determine desired placement of ball	A5	Golfer determines an inappropriate or unachievable end state	Shot selection is inappropriate	2.4.	M	H	Use training to enhance awareness of limitations
2.3.2. Consider shot type	A9	Golfer fails to fully consider full range of shot possibilities	Shot selection may be inappropriate as another shot type is more appropriate	2.4.	M	M	Specific training in shot selection strategies
2.3.3. Consider club	A6	Golfer considers and selects wrong club	Wrong club selection leads to over or under hit shot	2.4.	H	M	Use training to enhance awareness of limitations Specific training in club selection
2.3.3. Consider club	A9	Golfer fails to fully consider club options available	Wrong club selection leads to over or under hit shot	2.4.	M	M	Use training to enhance awareness of limitations Specific training in club selection
2.3.4. Consider power/swing strength	A5	Golfer considers/selects an inappropriate level of power/swing strength	Shot selection is inappropriate and leads to over or under hit shot	2.4.	H	M	Use training to enhance awareness of limitations Specific training in power/swing strength selection

TABLE 5.2 (CONTINUED)
Golf SHERPA Extract

Task Step	Error Mode	Description	Consequence	Recovery	P	C	Remedial Measures
2.3.5. Consider shot direction	A5	Golfer considers/selects an inappropriate direction (i.e., wrongly overcompensates for wind)	Shot selection is inappropriate and leads to wayward shot	2.4.	H	M	Specific training on influences on shot direction
2.3.6. Consider shot height	A5	Golfer considers/selects an inappropriate height for shot (i.e., wrongly overcompensates for wind or ability to stop ball on green)	Shot selection is inappropriate and leads to wayward shot	2.4.	H	M	Specific training on influences on shot height
2.3.7. Consider spin required	A4	Golfer considers/selects too much spin for shot	Level of spin selected for shot is inappropriate and ball may fall short/run off green	2.4.	M	M	Specific training in spin selection
2.3.8. Consider effects of weather on shot	A5	Golfer misunderstands effects of weather on shot	Shot selection is inappropriate and leads to wayward shot	2.4.	H	M	Specific training on influences on shot direction
2.4. Visualise shot	A5	Shot visualisation does not correctly take into account all of the effects applicable from 2.3.2.–2.3.8.	Shot visualisation is inappropriate and shot selection will be wrong or inappropriate for conditions	3.1.3. 3.1.4. 3.1.5.	H	M	Training in shot visualisation Training in correctly weighing scenario features and their impact on shots
2.4. Visualise shot	A9	Shot visualisation does not fully take into account all of the effects applicable from 2.3.2–2.3.8.	Shot visualisation is inappropriate and shot selection will be wrong or inappropriate for conditions	3.1.3. 3.1.4. 3.1.5.	H	M	Training in shot visualisation Training in correctly weighing scenario features and their impact on shots
2.5. Select shot	S1	Shot selected is inappropriate (e.g., wrong club, wrong placement, wrong shot type)	Shot selected is inappropriate for situation/conditions/hole	3.1.3. 3.1.4. 3.1.5.	H	H	Training in shot selection Training in correctly weighing scenario features and their impact on shots
3.1.1.1. Select tee	S1	Wrong tee is selected for conditions and chosen shot	Tee selected is inappropriate for shot and conditions (e.g., too long/high)	Immediate or 3.1.1.2. 3.1.1.3.	L	M	N/A

Human Error Identification and Analysis Methods 139

Task	Code	Error description	Consequence	Recovery	P	C	Remedy
3.1.1.2. Place tee in ground	A4	Tee is placed too far into ground	Ball position is too low for shot	Immediate or 3.1.1.3.	L	M	N/A
3.1.1.2. Place tee in ground	A4	Tee is not placed into ground sufficiently	Ball position is too high for shot	Immediate or 3.1.1.3.	L	M	N/A
3.1.1.3. Place ball on tee	A5	Golfer misaligns placement of ball on tee	Ball falls off tee	Immediate	L	L	N/A
3.1.2. Read green	R2	Golfer/caddie misreads green	Misunderstands the green and putt shot is inappropriate (e.g., too hard, too soft, too far right, too far left)	1.5.	H	H	Training in reading greens Putt practise
3.1.3. Make practice swing	A4	Swing is too strong	Golfer is not happy with swing and repeats	Immediate	H	L	N/A
3.1.3. Make practice swing	A4	Swing strength is too little	Golfer is not happy with swing and repeats	Immediate	H	L	N/A
3.1.3. Make practice swing	R2	Golfer wrongly assumes that practice swing is appropriate	Swing may be too light/too strong and shot execution will be inadequate	1.5.	H	H	Swing training
3.1.5. Set up swing	A5	Swing is not set up optimally	Shot execution may suffer due to swing set-up being inadequate	3.1.7	L	H	Swing set-up training
4.1. Make drive	A4	Golfer strikes ball with too much power in swing	Shot is over hit and overshoots desired location	Immediate	H	H	Driving training including reduced power driving
4.1. Make drive	A4	Golfer strikes ball with too little power in swing	Shot is under hit and overshoots desired location	Immediate	M	H	Driving training including reduced power driving
4.1. Make drive	A5	Swing is misaligned and golfer slices ball	Shot is sliced and ball travels from left to right and does not end up in desired location	Immediate	M	H	Driving training/practise
4.1. Make drive	A5	Swing is misaligned and golfer hooks ball	Shot is hooked and ball travels from right to left and does not end up in desired location	Immediate	M	H	Driving training/practise
4.1. Make drive	A5	Swing is misaligned and golfer shanks ball	Shot is shanked and ball travels sharply to the right	Immediate	L	H	Driving training/practise
4.1. Make drive	A5	Swing is misaligned and golfer strikes ball too high	Ball is thinned and ball does not travel distance required	Immediate	L	H	Driving training/practise

TABLE 5.2 (CONTINUED)
Golf SHERPA Extract

Task Step	Error Mode	Description	Consequence	Recovery	P	C	Remedial Measures
4.1. Make drive	A5	Swing is misaligned and golfer strikes ball too high	Ball is thinned and ball does not travel distance required	Immediate	L	H	Driving training/practise
4.1. Make drive	A5	Swing is misaligned and golfer "gets underneath" ball	Ball travels up and high and does not reach desired location	Immediate	L	H	Driving training/practise
4.1. Make drive	A8	Golfer hits a thin air shot	Golfer fails to make drive and incurs a shot penalty	Immediate	L	H	Driving training/practise
4.1. Make drive	A9	Swing incomplete	Backswing/downswing incomplete due to snagging on tree branch, rough, etc.	Immediate	L	H	Take drop
4.1. Make drive	A9	Pull out of swing due to external influence, e.g., spectator noise	Shot is not completed and penalty shot may be incurred	Immediate	L	H	N/A
4.1. Make drive	A6	Wrong club selected	Shot is over or under hit	5.	M	H	Club selection training Consult caddie
4.2–4.7. See 4.1.	A5 A8	4.2–4.7. See 4.1.	4.2–4.7. See 4.1.	4.2–4.7. See 4.1.			4.2–4.7. See 4.1.
4.8. Make putt	A4	Golfer over hits putt	Ball may overshoot hole or skip hole	Immediate	H	M	Putting practise
4.8. Make putt	A4	Golfer under hits putt	Ball does not reach hole	Immediate	H	M	Putting practise
4.8. Make putt	A5	Golfer misaligns putt	Ball misses hole	Immediate	H	M	Putting practise
4.8. Make putt	A6/A9	Golfer strikes green first	Putt does not have sufficient power	Immediate	L	H	Putting practise
4.8. Make putt	A5	Golfer mis-hits ball	Putt is not struck appropriately	Immediate	M	H	Putting practise

Human Error Identification and Analysis Methods

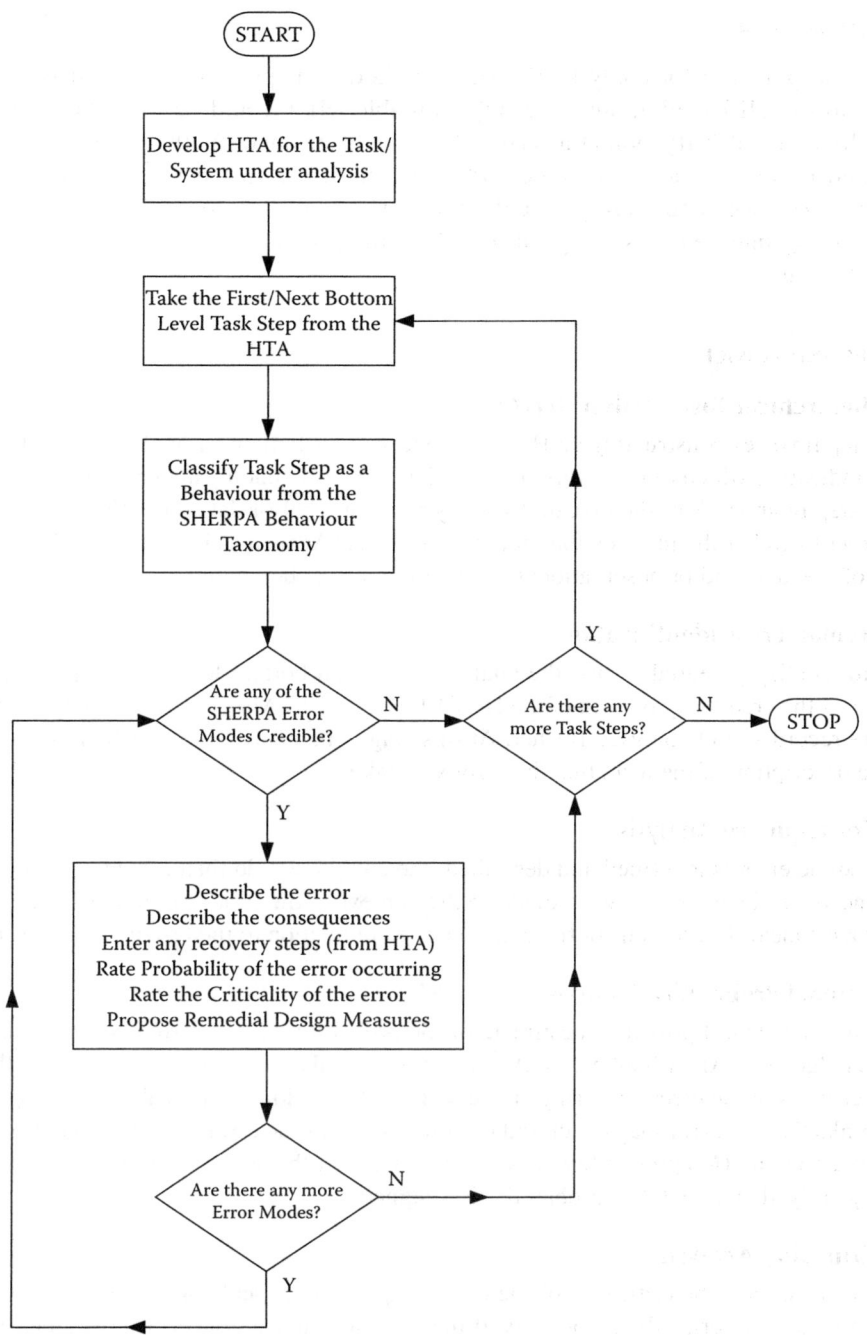

FLOWCHART 5.3 SHERPA flowchart.

Domain of Application

The HET method was originally developed for the aviation domain for use in the certification of flight deck technology. However, the HET EEM taxonomy is generic, allowing the method to be applied in any domain.

Application in Sport

There is great potential for applying the HET method in a sporting context. It is a simplistic and quick to use HEI method and is easily auditable. HET could be used during the design concept phases to identify potential errors that could be made with sports technology and devices, and it could be applied to identify, a priori, errors that performers may make during task performance. The findings could be used to propose remedial design measures or to develop training interventions designed to reduce the probability of error occurrence during future performances.

Procedure and Advice

Step 1: Hierarchical Task Analysis (HTA)

The first step involves constructing an HTA of the task or system under analysis. The HET method works by indicating which of the errors from the HET error taxonomy are credible at each bottom-level task step in an HTA of the task under analysis. A number of data collection methods may be used in order to gather the information required for the HTA, such as interviews with SMEs, walk-throughs of the task, and/or observational study of the task under analysis.

Step 2: Human Error Identification

In order to identify potential errors, the analyst takes each bottom-level task step from the HTA and considers the credibility of each of the HET EEMs. Any EEMs that are deemed credible by the analyst are recorded and analysed further. At this stage, the analyst ticks each credible EEM and provides a description of the form that the error will take.

Step 3: Consequence Analysis

Once a credible error is identified and described, the analyst should then consider and describe the consequence(s) of the error if it were made in the context of the task step in question. The analyst should provide clear descriptions of the consequences in relation to the task under analysis.

Step 4: Ordinal Probability Analysis

Next, the analyst should provide an estimate of the probability of the error occurring, based upon subjective judgement. An ordinal probability value is entered as low, medium, or high. If the analyst feels that chances of the error occurring are very small, then a low (L) probability is assigned. If the analyst thinks that the error may occur and has knowledge of the error occurring on previous occasions, then a medium (M) probability is assigned. Finally, if the analyst thinks that the error would occur frequently, then a high (H) probability is assigned.

Step 5: Criticality Analysis

Next, the analyst rates the criticality of the error in question. A scale of low, medium, and high is also used to rate error criticality. Normally, if the error would lead to a critical incident (in relation to the task in question), or if the error results in the task not being successfully performed, then it is rated as a highly critical error.

Step 6: Interface Analysis

If the analysis is focussing on the errors made using a sports device, such as a training watch or a cycle computer, then the final step involves determining whether or not the interface under analysis passes the certification procedure. The analyst assigns a "pass" or "fail" rating to the interface under analysis based upon the associated error probability and criticality ratings. If a high probability and a high criticality have been assigned, then the interface in question is classed as a "fail";

otherwise, any combination of probability and criticality and the interface in question is classed as a "pass."

Step 7: Review and Refine Analysis

HET is an iterative process and normally requires many passes before the analysis is completed to a satisfactory level. The analyst should therefore spend as much time as possible reviewing and refining the analysis, preferably in conjunction with SMEs for the task or scenario under analysis. The probability and criticality ratings in particular often require numerous revisions.

Advantages

1. The HET methodology is quick, simple to learn and use, and requires very little training.
2. HET provides a comprehensive EEM taxonomy, developed based upon a review of existing HEI EEM taxonomies.
3. HET is easily auditable as it comes in the form of an error pro-forma.
4. The HET taxonomy prompts the analyst for potential errors.
5. HET validation studies have generated encouraging reliability and validity data (Harris et al., 2005; Stanton et al., 2006).
6. Although the error modes in the HET EEM taxonomy were developed specifically for the aviation domain, they are generic, ensuring that the method can potentially be used in a wide range of different sporting domains.

Disadvantages

1. For large, complex tasks HET analyses may become overly time consuming and tedious to perform.
2. HET analyses do not provide remedial measures or countermeasures.
3. Extra work is required if the HTA is not already available.
4. HET does not deal with the cognitive component of errors.
5. In its present form HET does not consider performance-shaping factors, although these can be applied by the analyst.
6. HET only considers errors at the sharp end of system operation. The method does not consider system or organisational failures.

Related Methods

There are many taxonomic-based HEI methods available, including SHERPA (Embrey, 1986), Cognitive Reliability and Error Analysis Method (CREAM; Hollnagel, 1998), and TRACEr (Shorrock and Kirwan, 2002). An HET analysis requires an initial HTA (or some other specific task description) to be performed for the task in question. The data used in the development of the HTA may be collected through the application of a number of different methods, including observational study, interviews, and walkthrough analysis.

Approximate Training and Application Times

In HET validation studies Harris et al. (2005) reported that with non-Human Factors professionals, the approximate training time for the HET methodology is around 90 minutes. Application time is typically low, although it varies depending on the scenario being analysed. Harris et al. (2005), for example, reported a mean application time of 62 minutes based upon an analysis of the flight task, "Land aircraft X at New Orleans using the Autoland system."

Reliability and Validity

Harris et al. (2005) and Stanton et al. (2006) report sensitivity index ratings between 0.7 and 0.8 for participants using the HET methodology to predict potential design-induced pilot errors for an aviation approach and landing task. These figures represent a high level of accuracy of the error predictions (the closer to 1.0 the more accurate the error predictions are). It was also reported that participants using the HET method achieved higher SI ratings than those using the SHERPA, Human Error HAZOP, and HEIST methods to predict errors for the same task.

Tools Needed

HET can be carried out using the HET error pro-forma, an HTA of the task under analysis, functional diagrams of the interface under analysis, and a pen and paper.

Example

An HET analysis was conducted for the Garmin 305 forerunner device (see Chapter 9). The analysis focussed on the device programming and then running task "Use forerunner device to complete sub 5.30-minute mile 10 miles run." The HTA for this task is presented in Figure 5.10. An extract of the HET analysis is presented in Figure 5.7.

Flowchart

(See Flowchart 5.4.)

Scenario: *Use forerunner device to complete sub 5.30 minute mile 10 miles run*		Task Step: 4.2.3. Read distance ran	Interface Elements: Main display screen (distance)							
Error Mode	Description	Outcome	Likelihood			Criticality			Pass	Fail
			L	M	H	L	M	H		
Fail to execute	Fail to check distance ran	Runner may misunderstand distance ran and may inappropriately alter race strategy	✓				✓		✓	
Task execution incomplete										
Task execution in wrong direction										
Wrong task executed										
Task repeated										
Task executed on the wrong interface element	Runner checks pace or time display instead	Runner may misunderstand distance ran and may inappropriately alter race strategy		✓			✓		✓	
Task executed too early										
Task executed too late										
Task executed too much										
Task executed too little										
Misread information	Runner misreads distance reading	Runner may misunderstand distance ran and may inappropriately alter race strategy	✓				✓		✓	
Other	Runner cannot read display due to glare/shadow	Runner does not get distance ran information and has to use backlight feature		✓	✓				✓	

FIGURE 5.7 HET analysis extract.

Human Error Identification and Analysis Methods

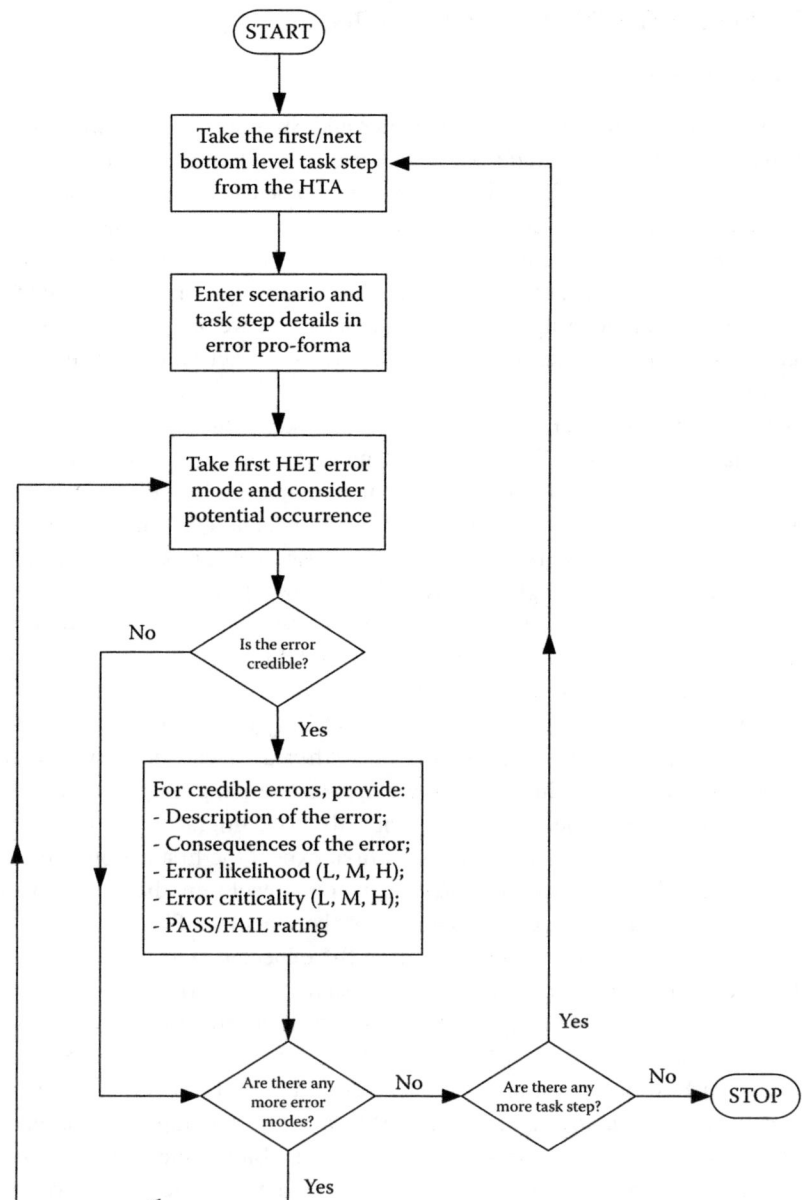

FLOWCHART 5.4 HET flowchart.

Recommended Texts

Stanton, N., Harris, D., Salmon, P. M., Demagalski, J. M., Marshall, A., Young, M. S., Dekker, S. W. A., and Waldmann, T. (2006). Predicting design induced pilot error using HET (Human Error Template)—A new formal human error identification method for flight decks. *Journal of Aeronautical Sciences* February: 107–115.

TASK ANALYSIS FOR ERROR IDENTIFICATION

Background and Applications

Task Analysis for Error Identification (TAFEI; Baber and Stanton, 2002) is a method that enables analysts to predict the errors associated with device use by modelling the interaction between the user and the device under analysis. TAFEI assumes that people use devices in a purposeful manner, such that the interaction may be described as a "cooperative endeavour," and it is by this process that problems arise. Furthermore, the method makes the assumption that actions are constrained by the state of the product at any particular point in the interaction, and that the device offers information to the user about its functionality. Thus, the interaction between users and devices progresses through a sequence of states. At each state, the user selects the action most relevant to their goal, based on the system image.

The foundation for the approach is based on general systems theory. This theory is potentially useful in addressing the interaction between sub-components in systems (i.e., the human and the device). It also assumes a hierarchical order of system components, i.e., all structures and functions are ordered by their relation to other structures and functions, and any particular object or event is comprised of lesser objects and events. Information regarding the status of the machine is received by the human part of the system through sensory and perceptual processes and converted to physical activity in the form of input to the machine. The input modifies the internal state of the machine and feedback is provided to the human in the form of output. Of particular interest here is the boundary between humans and machines, as this is where errors become apparent. It is believed that it is essential for a method of error prediction to examine explicitly the nature of the interaction.

The method draws upon the ideas of scripts and schema. It can be imagined that a person approaching a ticket-vending machine might draw upon a "vending machine" or a "ticket kiosk" script when using a ticket machine. From one script, the user might expect the first action to be "Insert Money," but from the other script, the user might expect the first action to be "Select Item." The success, or failure, of the interaction would depend on how closely they were able to determine a match between the script and the actual operation of the machine. The role of the comparator is vital in this interaction. If it detects differences from the expected states, then it is able to modify the routines. Failure to detect any differences is likely to result in errors.

Examples of applications of TAFEI include the prediction of errors when boiling a kettle (Baber and Stanton, 1994; Stanton and Baber, 1998), using word processing packages (Stanton and Baber, 1996), withdrawing cash from automatic teller machines (Burford, 1993), using medical applications (Baber and Stanton, 1999; Yamaoka and Baber, 2000), recording on tape-to-tape machines (Baber and Stanton, 1994), programming video-cassette recorders (Baber and Stanton, 1994), operating radio-cassette machines (Stanton and Young, 1999), retrieving a phone number on mobile phones (Baber and Stanton, 2002), buying a rail ticket on the ticket machines on the London Underground (Baber and Stanton, 1996a), and operating high-voltage switchgear in substations (Glendon and McKenna, 1995).

Domain of Application

The method was originally conceived as a way of identifying errors during product design; however, the procedure is generic and can be applied in any domain.

Application in Sport

There is great scope to apply TAFEI in a sporting context during the design and evaluation of sports products, equipment, and technology, such as training (e.g., training watches) and performance aids

Human Error Identification and Analysis Methods

(e.g., golf GPS devices). In this context, TAFEI would enable analysts to predict errors associated with device use by modelling the interaction between the user and the device under analysis.

PROCEDURE AND ADVICE

Step 1: Construct HTA for Device under Analysis

For illustrative purposes of how to conduct the method, a simple, manually operated electric kettle is used. The first step in a TAFEI analysis is to obtain an appropriate HTA for the device, as shown in Figure 5.8. As TAFEI is best applied to scenario analyses, it is wise to consider just one specific goal, as described by the HTA (e.g., a specific, closed-loop task of interest) rather than the whole design. Once this goal has been selected, the analysis proceeds to constructing State Space Diagrams (SSDs) for device operation.

Step 2: Construct State Space Diagrams

Next, SSDs are constructed to represent the behaviour of the artefact. An SSD essentially consists of a series of states through which the device passes from a starting state to the goal state. For each series of states, there will be a current state, and a set of possible exits to other states. At a basic level, the current state might be "off," with the exit condition "switch on" taking the device to the state "on." Thus, when the device is "off" it is "waiting to ..." an action (or set of actions) that will take it to the state "on." Upon completion of the SSD, the analyst should have an exhaustive set of states for the device under analysis. Numbered plans from the HTA are then mapped onto the SSD, indicating which human actions take the device from one state to another. Thus the plans are mapped onto the state transitions (if a transition is activated by the machine, this is also indicated on the SSD, using the letter "M" on the TAFEI diagram). This results in a TAFEI diagram, as shown in Figure 5.9. Potential state-dependant hazards have also been identified.

Step 3: Create Transition Matrix

Finally, a transition matrix is devised to display state transitions during device use. TAFEI aims to assist the design of artefacts by illustrating when a state transition is possible but undesirable (i.e., illegal). Making all illegal transitions impossible should facilitate the cooperative endeavour of device use. All possible states are entered as headers on a matrix—see Table 5.3. The cells represent state transitions (e.g., the cell at row 1, column 2 represents the transition between state 1 and state 2), and are filled in one of three ways. If a transition is deemed impossible (i.e., you simply cannot go from this state to that one), "-------" is entered into the cell. If a transition is deemed possible

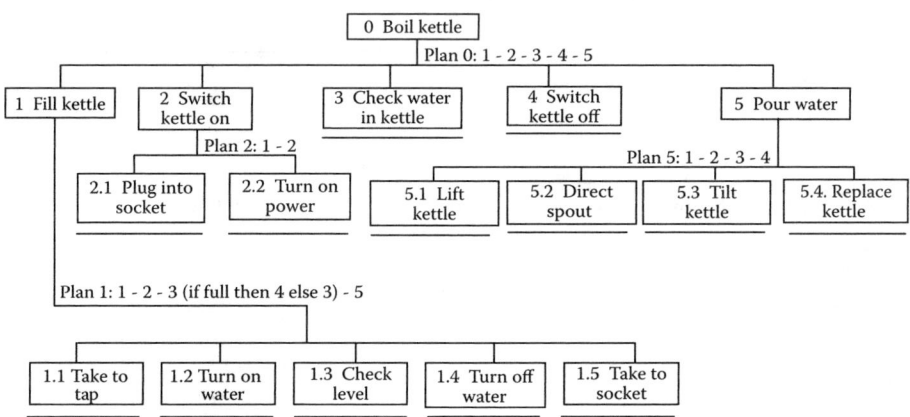

FIGURE 5.8 HTA for boil kettle task.

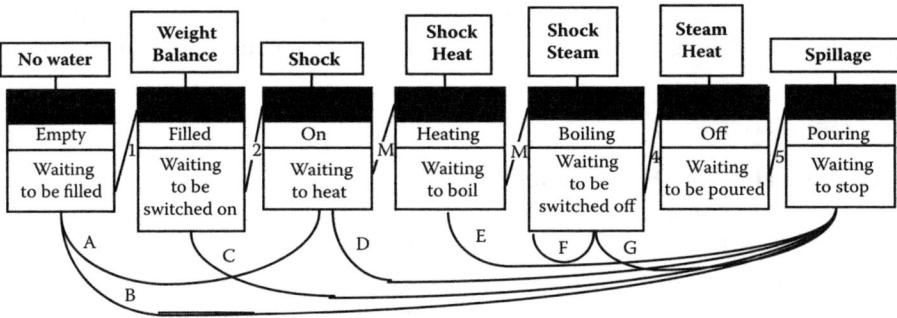

FIGURE 5.9 State space TAFEI diagram.

and desirable (i.e., it progresses the user towards the goal state—a correct action), this is a legal transition and "L" is entered into the cell. If, however, a transition is both possible but undesirable (a deviation from the intended path—an error), this is termed illegal and the cell is filled with an "I." The idea behind TAFEI is that usability may be improved by making all illegal transitions (errors) impossible, thereby limiting the user to only performing desirable actions. It is up to the analyst to conceive of design solutions to achieve this.

The states are normally numbered, but in this example the text description is used. The character "L" denotes all of the error-free transitions and the character "I" denotes all of the errors. Each error has an associated character (i.e., A to G), for the purposes of this example and so that it can be described Table 5.4.

Obviously the design solutions presented in Table 5.4 are illustrative and would need to be formally assessed for their feasibility and cost. What TAFEI does best is enable the analysis to model the interaction between human action and system states. This can be used to identify potential errors and consider the task flow in a goal-oriented scenario. Potential conflicts and contradictions in task flow should become known.

Advantages

1. TAFEI offers a structured and thorough procedure for predicting errors that are likely to arise during human-device interaction.
2. TAFEI enables the analyst to model the interaction between human action and system states.
3. TAFEI has a sound theoretical underpinning.

TABLE 5.3
Transition Matrix

		To State						
		Empty	Filled	On	Heating	Boiling	Off	Pouring
From State	Empty	---------	L (1)	I (A)	---------	---------	---------	I (B)
	Filled		---------	L (2)	---------	---------	---------	I (C)
	On			---------	L (M)	---------	---------	I (D)
	Heating					L (M)	---------	I (E)
	Boiling					I (F)	L (4)	I (G)
	Off							L (5)
	Pouring							

TABLE 5.4
Error Descriptions and Design Solutions

Error	Transition	Error Description	Design Solution
A	1 to 3	Switch empty kettle on	Transparent kettle walls and/or link to water supply
B	1 to 7	Pour empty kettle	Transparent kettle walls and/or link to water supply
C	2 to 7	Pour cold water	Constant hot water or autoheat when kettle placed on base after filling
D	3 to 7	Pour kettle before boiled	Kettle status indicator showing water temperature
E	4 to 7	Pour kettle before boiled	Kettle status indicator showing water temperature
F	5 to 5	Fail to turn off boiling kettle	Auto cut-off switch when kettle boiling
G	5 to 7	Pour boiling water before turning kettle off	Auto cut-off switch when kettle boiling

4. TAFEI is a flexible and generic methodology that can be applied in any domain in which humans interact with technology.
5. TAFEI can include error reduction proposals and can be used to inform product design/redesign.
6. It can be used for identifying human errors which can be used to inform the design of anything from kettles to railway ticket machines (Baber and Stanton, 1994).
7. It can be applied to functional drawings and design concepts throughout the design lifecycle.

Disadvantages

1. Due to the requirement for HTA and SSDs, TAFEI can be time consuming to apply compared to other HEI approaches. Kirwan (1998a), for example, reports that TAFEI is a highly resource intensive method.
2. For complex systems the analysis is likely to be complex and the outputs (i.e., SSDs and transition matrix) large and unwieldy.
3. Depending on the HTA used, only a limited subset of possible errors may be identified.
4. For complex devices the SSDs may be difficult to acquire or develop and the number of states that the device can potentially be in may overwhelm the analyst.
5. Any remedial measures are based on the judgement of the analysts involved.
6. TAFEI requires some skill to perform effectively and is difficult to grasp initially.
7. TAFEI is limited to goal-directed behaviour.
8. TAFEI may be difficult to learn and time consuming to train.

Related Methods

TAFEI uses HTA and SSDs to predict illegal transactions. The data used in the development of the HTA and SSDs may be collected through the application of a number of different methods, including observational study, interviews, and walkthrough analysis, and also review of relevant documentation, such as user and training manuals.

Approximate Training and Application Times

Analysts with little or no knowledge of schema theory may require significant training in the TAFEI method. The application time is dependent upon the device; however, for complex devices with many different states it may be high. Stanton and Young (1999) report that TAFEI is relatively quick to train and apply. For example, in their study of radio-cassette machines, training in the TAFEI method took approximately 3 hours. In the application of the method by recently trained people, it took approximately 3 hours in the radio-cassette study to predict the errors.

Reliability and Validity

Provided the HTA and SSDs used are accurate, the reliability and validity of the method should be high. Stanton and Baber (2002) and Baber and Stanton (1996a) report promising performance of the TAFEI method in validation studies.

Tools Needed

TAFEI is normally conducted using pen and paper; however, drawing software packages such as Microsoft Visio are typically used to produce TAFEI output diagrams.

Example

A TAFEI analysis of a Garmin Forerunner 305 training watch was conducted. The first step involved constructing an HTA for the Garmin device. In order to make the analysis manageable, particularly the SSD construction phase, the analysis focussed on the following specific goal-related scenario: "Use Forerunner device to complete sub 5.30-minute mile 10-mile run." The scenario involved using the Forerunner device to set up a sub 5.30-minute mile pace alert system and then completing a 10-mile run using the device and pace alert system to maintain a sub 5.30-minute mile pace. The purpose of restricting the task in this way was to limit the analysis in terms of size and time taken. The HTA for the scenario "Use Forerunner device to complete sub 5.30-minute mile 10-mile run" is presented in Figure 5.10. Next, an SSD for the Garmin 305 Forerunner device was constructed. The SSD is presented in Figure 5.11. The transition matrix is presented in Table 5.5.

For this scenario, 27 of the possible transitions are defined as illegal. These can be reduced to the following basic error types:

1. Inadvertently switching the device off
2. Inadvertently starting the timer while programming the device
3. Inadvertently stopping the timer during the run
4. Inadvertently selecting/going back to the wrong mode
5. Inadvertently starting the timer once the run has been completed

Flowchart

(See Flowchart 5.5.)

Recommended Texts

Baber, C., and Stanton, N. A. (2002). Task analysis for error identification: Theory, method and validation. *Theoretical Issues in Ergonomics Science* 3(2):212–27.
Stanton, N. A., and Baber, C. (2005). Validating task analysis for error identification: Reliability and validity of a human error prediction technique. *Ergonomics* 48(9):1097–1113.

Human Error Identification and Analysis Methods 151

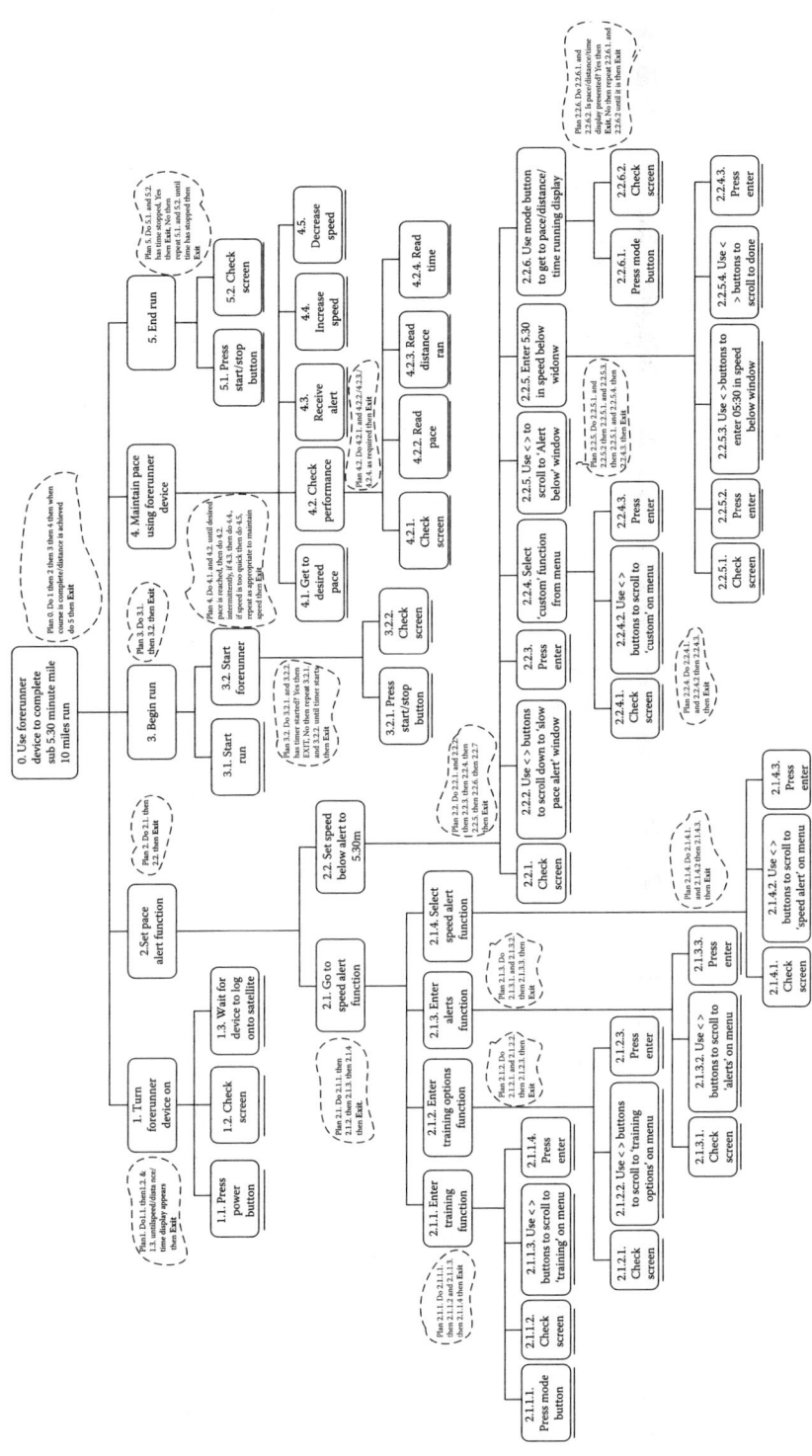

FIGURE 5.10 HTA for "Use Forerunner device to complete sub 5.30-minute mile 10-mile run" scenario.

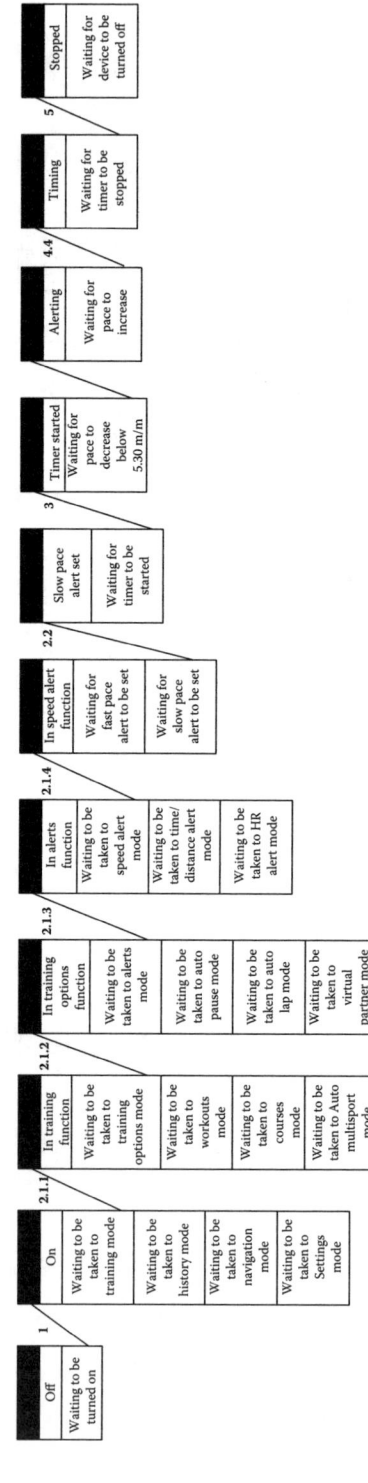

FIGURE 5.11 Garmin 305 Forerunner SSD.

TABLE 5.5
Transition Matrix

	Off	On	In Training Function	In Training Options	In Alerts Function	In Speed Alert Function	Slow Pace Alert Set	Timer Stalled	Alerting	Timing	Stopped
Off	—	L	—	—	—	—	—	—	—	—	—
On	L	—	L	—	—	—	—	I	—	I	—
In training function	I	—	—	L	—	—	—	I	—	I	—
In training options function	I	—	I	—	L	—	—	I	—	I	—
In alerts function	I	—	—	I	—	L	—	I	—	I	—
In speed alert function	I	—	—	—	I	—	L	I	—	I	—
Slow pace alert set	I	—	—	—	—	I	—	L	L	L	I
Timer started	I	—	—	—	—	—	—	—	L	L	I
Alerting	I	—	—	—	—	—	—	—	—	L	I
Timing	I	—	—	—	—	—	—	—	L	—	I
Stopped	L	—	—	—	—	—	—	—	—	I	—

TECHNIQUE FOR HUMAN ERROR ASSESSMENT

BACKGROUND AND APPLICATIONS

The Technique for Human Error Assessment (THEA; Pocock et al., 2001) was developed to aid designers and engineers in the identification of potential user interaction problems in the early stages of interface design. The impetus for the development of THEA was the requirement for an HEI tool that could be used effectively and easily by non-Human Factors specialists. To that end, it is suggested by the creators that the method is more suggestive and much easier to apply than typical HEI methods. The method itself is based upon Norman's model of action execution (Norman, 1988) and uses a series of questions or error identifier prompts in a checklist-style approach based upon goals, plans, performing actions and perception, evaluation, and interpretation. THEA also utilises a scenario-based analysis, whereby the analyst exhaustively describes the scenario under analysis before any error analysis is performed.

DOMAIN OF APPLICATION

The THEA method is generic and can be applied in any domain.

APPLICATION IN SPORT

In a similar fashion to the TAFEI approach, there is scope to apply THEA in a sporting context during the design of sports equipment and technology, such as training (e.g., training watches) and performance aids (e.g., golf GPS devices). In this context THEA would enable analysts to predict errors associated with a device, throughout the design life cycle (e.g., concept, mock-up, prototype), by modelling the interaction between the user and the device under analysis.

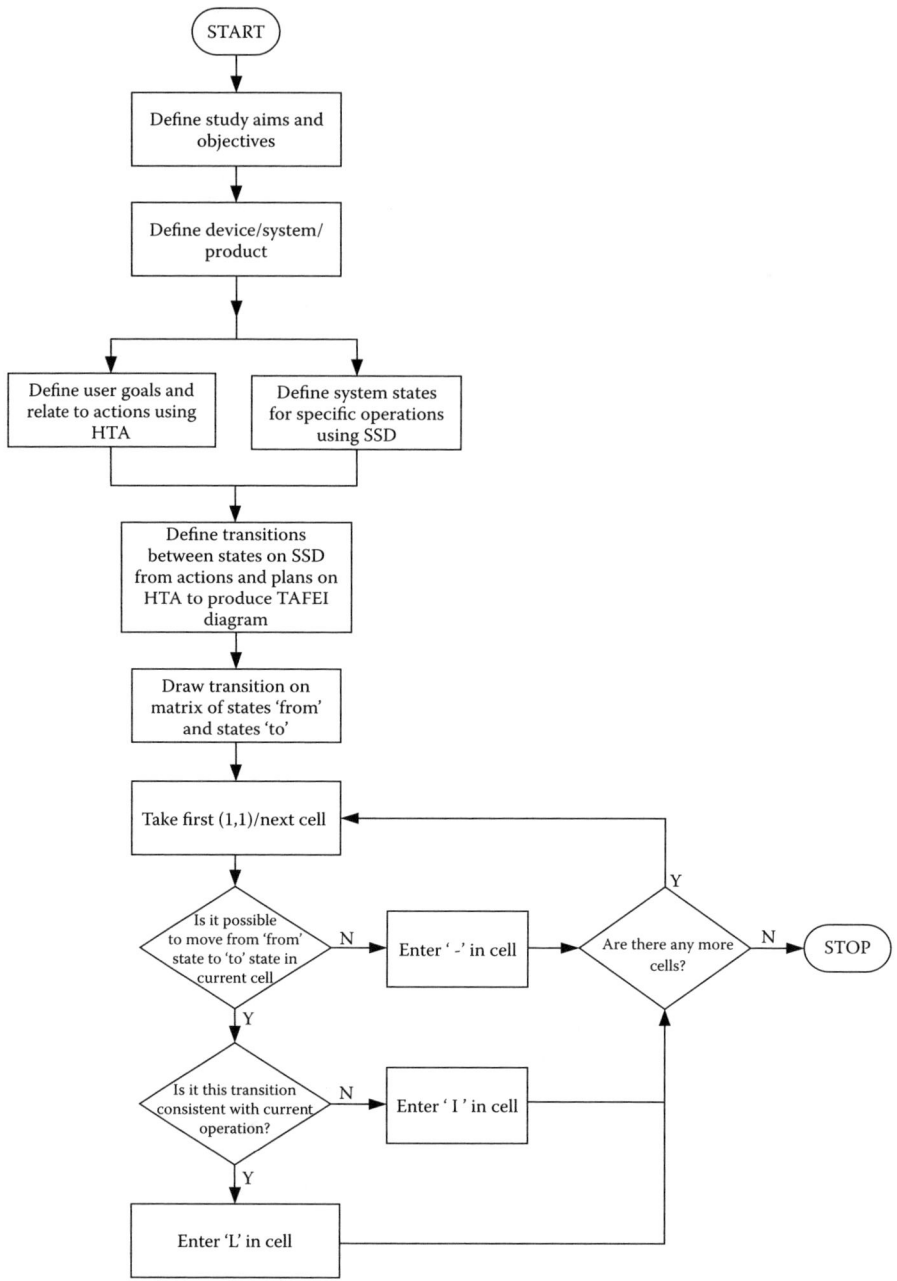

FLOWCHART 5.5 TAFEI flowchart.

Procedure and Advice

Step 1: System Description
Initially, a THEA analysis requires a formal description of the system and task or scenario under analysis. This system description should include details regarding the specification of the systems functionality and interface, and also if and how it interacts with any other systems (Pocock, Fields, Harrison, and Wright, 2001).

Step 2: Scenario Description

Next, the analyst should provide a description of the type of scenario under analysis. The creators have developed a scenario template that assists the analyst in developing the scenario description. The scenario description is conducted in order to give the analyst a thorough description of the scenario under analysis, including information such as actions and any contextual factors that may provide error potential. The scenario description template is presented in Table 5.6.

Step 3: Task Description and Goal Decomposition

A description of the goals and tasks that the operator or user would perform in the scenario is also required. This should describe goals, plans, and intended actions. It is recommended that an HTA of the task under analysis is conducted for this purpose. The HTA should then be used for decomposing the task goals into operations.

TABLE 5.6
Scenario Description Template

Agents
The human agents involved and their organisations
The roles played by the humans, together with their goals and responsibilities

Rationale
Why is this scenario an interesting or useful one to have picked?

Situation and Environment
The physical situation in which the scenario takes place
External and environmental triggers, problems, and events that occur in this scenario

Task Context
What tasks are performed?
Which procedures exist, and will they be followed as prescribed?

System Context
What devices and technology are involved?
What usability problems might participants have?
What effects can users have?

Action
How are the tasks carried out in context?
How do the activities overlap?
Which goals do actions correspond to?

Exceptional Circumstances
How might the scenario evolve differently, either as a result of uncertainty in the environment or because of variations in agents, situation, design options, system, and task context?

Assumptions
What, if any, assumptions have been made that will affect this scenario?

Source: Pocock, S., Fields, R. E., Harrison, M. D., and Wright, P. C. (2001). THEA—A reference guide. University of York Technical Report.

Step 4: Error Analysis

Next, the analyst has to identify and explain any human error that may arise during task performance. THEA provides a structured questionnaire or checklist-style approach in order to aid the analyst in identifying any possible errors. The analyst simply asks questions (from THEA) about the scenario under analysis in order to identify potential errors. For any credible errors, the analyst should record the error, its causes, and its consequences. The questions are normally asked about each goal or task in the HTA, or alternatively, the analyst can select parts of the HTA where problems are anticipated. The THEA error analysis questions comprise the following four categories:

- Goals
- Plans
- Performing actions
- Perception, interpretation, and evaluation

Examples of the THEA error analysis questions for each of the four categories are presented in Table 5.7.

Step 5: Design Implications/Recommendations

Once the analyst has identified any potential errors, the final step of the THEA analysis is to offer any design remedies for each error identified. This is based primarily upon the analyst's subjective judgement. However, the design issues section of the THEA questions also prompt the analyst for design remedies.

ADVANTAGES

1. THEA offers a structured approach to HEI.
2. The THEA method is easy to learn and use and can be used by non-Human Factors professionals.
3. As it is recommended that THEA be used very early in the system life cycle, potential interface problems can be identified and eradicated very early in the design process.
4. THEA error prompt questions are based on sound underpinning theory (Norman's action execution model).
5. THEA's error prompt questions aid the analyst in the identification of potential errors.
6. According to the method's creators, THEA is more suggestive and easier to apply than typical human reliability analysis methods (Pocock et al., 2001).
7. Each error question has associated consequences and design issues to aid the analyst.
8. THEA is a generic method, allowing it to be applied in any domain.

DISADVANTAGES

1. Although error questions prompt the analyst for potential errors, THEA does not use any error modes and so the analyst may be unclear on the types of errors that may occur.
2. THEA is a time-consuming approach and may be difficult to grasp for analysts with little experience of human error theory.
3. Error consequences and design issues provided by THEA are generic and limited.
4. There is little evidence of validation or uptake of the method in the academic literature.
5. HTA, task decomposition, and scenario description create additional work for the analyst.
6. For a method that is supposed to be usable by non-Human Factors professionals, the terminology used in the error analysis questions section may be confusing and hard to decipher.

TABLE 5.7
Example THEA Error Analysis Questions

Questions	Consequences	Design Issues
Goals		
G1—Are items triggered by stimuli in the interface, environment, or task?	If not, goals (and the tasks that achieve them) may be lost, forgotten, or not activated, resulting in omission errors.	Are triggers clear and meaningful? Does the user need to remember all of the goals?
G2—Does the user interface "evoke" or "suggest" goals?	If not, goals may not be activated, resulting in omission errors. If the interface does "suggest" goals, they may not always be the right ones, resulting in the wrong goal being addressed.	Example: graphical display of flight plan shows predetermined goals as well as current progress
Plans		
P1—Can actions be selected in situ, or is pre-planning required?	If the correct action can only be taken by planning in advance, then the cognitive work may be harder. However, when possible, planning ahead often leads to less error-prone behaviour and fewer blind alleys.	
P2—Are there well practised and predetermined plans?	If a plan isn't well known or practised, then it may be prone to being forgotten or remembered incorrectly. If plans aren't predetermined, and must be constructed by the user, then their success depends heavily on the user possessing enough knowledge about their goals and the interface to construct a plan. If predetermined plans do exist and are familiar, then they might be followed inappropriately, not taking account of the peculiarities of the current context.	
Performing Actions		
A1—Is there physical or mental difficulty in executing the actions?	Difficult, complex, or fiddly actions are prone to being carried out incorrectly.	
A2—Are some actions made unavailable at certain times?		
Perception, Interpretation, and Evaluation		
I1—Are changes in the system resulting from user action clearly perceivable?	If there is no feedback that an action has been taken, the user may repeat actions, with potentially undesirable effects.	
I2—Are the effects of user actions perceivable immediately?	If feedback is delayed, the user may become confused about the system state, potentially leading up to a supplemental (perhaps inappropriate) action being taken.	

Source: Pocock, S., Fields, R. E., Harrison, M. D., and Wright, P. C. (2001). THEA—A reference guide. University of York Technical Report.

Related Methods

THEA is similar to HEIST (Kirwan, 1994) in that it uses error prompt questions to aid the analysis. A THEA analysis should be conducted on an initial HTA of the task under analysis; also, various data collection methods may be used, including observational study, interviews, and walkthrough analysis.

Approximate Training and Application Times

Although no training and application time data are offered in the literature, it is apparent that the amount of training time would be minimal. The application time, however, would be high, especially for large, complex tasks.

Reliability and Validity

No data regarding reliability and validity are presented in the literature. The reliability of this approach may be questionable; however, the use of error prompts potentially enhances the method's reliability.

Tools Needed

To conduct a THEA analysis, pen and paper is required. The analyst would also require functional diagrams of the system/interface under analysis and the THEA error analysis questions.

Flowchart

(See Flowchart 5.6.)

Recommended Texts

Pocock, S., Harrison, M., Wright, P., and Johnson, P. (2001). THEA—A technique for human error assessment early in design. In *Human-computer interaction: INTERACT'01*, ed. M. Hirose, 247–54. IOS Press.

Pocock, S., Fields, R. E., Harrison, M. D., and Wright, P. C. (2001). *THEA—A reference guide*. University of York Technical Report.

Human Error Identification and Analysis Methods

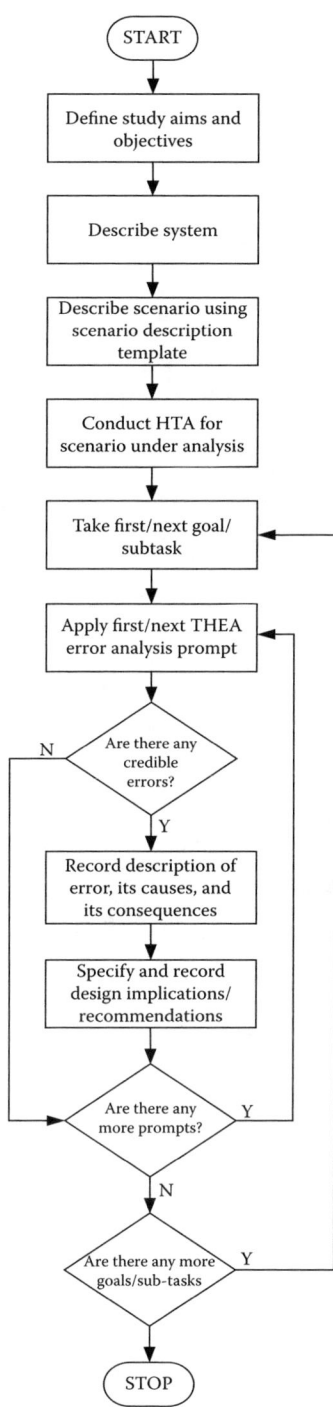

FLOWCHART 5.6 THEA flowchart.

6 Situation Awareness Assessment Methods

INTRODUCTION

Situation Awareness (SA) is the term that is used within Human Factors circles to describe the level of awareness that operators have of the situation that they are engaged in; it focuses on how operators develop and maintain a sufficient understanding of "what is going on" (Endsley, 1995b) in order to achieve success in task performance. A critical commodity in the safety critical domains, SA is now a key consideration in system design and evaluation (Endsley, Bolte, and Jones, 2003; Salmon et al., 2009). Accordingly, various theoretical models have been postulated in relation to individuals (e.g., Endsley, 1995b; Smith and Hancock, 1995), teams (e.g., Endsley and Robertson, 2000; Salas, Prince, Baker, and Shrestha, 1995), and systems (e.g., Artman and Garbis, 1998; Stanton et al., 2006). Further, various measurement approaches have been developed and applied in a range of domains (e.g., Endsley, 1995a; Hogg, Folleso, Strand-Volden, and Torralba, 1995; Stanton et al., 2006).

SITUATION AWARENESS THEORY

The concept first emerged as a topic of interest within the military aviation domain when it was identified as a critical asset for military aircraft crews during the first World War (Press, 1986; cited in Endsley, 1995b). Despite this, it did not begin to receive attention in academic circles until the late 1980s (Stanton and Young, 2000), when SA-related research began to emerge within the aviation and air traffic control domains (e.g., Endsley, 1989, 1993). Following a seminal special issue of the *Human Factors* journal on the subject in 1995, SA became a topic of considerable interest, and it has since evolved into a core theme within system design and evaluation. For example, SA-related research is currently prominent in the military (e.g., Salmon, Stanton, Walker, and Jenkins, 2009), civil aviation and air traffic control (e.g., Kaber, Perry, Segall, McClernon, and Prinzel, 2006), road transport (e.g., Ma and Kaber, 2007; Walker, Stanton, Kazi, Salmon, and Jenkins, in press), energy distribution (Salmon, Stanton, Walker, Jenkins, Baber, and McMaster, 2008), rail (Walker et al., 2006), naval (e.g., Stanton, Salmon, Walker, and Jenkins, 2009), sports (James and Patrick, 2004), healthcare and medicine (Hazlehurst, McMullen, and Gorman, 2007), and emergency services domains (e.g., Blandford and Wong, 2004).

Various definitions of SA are presented in the academic literature (e.g., Adams, Tenney, and Pew, 1995; Bedny and Meister, 1999; Billings, 1995; Dominguez, 1994; Fracker, 1991; Sarter and Woods, 1991; Taylor, R. M., 1990). The most prominent is that offered by Endsley (1995b), who defines SA as a cognitive product (resulting from a separate process labelled situation assessment) comprising "the perception of the elements in the environment within a volume of time and space, the comprehension of their meaning, and the projection of their status in the near future" (Endsley, 1995b, p. 36). Despite its popularity, Endsley's definition is by no means universally accepted. We prefer Smith and Hancock's (1995) view, which describes the construct as "externally directed consciousness" and suggests that SA is "the invariant in the agent-environment system that generates the momentary knowledge and behaviour required to attain the goals specified by an arbiter of performance in the environment" (Smith and Hancock, 1995, p. 145), with deviations between an individual's knowledge and the state of the environment being the variable that directs situation assessment behaviour and the subsequent acquisition of data from the environment.

The main incongruence between definitions lies in the reference to SA as either the process of gaining awareness (e.g., Fracker, 1991), as the product of awareness (e.g., Endsley, 1995b), or as a combination of the two (e.g., Smith and Hancock, 1995). This is a debate that will no doubt continue unabated; however, we feel that in order to fully appreciate the construct, an understanding of both the process and the product is required (Stanton, Chambers, and Piggott, 2001).

Individual Models of Situation Awareness

Inaugural SA models were, in the main, focussed on how individual operators develop and maintain SA while undertaking activity within complex systems (e.g., Adams, Tenney, and Pew, 1995; Endsley, 1995b; Smith and Hancock, 1995). Indeed, the majority of the models presented in the literature are individual focussed models, such as Endsley's three-level model (Endsley, 1995b), Smith and Hancock's perceptual cycle model (Smith and Hancock, 1995), and Bedny and Meister's activity theory model (Bedny and Meister, 1999). As well as being divided by the process versus product debate, SA models also differ in terms of their underpinning psychological approach. For example, the three-level model (Endsley, 1995b) is a cognitive theory that uses an information processing approach; Smith and Hancock's (1995) model is an ecological approach underpinned by Neisser's perceptual cycle model (Neisser, 1976); and Bedny and Meister's (1999) model uses an activity theory model to describe SA.

Endsley's three-level (Endsley, 1995b) model has undoubtedly received the most attention. The three-level model describes SA as an internally held cognitive product comprising three hierarchical levels that is separate from the processes (termed situation assessment) used to achieve it. Endsley's information processing–based model of SA is presented in Figure 6.1.

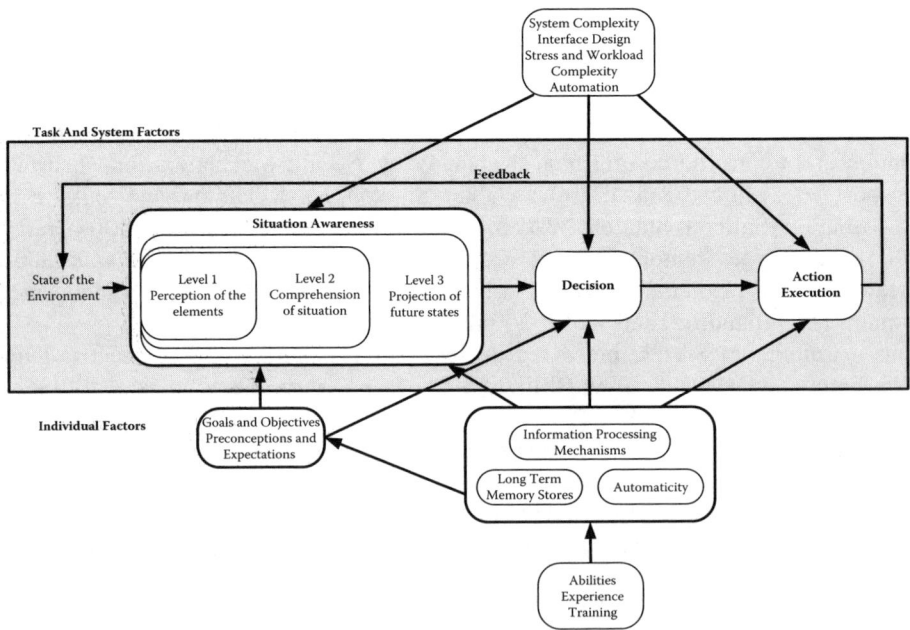

FIGURE 6.1 Endsley's three-level model of situation awareness. From Endsley, Human Factors, 1995. With permission.

The model depicts SA as a component of the information processing chain that follows perception and leads to decision making and action execution. According to the model, SA acquisition and maintenance is influenced by various factors including individual factors (e.g., experience, training, workload, etc.), task factors (e.g., complexity), and systemic factors (e.g., interface design) (Endsley, 1995b).

Endsley's account focuses on the individual as a passive information receptor and divides SA into three hierarchical levels. The first step involves perceiving the status, attributes, and dynamics of task-related elements in the surrounding environment (Endsley, 1995b). At this stage, the data are merely perceived and no further processing takes place. The data perceived are dependent on a range of factors, including the task being performed; the operator's goals, experience, and expectations; and systemic factors such as system capability, interface design, level of complexity, and automation. Level 2 SA involves the interpretation of level 1 data in a way that allows an individual to comprehend or understand its relevance in relation to their task and goals. During the acquisition of level 2 SA "the decision maker forms a holistic picture of the environment, comprehending the significance of objects and events" (Endsley, 1995b, p. 37). Similar to level 1, the interpretation and comprehension of SA-related data is influenced by an individual's goals, expectations, experience in the form of mental models, and preconceptions regarding the situation. The highest level of SA, according to Endsley, involves prognosticating future system states. Using a combination of level 1 and 2 SA-related knowledge, and experience in the form of mental models, individuals can forecast likely future states in the situation. For example, a soccer player might forecast, based on level 1– and level 2–related information, an opposing soccer team player's next move, which may be a pass to a certain player, an attempt to dribble round his or her marker, or a shot on goal. The player can do this through perceiving elements such as the locations of opposing players, player posture, or positioning and gestures, comprehending what they mean and then comparing this to experience to derive what might happen next. Features in the environment are mapped to mental models in the operator's mind, and the models are then used to facilitate the development of SA (Endsley, 1995b). Mental models are therefore used to facilitate the achievement of SA by directing attention to critical elements in the environment (level 1), integrating the elements to aid understanding of their meaning (level 2), and generating possible future states and events (level 3).

Smith and Hancock's (1995) ecological approach takes a more holistic stance, viewing SA as a "generative process of knowledge creation and informed action taking" (p. 138). Their description is based upon Neisser's (1976) perceptual cycle model, which describes an individual's interaction with the world and the influential role of schemata in these interactions. According to the perceptual cycle model, our interaction with the world (termed explorations) is directed by internally held schemata. The outcome of interaction modifies the original schemata, which in turn directs further exploration. This process of directed interaction and modification continues in an infinite cyclical nature. Using this model, Smith and Hancock (1995) suggest that SA is neither resident in the world nor in the person, but resides through the interaction of the person with the world. Smith and Hancock (1995, p. 138) describe SA as "externally directed consciousness," that is, an "invariant component in an adaptive cycle of knowledge, action and information." Smith and Hancock (1995) argue that the process of achieving and maintaining SA revolves around internally held schema, which contain information regarding certain situations. These schema facilitate the anticipation of situational events, directing an individual's attention to cues in the environment and directing his or her eventual course of action. An individual then conducts checks to confirm that the evolving situation conforms to his or her expectations. Any unexpected events serve to prompt further search and explanation, which in turn modifies the operator's existing model. The perceptual cycle model of SA is presented in Figure 6.2.

Smith and Hancock (1995) identify SA as a subset of the content of working memory in the mind of the individual (in one sense it is a product). However, they emphasise that attention is externally directed rather than introspective (and thus is contextually linked and dynamic). Unlike the three-level model, which depicts SA as a product separate from the processes used to achieve it, SA is therefore viewed as both process and product. Smith and Hancock's (1995) complete model therefore

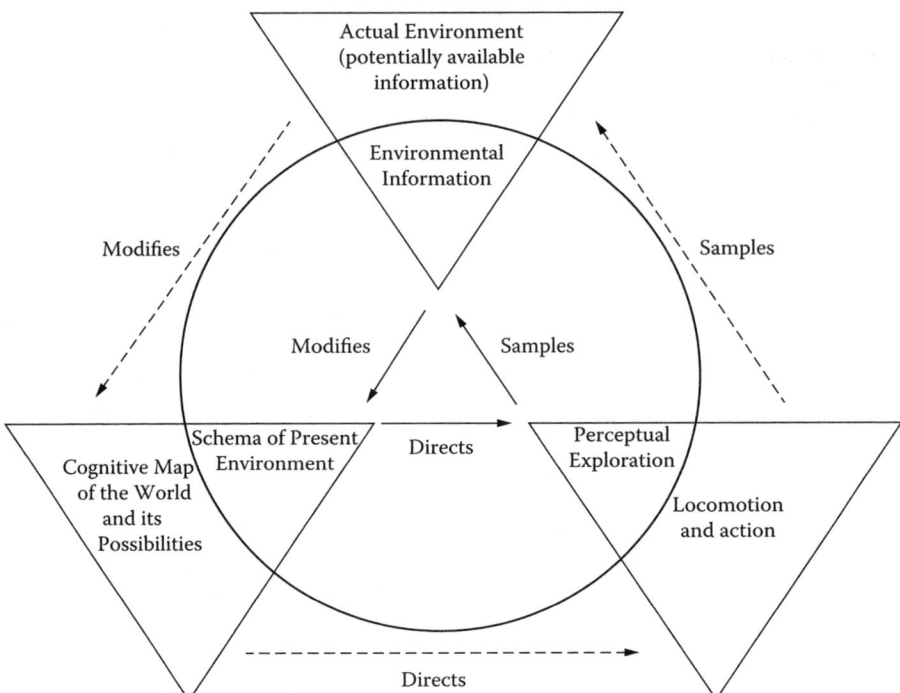

FIGURE 6.2 Smith and Hancock's perceptual cycle model of situation awareness. *Source*: Adapted from Smith, K. and Hancock, P. A. (1995). With permission.

views SA as more of a holistic process that influences the generation of situational representations. For example, in reference to air traffic controllers "losing the picture," Smith and Hancock suggest, "SA is not the controller's picture. Rather it is the controller's SA that builds the picture and that enables them to know that what they know is insufficient for the increasing demands" (Smith and Hancock, 1995, p. 142). The model has sound underpinning theory (Neisser, 1976) and is complete in that it refers to the continuous cycle of SA acquisition and maintenance, including both the process (the continuous sampling of the environment) and the product (the continually updated schema) of SA. Their description also caters to the dynamic nature of SA and more clearly describes an individual's interaction with the world in order to achieve and maintain SA, whereas Endsley's model seems to place the individual as a passive information receiver. The model therefore considers the individual, the situation, and the interactions between the two.

Team Models of Situation Awareness

Team SA is indubitably more complex than individual SA. Salas et al. (1995) point out that there is a lot more to team SA than merely combining individual team member SA. Various models of team SA have been proposed (e.g., Salas et al, 1995; Shu and Furuta, 2005; Stanton et al., 2006; 2009; Wellens, 1993). The most commonly used approach has involved the scaling up of Endsley's (1995b) three-level model to cater for team SA, using the related but distinct concepts of team and shared SA (e.g., Endsley and Jones, 1997; Endsley and Robertson, 2000). Shared SA is defined as "the degree to which team members have the same SA on shared SA requirements" (Endsley and Jones, 1997). Team SA, on the other hand, is defined as "the degree to which every team member possesses the SA required for his or her responsibilities" (Endsley, 1989). Shared SA models suggest that during team activities SA overlaps between team members, in that individuals need to perceive, comprehend, and project SA elements that are specifically related to their role within the team, but also elements that are required by themselves and by other members of the team. Successful team performance therefore requires that individual team members have good SA on their specific elements and the same SA for shared SA elements (Endsley and Robertson, 2000). Again using soccer as an example, different players within the team have their own distinct SA requirements related to their role and position within the team, but also have shared SA requirements that are common across all players. For example, a midfielder whose role it is to man mark a specific opposition player would have very different SA requirements to a striker whose job it is to create space in the attacking third and score goals; however, they would also possess shared SA requirements, such as the ball, time, score, rules, overall team tactics, and the referee.

Much like individual SA, the concept of team SA is plagued by contention. For example, many have expressed concern over the use of Endsley's individual operator three-level model to describe team SA (Artman and Garbis, 1998; Gorman, Cooke, and Winner, 2006; Patrick, James, Ahmed, and Halliday, 2006; Salmon, Stanton, Walker, and Green, 2006; Salmon et al., 2008; Stanton et al., 2009; Shu and Furuta, 2005; Siemieniuch and Sinclair, 2006; Sonnenwald, Maglaughlin and Whitton, 2004), and also regarding the relatively blunt characterisation of shared SA (e.g., Stanton et al., 2009). Describing team SA using the three-level model is problematic for a number of reasons, not least because it was and is primarily an individual operator model. Further, the concept of shared SA remains ambiguous and leaves key questions unanswered. For example, does "shared" SA mean that team members understand elements of a situation in exactly the same manner, or is SA shared in the sense that different agents each have a different piece of it? Can team members with access to the same information really arrive at the same situational picture? How do different team member roles, tasks, experience, and schema map onto the idea of shared SA? As a consequence of existing team SA models with individualistic origins, these are all significant issues that have not been satisfactorily dealt with by existing team and shared SA approaches.

In response to these problems, recent distributed cognition-based Distributed Situation Awareness (DSA) models have attempted to clarify SA in collaborative environments (e.g., Salmon et al., 2009; Stanton et al., 2006; Stanton et al., 2009). Distributed cognition (e.g., Hutchins, 1995) moves the focus on cognition out of the heads of individual operators and onto the overall system consisting of human and technological agents; cognition transcends the boundaries of individual actors and "systemic" cognition is achieved by the transmission of representational states throughout the system (Hutchins, 1995). Artman and Garbis (1998) first called for a systems perspective model of SA, suggesting that the predominant individualistic models of that time were inadequate for studying SA during teamwork. Instead they urged a focus on the joint cognitive system as a whole, and subsequently defined team SA as "the active construction of a model of a situation partly shared and partly distributed between two or more agents, from which one can anticipate important future states in the near future" (Artman and Garbis, 1998, p. 2). Following this, the foundations for a theory of DSA in complex systems, laid by Stanton et al. (2006) were built upon by Stanton et al. (2009) who outlined a model of DSA, developed based on applied research in a range of military and civilian command and control environments.

Briefly, the model of Stanton et al. is underpinned by four theoretical concepts: schema theory (e.g., Bartlett, 1932), genotype and phenotype schema, Neisser's (1976) perceptual cycle model of cognition, and, of course, Hutchins's (1995) distributed cognition approach. Following Hutchins (1995) and Artman and Garbis (1998), the model takes a systems perspective approach on SA and views SA as an emergent property of collaborative systems; SA therefore arises from the interactions between agents. At a systemic level, awareness is distributed across the different human and technological agents involved in collaborative endeavour. Stanton et al. (2006; 2009) suggest that a system's awareness comprises a network of information on which different components of the system have distinct views and ownership of information. Scaling the model down to individual team members, it is suggested that team member SA represents the state of their perceptual cycle (Neisser, 1976); individuals possess genotype schema that are triggered by the task-relevant nature of task performance, and during task performance the phenotype schema comes to the fore. It is this task- and schema-driven content of team member SA that brings the ubiquitous shared SA (e.g., Endsley and Robertson, 2000) notion into question. Rather than possess shared SA (which suggests that team members understand a situation or elements of a situation in the same manner), the model instead suggests that team members possess unique, but compatible, portions of awareness. Team members experience a situation in different ways as defined by their own personal experience, goals, roles, tasks, training, skills, schema, and so on. Compatible awareness is therefore the phenomenon that holds distributed systems together (Stanton et al., 2006; Stanton et al., 2009). Each team member has their own awareness, related to the goals that they are working toward. This is not the same as that of other team members, but is such that it enables them to work with adjacent team members. Although different team members may have access to the same information, differences in goals, roles, the tasks being performed, experience, and their schema mean that their resultant awareness of it is not shared; instead the situation is viewed differently based on these factors. Each team members' SA is, however, compatible since it is different in content but is collectively required for the system to perform collaborative tasks optimally.

Of course, the compatible SA view does not discount the sharing of information, nor does it discount the notion that different team members have access to the same information; this is where the concept of SA "transactions" applies. Transactive SA describes the notion that DSA is acquired and maintained through transactions in awareness that arise from communications and sharing of information. A transaction in this case represents an exchange of SA between one agent and another (where agent refers to humans and artefacts). Agents receive information; it is integrated with other information and acted on; and then agents pass it on to other agents. The interpretation of that information changes per team member. The exchange of information between team members leads to transactions in SA; for example, when a soccer coach provides instruction to a soccer team during the game, the resultant transaction in SA for each player is different, depending on their role in the team. Each player is using the information for their own ends, integrated into their own schemata, and reaching an individual interpretation. Thus, the transaction is an exchange rather than a sharing of awareness. Each agent's SA (and so the overall DSA) is therefore updated via SA transactions. Transactive SA elements from one model of a situation can form an interacting part of another without any necessary requirement for parity of meaning or purpose; it is the systemic transformation of situational elements as they cross the system boundary from one team member to another that bestows upon team SA an emergent behaviour.

SITUATION AWARENESS AND SPORT

The concept of SA is highly applicable in sport. James and Patrick (2004), for example, point out that within various sports, expert performers are defined by an ability to be aware of key components of the game and to make more appropriate decisions when faced with multiple options under time constraints (Ward and Williams, 2003; cited in James and Patrick, 2004). James and Patrick (2004) go on to emphasise the importance of investigating the nature of SA in a sporting context. Surprisingly then perhaps, a review of the literature reveals that there has been scant reference to the concept within Sports Science journals. Further investigation reveals, however, that this may be down to semantics more than anything else, since there are close similarities between the SA concept and other concepts that have been investigated by sports scientists. Fiore and Salas (2008), for example, suggest that research on SA is very similar to anticipation, perceptual-cognitive skills, and tactical decision-making research undertaken by sports psychologists. They also suggest that many sports psychology studies have generated significant findings with regard to gaining and maintaining SA in fast, externally paced environments. James and Patrick (2004) also identify various sports-based research which is applicable to SA, including the use of occlusion methods, eye tracking devices, and verbal reports to identify the cues used by expert sports performers and also the identification and specification of SA requirements in both individual and team sports.

Within individual and team sports, having an accurate awareness of the situation is obviously a critical commodity. In individual sports, such as tennis, running, and cycling, being aware of what is going on and what is likely to happen are key to successful performance. In tennis, for example, being aware of where the ball and opponent are, where the ball and opponent are going, and how the ball is going to behave are key aspects of performance, as are being aware of one's own position on the court, condition, and current score. In particular, being able to comprehend aspects of the situation, relate them to stored knowledge and previous experiences, project future states, and then behave accordingly is paramount. For example, in tennis, forecasting, based on posture, stance, grip, previous behaviour, ability, and countless other factors, which shot the opponent is about to play and where the ball is likely to end up, is critical. Further, when making a shot, awareness of the flight of the ball, one's own position on court, the opponent's position, and likely reactions are also key aspects.

The concept is equally applicable to team sports. For example, in soccer, knowing what is going on and what is going to happen next is critical. When in possession, being aware of where one is and where team-mates and opposition players are and are likely to be in the near future, what tactics are applicable to the situation, and which team-mates want the ball and are making runs in order to receive the ball are examples of key SA requirements. When a team-mate is in possession, knowing what he or she is likely to do next is also important. Further, when the opposition is in possession, being aware of who to mark and what opposition players are going to do next are also key elements of performance. The identification of SA requirements for different sports is of particular interest; for example, the differences in SA requirements across positions, identifying which SA requirements are compatible across team members, and also analysing the processes that are used by the team to build and maintain team SA are all pertinent lines of enquiry.

As pointed out above, evidence of studies focussing on the concept of SA is sparse. Pederson and Cooke (2006) do highlight the importance of SA in American football and speculate on ways to improve American football team SA, such as enhancing coordination through Internet video games or expert instruction; however, this aside we found no studies making a specific reference to the concept of SA within Sports Science journals. Ostensibly, much of this is down to semantics; however, James and Patrick (2004) also highlighted the difficulty associated with the assessment of SA in a sporting context and the lack of ecologically valid methods that also capture the diversity of the sporting situations under analysis. Certainly, there is no doubt that studying the SA concept is likely to lead to theoretical and performance gains. From a sporting point of view, identifying SA requirements, and instances where players have degraded or erroneous levels of SA, can be used to develop coaching interventions and performer strategies designed to enhance SA. In addition, new tactics, formations, sports technology, or performance aids require testing for their affect on the SA levels of the performers using them. The differences between the nature and level of SA acquired by elite and novice performers is also likely to be of interest, as is the differences in SA requirements between different sports.

MEASURING SITUATION AWARENESS

Elaborating on the nature of SA during sporting activities requires that valid measures be in place for describing and assessing the concept. Various methods of SA measurement exist (for detailed reviews see Salmon et al., 2006; Salmon et al., 2009). Existing SA measures can be broadly categorised into the following types (Salmon et al., 2006): SA requirements analysis, freeze probe recall methods, real-time probe methods, post-trial subjective rating methods, observer rating methods, and process indices. SA requirements analysis forms the first step in an SA assessment effort and is used to identify what exactly it is that comprises SA in the scenario and environment in question. Endsley (2001) defines SA requirements as "those dynamic information needs associated with the major goals or sub-goals of the operator in performing his or her job" (p. 8). According to Endsley (2001), they concern not only the data that operators need, but also how the data are integrated to address decisions. Matthews, Strater, and Endsley (2004) highlight the importance of conducting SA requirements analysis when developing reliable and valid SA metrics. By far the most popular approach for assessing SA, freeze probe methods involve the administration of SA-related queries on-line (i.e., during task performance) during "freezes" in a simulation of the task under analysis. Participant responses are compared to the state of the system at the point of the freeze and an overall SA score is calculated at the end of the trial. Developed to limit the high level of task intrusion imposed by freeze probe methods, real-time probe methods involve the administration of SA-related queries on-line, but with no freeze of the task under analysis (although the reduction in intrusion is questionable). Self-rating methods involve the use of rating scales to elicit subjective assessments of participant SA; they reflect how aware participants perceived themselves to be during task performance. Observer rating methods typically involve SMEs observing participants during task performance and then providing a rating of participant SA based upon pre-defined observable SA-related behaviours exhibited by those being assessed. Using performance measures to assess SA involves measuring relevant aspects of participant performance during the task under analysis; depending upon the task, certain aspects of performance are recorded in order to determine an indirect measure of SA. Finally, process indices involve analysing the processes that participants use in order to develop SA during task performance. Examples of SA-related process indices include the use of eye tracking devices to measure participant eye movements during task performance (e.g., Smolensky, 1993), the results of which are used to determine how the participant's attention was allocated during task performance, and concurrent verbal protocol analysis, which involves creating a written transcript of operator behaviour as they perform the task under analysis (see Chapter 3).

This chapter focuses on the following six SA assessment methods: SA requirements analysis (Endsley, 1993); the Situation Awareness Global Assessment Technique (SAGAT; Endsley, 1995a); the Situation Present Assessment Method (SPAM; Durso, Hackworth, Truitt, Crutchfield, and Manning, 1998); the Situation Awareness Rating Technique (SART; Taylor, R. M., 1990); the SA-Subjective Workload Dominance method (SA-SWORD; Vidulich and Hughes, 1991); and the propositional network method (Salmon et al., 2009; Stanton et al., 2006; Stanton et al., 2009). VPA, a process index that is also used to assess SA, is described in Chapter 3.

The SA requirements analysis method (Endsley, 1993) uses SME interviews and goal-directed task analysis to identify the SA requirements for a particular task or scenario. SAGAT (Endsley, 1995a) is the most popular freeze probe method and was developed to assess pilot SA based on the three levels of SA postulated in Endsley's three-level model. The SPAM method (Durso at al., 1998) is a real-time probe method that was developed for use in the assessment of air traffic controllers' SA and uses on-line real-time probes to participant SA. SART (Taylor, R. M., 1990) is a subjective rating method that uses 10 dimensions to measure operator SA. Participant ratings on each dimension are combined in order to calculate a measure of participant SA. SA-SWORD is a subjective rating method that is used to compare the levels of SA afforded by two different displays, devices, or interfaces. The propositional network method (Salmon et al., 2009) is used to model DSA in collaborative environments and represents SA as a network of information elements on which different team members have differing views. A summary of the SA assessment methods described is presented in Table 6.1.

SITUATION AWARENESS REQUIREMENTS ANALYSIS

BACKGROUND AND APPLICATIONS

SA requirements analysis is used to identify exactly what it is that comprises SA in the scenario and environment under analysis. Endsley (2001) defines SA requirements as "those dynamic information needs associated with the major goals or sub-goals of the operator in performing his or her job" (p. 8). According to Endsley (2001), SA requirements concern not only the data that operators need, but also how the data are integrated to address decisions. Matthews, Strater, and Endsley

Situation Awareness Assessment Methods

(2004) suggest that a fundamental step in developing reliable and valid SA metrics is to identify the SA requirements of a given task. Further, Matthews, Strater, and Endsley (2004) point out that that knowing what the SA requirements are for a given domain provides engineers and technology developers a basis to develop optimal system designs that maximise human performance rather than overloading workers and degrading their performance.

Endsley (1993) and Matthews, Strater, and Endsley (2004) describe a generic procedure for conducting SA requirements analyses that uses unstructured interviews with SMEs, goal-directed task analysis, and questionnaires in order to determine SA requirements for a particular task or system. Endsley's methodology focuses on SA requirements across the three levels of SA specified in her information processing–based model of SA (level 1—perception of elements, level 2—comprehension of meaning, level 3—projection of future states).

Domain of Application

The SA requirements analysis procedure is generic and has been applied in various domains, including the military (Bolstad, Riley, Jones, and Endsley, 2002; Matthews, Strater, and Endsley, 2004) and air traffic control (Endsley, 1993).

Application in Sport

Endsley's methodology provides a useful means of identifying the SA requirements extant in different sports. The SA requirements of different team members in sports teams are also a pertinent line of enquiry. This information can then be used to inform the development of SA measures for different sports, the design and development of coaching and training interventions, and the design and evaluation of sports products, technology, and performance aids.

Procedure and Advice

Step 1: Define the Task(s) under Analysis

The first step in an SA requirements analysis is to clearly define the task or scenario under analysis. It is recommended that the task be described clearly, including the different actors involved, the task goals, and the environment within which the task is to take place. An SA requirements analysis requires that the task be defined explicitly in order to ensure that the appropriate SA requirements are comprehensively assessed.

Step 2: Select Appropriate SMEs

The SA requirements analysis procedure is based upon eliciting SA-related knowledge from SMEs. Therefore, the analyst should next select a set of appropriate SMEs. The more experienced the SMEs are in the task environment under analysis the better, and the analyst should strive to use as many SMEs as possible to ensure comprehensiveness.

Step 3: Conduct SME Interviews

Once the task under analysis is defined clearly and appropriate SMEs are identified, a series of unstructured interviews with the SMEs should be conducted. First, participants should be briefed on the topic of SA and the concept of SA requirements analysis. Following this, Endsley (1993) suggests that each SME should be asked to describe, in their own words, what they feel comprises "good" SA for the task in question. They should then be asked what they would want to know in order to achieve perfect SA. Finally, the SME should be asked to describe what each of the SA elements identified is used for during the task under analysis, e.g., decision making, planning, actions. Endsley (1993) also suggests that once the interviewer has exhausted the SME's knowledge, they

TABLE 6.1
Situation Awareness Assessment Methods Summary Table

Name	Domain	Application in Sport	Training Time	App. Time	Input Methods	Tools Needed	Main Advantages	Main Disadvantages	Outputs
Situation Awareness Requirements Analysis (Matthews, Strater, and Endsley, 2004)	Generic	Identification of situation awareness requirements	Low	High	Task analysis Interviews Questionnaire	Audio recording device Pen and paper Microsoft Word	1. Provides a structured approach for identifying the SA requirements associated with a particular task or scenario 2. The output tells us exactly what it is that needs to be known by different actors during task performance 3. The output has many uses, including for developing SA measures or to inform the design of coaching and training interventions, procedures, or new technology	1. Highly time consuming to apply 2. Requires a high level of access to SMEs 3. Identifying SA elements and the relationships between them requires significant skill and knowledge of SA on behalf of the analyst involved	List of SA requirements for the task or scenario under analysis across Endsley's three levels: 1. Perception of the elements 2. Comprehension of their meaning 3. Projection of future states
Situation Awareness Global Assessment Technique (SAGAT; Endsley, 1995a)	Generic	Situation awareness measurement Analysis of SA levels afforded by different tactics, positions, and sports products/equipment Analysis of SA levels achieved during different events	Low	Low	SA requirements analysis	Simulator	1. Avoids the problems associated with collecting SA data post task, such as memory degradation and forgetting low SA periods of the task 2. SA scores can be viewed in total and also across the three levels specified by Endsley's model 3. By far the most popular approach for measuring SA and has the most validation evidence associated with it	1. Highly intrusive and cannot be applied during real-world tasks 2. Various preparatory activities are required, making the total application time high 3. Typically requires expensive high fidelity simulators and computers	SA scores for participants across Endsley's three levels: 1. Perception of the elements 2. Comprehension of their meaning 3. Projection of future states

Situation Awareness Assessment Methods

Method	Type	Purpose	Level	Equipment/Analysis	Tools	Advantages	Disadvantages	Output
Situation Present Assessment Method (SPAM; Durso et al., 1998)	Generic	Situation awareness measurement. Analysis of SA levels afforded by different tactics, positions, and sports products/equipment. Analysis of SA levels achieved during different events	Low	SA requirements analysis	Simulator	1. On-line administration removes the need for task freezes 2. On-line administration removes the various problems associated with collecting SA data post trial 3. Requires little training	1. Highly intrusive and cannot be applied during real-world tasks 2. Probes could alert participants to aspects of the task that they are not aware of 3. Various preparatory activities are required, including the conduct of SA requirements analysis and the generation of queries	SA scores for participants across Endsley's three levels: 1. Perception of the elements 2. Comprehension of their meaning 3. Projection of future states
Situation Awareness Rating Technique (SART; Taylor, R. M., 1990)	Generic	Situation awareness measurement. Analysis of SA levels afforded by different tactics, positions, and sports products/equipment. Analysis of SA levels achieved during different events	Low	Observational study	Pen and paper	1. Quick and easy to apply and requires very little training 2. Data obtained are quick and easy to analyse 3. The SART dimensions are generic, allowing it to be applied in any domain	1. There are various problems associated with collecting subjective SA data post trial 2. It is questionable whether the SART dimensions actually represent SA or not 3. SART has performed poorly in various validation studies	Subjective rating of total SA and across three dimensions: 1. Demand 2. Supply 3. Understanding
SA-SWORD (Situation Awareness Subjective Workload Dominance; Vidulich, and Hughes, 1991)	Generic	Comparison of SA levels afforded by different device, tactics, position, etc.	Low	Observational study	Pen and paper	1. Quick and easy to use and requires only minimal training 2. Generic; can be applied in any domain 3. Useful when comparing two different devices and their affect on operator SA	1. What provides good SA for one operator may provide poor SA for another 2. Does not provide a direct measure of SA 3. There are various problems associated with collecting subjective SA data post trial	Rating of one task/device's SA dominance over another
Propositional networks (Salmon et al., 2009)	Generic	Model of systems awareness during task performance	High	Observational study. Verbal protocol analysis. Hierarchical task analysis	Pen and paper. Microsoft Visio. WESTT	1. Depicts the information elements underlying the system's SA and the relationships between them 2. Also depicts the awareness of individuals and sub teams working within the system 3. Avoids most of the flaws associated with measuring/modelling SA	1. More of a modelling approach than a measure 2. Can be highly time consuming and laborious 3. It is difficult to present larger networks within articles, reports, or presentations	Network model of system's SA including information elements and the relationships between them

should offer their own suggestions regarding SA requirements, and discuss their relevance. It is recommended that each interview be recorded using either video or audio recording equipment. Following completion of the interviews, all data should be transcribed.

Step 4: Conduct Goal-Directed Task Analysis

Once the interview phase is complete, a goal-directed task analysis should be conducted for the task or scenario under analysis. Endsley (1993) prescribes her own goal-directed task analysis method; however, it is also possible to use HTA (Annett et al., 1971) for this purpose, since it focuses on goals and their decomposition. For this purpose, the HTA procedure presented in Chapter 3 should be used. Once the HTA is complete, the SA elements required for the completion of each step in the HTA should be added. This step is intended to ensure that the list of SA requirements identified during the interview phase is comprehensive. Upon completion, the task analysis output should be reviewed and refined using the SMEs used during the interview phase.

Step 5: Compile List of SA Requirements Identified

The outputs from the SME interview and goal-directed task analysis phases should then be used to compile a list of SA requirements for the different actors involved in the task under analysis.

Step 6: Rate SA Requirements

Endsley's method uses a rating system to sort the SA requirements identified based on importance. The SA elements identified should be compiled into a rating type questionnaire, along with any others that the analyst(s) feels are pertinent. Appropriate SMEs should then be asked to rate the criticality of each of the SA elements identified in relation to the task under analysis. Items should be rated as not important (1), somewhat important (2), or very important (3). The ratings provided should then be averaged across subjects for each item.

Step 7: Determine SA Requirements

Once the questionnaires have been collected and scored, the analyst(s) should use them to determine the SA elements for the task or scenario under analysis. How this is done is dependent upon the analyst(s') judgement. It may be that the elements specified in the questionnaire are presented as SA requirements, along with a classification in terms of importance (e.g., not important, somewhat important, or very important).

Step 8: Create SA Requirements Specification

The final stage involves creating an SA requirements specification that can be used by other practitioners (e.g., system designers, methods developers). The SA requirements should be listed for each actor involved in the task or scenario under analysis. Endsley (1993) and Matthews, Strater, and Endsley (2004) demonstrate how the SA requirements can be categorised across the three levels of SA as outlined by the three-level model; however, this may not be necessary depending on the specification requirements. It is recommended that SA requirements should be listed in terms of what it is that needs to be known, what information is required, how this information is used (i.e., what the linked goals and decisions are), and what the relationships between the different pieces of information actually are, i.e., how they are integrated and used by different actors. Once the SA requirements are identified for each actor in question, a list should be compiled, including tasks, SA elements, the relationships between them, and the goals and decisions associated with them.

ADVANTAGES

1. SA requirements analysis provides a structured approach for identifying the SA requirements associated with a particular task or scenario.
2. The output tells exactly what it is that needs to be known by different actors during task performance.
3. The output has many uses, including for developing SA measures or to inform the design of coaching and training interventions, procedures, or new technology.
4. If conducted properly the method has the potential to be exhaustive.
5. The procedure is generic and can be used to identify the SA requirements associated with any task in any domain.
6. It has great potential to be used as a method for identifying the SA requirements associated with different sports, and also with different positions in sporting teams, the outputs of which can be used to inform the design of training interventions, tactics, performance aids, and sports technology.
7. It has been used to identify the SA requirements associated with various roles, including infantry officers (Matthews, Strater, and Endsley 2004), pilots (Endsley, 1993), aircraft maintenance teams (Endsley and Robertson, 2000), and air traffic controllers (Endsley and Rogers, 1994).

DISADVANTAGES

1. Due to the use of interviews and task analysis methods, the method is highly time consuming to apply.
2. It requires a high level of access to SMEs for the task under analysis.
3. Identifying SA elements and the relationships between them requires significant skill on behalf of the analysts involved.
4. Analyses may become large and unwieldy for complex collaborative systems.
5. Analysts require an in-depth understanding of the SA concept.
6. The output does not directly inform design.

RELATED METHODS

The SA requirements analysis procedure outlined by Endsley (1993) was originally conceived as a way of identifying the SA elements to be tested using the SAGAT freeze probe recall method. The SA requirements analysis method itself uses interviews with SMEs and also goal-directed task analysis, which is similar to the HTA approach.

APPROXIMATE TRAINING AND APPLICATION TIMES

Provided analysts have significant experience of the SA concept, interviews, and task analysis methods, the training time for the SA requirements analysis method is low; however, for novice analysts new to the area and without experience in interview and task analysis methods, the training time required is high. The application time for the SA requirements analysis method is high, including the conduct of interviews, transcription of interview data, conduct of task analysis for the task in question, the identification and rating of SA elements, and finally the compilation of SA requirements and the relationships between them.

Reliability and Validity

The reliability and validity of the SA requirements method is difficult to assess. Provided appropriate SMEs are used throughout the process, the validity should be high; however, the methods reliability may be questionable.

Tools Needed

At its most basic, the SA requirements analysis procedure can be conducted using pen and paper; however, in order to make the analysis as simple and as comprehensive as possible, it is recommended that audio recording equipment is used to record the interviews, and that a computer with a word processing package (such as Microsoft Word) and SPSS are used during the design and analysis of the questionnaire. A drawing package such as Microsoft Visio is also useful when producing the task analysis and SA requirements analysis outputs.

Example

For example purposes, the soccer HTA presented in Chapter 3 was used to generate the SA requirements extant during a soccer game. The example extract presented focuses on the sub-goal decomposition for the task "make pass." The SA requirements were derived for each of the "make pass" sub-goals. The HTA sub-goal decomposition is presented in Figure 6.3. The associated SA requirements are presented in Figure 6.4.

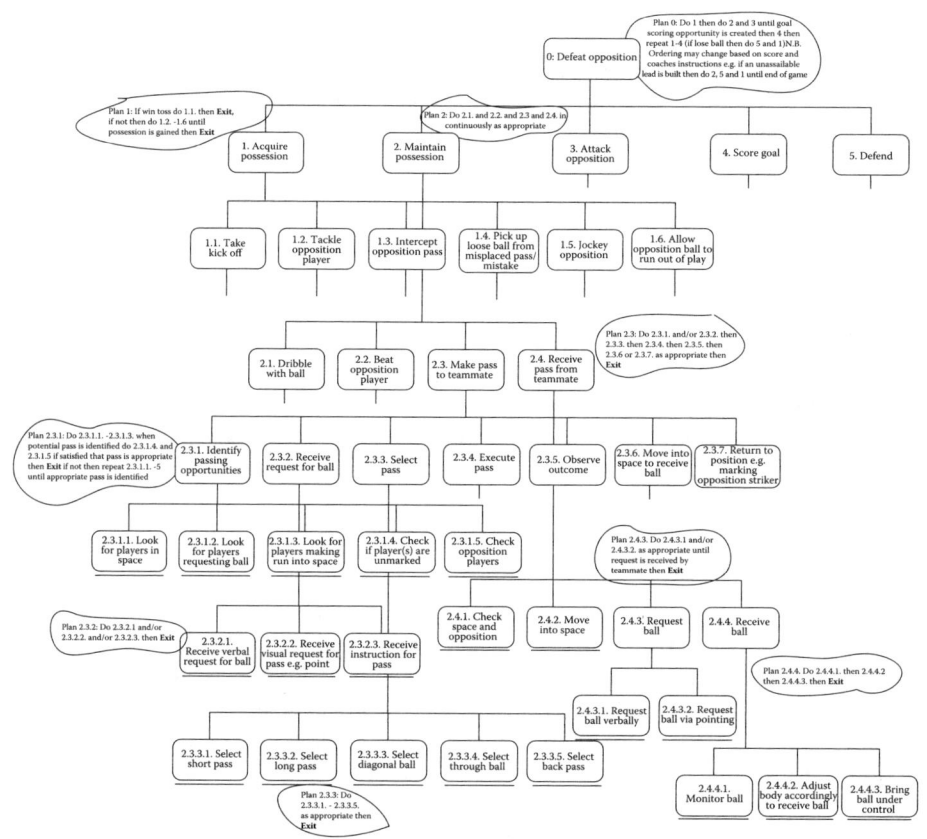

FIGURE 6.3 HTA passing sub-goal decomposition.

Situation Awareness Assessment Methods

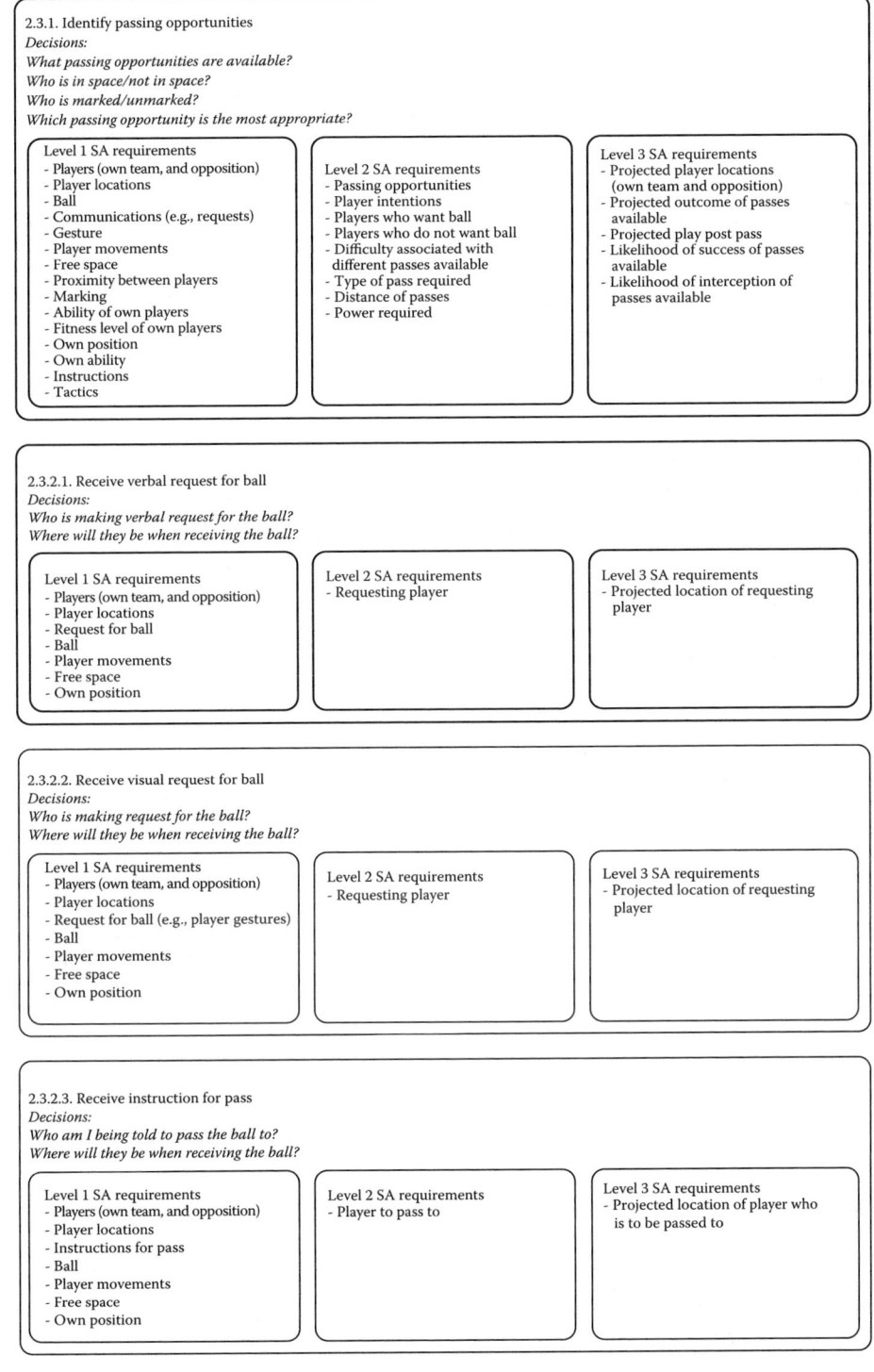

FIGURE 6.4 SA requirements extract for "make pass" sub-goals.

2.3.3. Select pass type
Decisions:
What type of pass is required for the intended pass?

Level 1 SA requirements	Level 2 SA requirements	Level 3 SA requirements
- Players (own team, and opposition) - Intended recipient - Location of intended recipient - Location of other players - Ball - Player movements (i.e., runs) - Player comms (i.e., gestures) - Free space - Proximity between players - Marking - Own position - Own ability	- Distance of pass - Power level required - Type of passes available - Type of pass required - Most appropriate pass available	- Projected recipient location - Projected locations of other players (own team and opposition) - Outcome of passes available - Likelihood of success of different pass types - Likelihood of interception of different pass types

2.3.4. Execute pass
Decisions:
Is chosen pass still on?

Level 1 SA requirements	Level 2 SA requirements	Level 3 SA requirements
- Players (own team, and opposition) - Player locations - Ball - Communications (e.g., requests) - Player movements - Free space - Proximity between players - Marking - Own position - Own ability - Posture	- Type of pass required - Distance of passes - Power required - Action required - Viability of chosen pass	- Projected player locations (own team and opposition) - Projected player actions (i.e., attempt to intercept pass)

2.3.5. Observe outcome
Decisions:
Did pass make it to intended recipient?
Are my own team still in possession of the ball?

Level 1 SA requirements	Level 2 SA requirements	Level 3 SA requirements
- Players (own team, and opposition) - Player locations - Ball - Possession of ball - Player movements	- Outcome of pass	- N/A

FIGURE 6.4 (Continued.)

Flowchart

(See Flowchart 6.1.)

Recommended Texts

Endsley, M. R., Bolte, B., and Jones, D. G. (2003). *Designing for situation awareness: An approach to user-centred design.* London: Taylor & Francis.

Matthews, M. D., Strater, L. D., and Endsley, M. R. (2004). Situation awareness requirements for infantry platoon leaders. *Military Psychology* 16:149–61.

Situation Awareness Assessment Methods

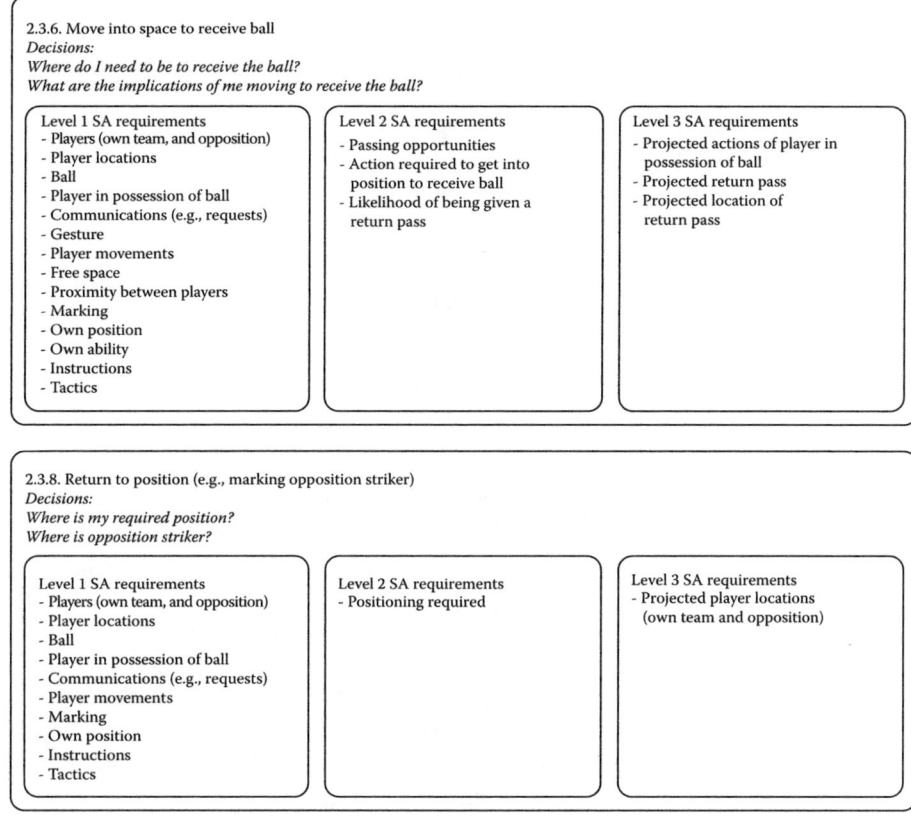

FIGURE 6.4 (Continued.)

SITUATION AWARENESS GLOBAL ASSESSMENT TECHNIQUE

BACKGROUND AND APPLICATIONS

The Situation Awareness Global Assessment Technique (SAGAT; Endsley, 1995b) is the most popular freeze probe recall method and was developed to assess pilot SA based on the three levels of SA postulated by Endsley's three-level model (level 1, perception of the elements; level 2, comprehension of their meaning; and level 3, projection of future states). SAGAT is simulator based, and involves querying participants for their knowledge of relevant SA elements during random freezes in a simulation of the task under analysis. During the freezes, all simulation screens and displays are blanked, and relevant SA queries for that point of the task or scenario are administered.

DOMAIN OF APPLICATION

SAGAT was originally developed for use in the military aviation domain; however, numerous variations of the method have since been applied in other domains, including an air-to-air tactical aircraft version (Endsley, 1990), an advanced bomber aircraft version (Endsley, 1989), and an air traffic control version (Endsley and Kiris, 1995). SAGAT-style approaches can be applied in any domain provided the queries are developed based on an SA requirements analysis carried out in the domain in question.

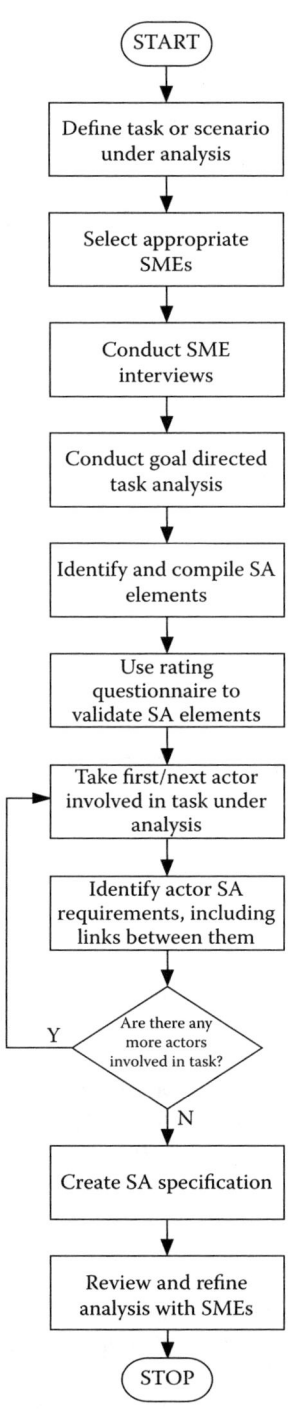

FLOWCHART 6.1 SA requirements analysis flowchart.

APPLICATION IN SPORT

The SAGAT method can be applied in a sporting context provided simulations of the task or scenario under analysis exist. This may be for performance aid or sports technology evaluation in order to determine the effect that a new device has on SA. Further, comparisons of the SA levels achieved by novice and expert performers could also be made using SAGAT type analyses.

PROCEDURE AND ADVICE

Step 1: Define Analysis Aims

First, the aims of the analysis should be clearly defined, since this affects the scenarios used and the types of SAGAT queries administered. For example, the aims of the analysis may be to evaluate the impact that a new performance aid or technological device has on SA during task performance, or it may be to compare novice and expert performer SA during a particular task.

Step 2: Define Task(s) or Scenario under Analysis

The next step involves clearly defining the task or scenario under analysis. It is recommended that the task be described clearly, including the different actors involved, the task goals, and the environment within which the task is to take place.

Step 3: Conduct SA Requirements Analysis and Generate SAGAT Queries

Once the task under analysis is determined, the SA requirements analysis procedure should be undertaken and a series of appropriate queries should be developed based on the SA requirements analysis outputs. Jones and Kaber (2004) highlight the importance of this phase, suggesting that the foundation of successful SAGAT data collection efforts rests solely on the efficacy of the queries used. The queries generated should cover the three levels of SA as prescribed by Endsley's model (i.e., perception, comprehension, and projection). Jones and Kaber (2004) stress that the wording of the queries should be compatible with the operator's frame of reference and appropriate to the language typically used in the domain under analysis.

Step 4: Brief Participants

Once appropriate participants have been recruited based on the analysis requirements, the data collection phase can begin. First, however, it is important to brief the participants involved. This should include an introduction to the area of SA and a description and demonstration of the SAGAT methodology. At this stage participants should also be briefed on what the aims of the study are and what is required of them as participants.

Step 5: Conduct Pilot Run(s)

Before the data collection process begins, it is recommended that a number of pilot runs of the SAGAT data collection procedure be undertaken. A number of small test scenarios should be used to iron out any problems with the data collection procedure, and the participants should be encouraged to ask any questions. The pilot runs should include multiple SAGAT freezes and query administrations. Once the participant is familiar with the procedure and is comfortable with his or her role, the "real" data collection process can begin.

Step 6: Begin SAGAT Data Collection

Once the participants fully understand the SAGAT method and the data collection procedure involved, the SAGAT data collection phase can begin. This begins by instructing the participant(s) to undertake the task under analysis.

Step 7: Freeze the Simulation

SAGAT works by temporarily freezing the simulation at predetermined random points and blanking all displays or interfaces (Jones and Kaber, 2004). According to Jones and Kaber (2004) the following guidelines are useful:

- The timing of freezes should be randomly determined.
- SAGAT freezes should not occur within the first 3 to 5 minutes of the trial.
- SAGAT freezes should not occur within one minute of each other.
- Multiple SAGAT freezes should be used.

Step 8: Administer SAGAT Queries

Once the simulation is frozen at the appropriate point, the analyst should probe the participant's SA using the pre-defined SA queries. These queries are designed to allow the analyst to gain a measure of the participant's knowledge of the situation at that exact point in time and should be directly related to the participant's SA at the point of the freeze. A computer programmed with the SA queries is normally used to administer the queries; however, queries can also be administered using pen and paper. To stop any overloading of the participants, not all SA queries are administrated in any one freeze, and only a randomly selected portion of the SA queries is administrated at any one time. Jones and Kaber (2004) recommend that no outside information should be available to the participants during query administration. For evaluation purposes, the correct answers to the queries should also be recorded; this can be done automatically by sophisticated computers/simulators or manually by an analyst. Once all queries are completed, the simulation should resume from the exact point at which it was frozen (Jones and Kaber, 2004). Steps 7 and 8 are repeated throughout the task until sufficient data are obtained.

Step 9: Query Response Evaluation and SAGAT Score Calculation

Upon completion of the simulator trial, participant query responses are compared to what was actually happening in the situation at the time of query administration. To achieve this, query responses are compared to the data recorded by the simulation computers or analysts involved. Endsley (1995a) suggests that this comparison of the real and perceived situation provides an objective measure of participant SA. Typically, responses are scored as either correct (1) or incorrect (0) and a SAGAT score is calculated for each participant. Additional measures or variations on the SAGAT score can be taken depending upon study requirements, such as time taken to answer queries.

Step 10: Analyse SAGAT Data

SAGAT data are typically analysed across conditions (e.g., trial 1 vs. trial 2) and the three SA levels specified by Endsley's three-level model. This allows query responses to be compared across conditions and also levels of SA to be compared across participants.

ADVANTAGES

1. SAGAT provides an on-line, objective measure of SA, removing the problems associated with collecting subjective SA data (e.g., a correlation between SA ratings and task performance).
2. Since queries are administered on-line during task performance SAGAT avoids the problems associated with collecting SA data post task, such as memory degradation and forgetting low SA periods of the task.
3. SA scores can be viewed in total and also across the three levels specified by Endsley's model.
4. SAGAT is the most popular approach for measuring SA and has the most validation evidence associated with it (Jones and Endsley, 2000, Durso et al., 1998).

5. The specification of SA scores across Endsley's three levels is useful for designers and easy to understand.
6. Evidence suggests that SAGAT is a valid metric of SA (Jones and Kaber, 2004).
7. The method is generic and can be applied in any domain.

Disadvantages

1. Various preparatory activities are required, including the conduct of SA requirements analysis and the generation of numerous SAGAT queries.
2. The total application time for the whole procedure (i.e., including SA requirements analysis and query development) is high.
3. Using the SAGAT method typically requires expensive high fidelity simulators and computers.
4. The use of task freezes and on-line queries is highly intrusive.
5. SAGAT cannot be applied during real-world and/or collaborative tasks.
6. SAGAT does not account for distributed cognition or distributed situation awareness theory (Salmon et al., 2009). For example, in a joint cognitive system it may be that operators do not need to be aware of certain elements as they are held by displays and devices. In this case SAGAT would score participant SA as low, even though the system has optimum SA.
7. SAGAT is based upon the three-level model of SA (Endsley, 1995b), which has various flaws (Salmon et al., 2009).
8. Participants may be directed to elements of the task that they are unaware of.
9. In order to use this approach one has to be able to determine what SA consists of a priori. This might be particularly difficult, if not impossible, for some scenarios.

Related Methods

SAGAT queries are generated based on an initial SA requirements analysis conducted for the task in question. Various versions of SAGAT have been applied, including an air-to-air tactical aircraft version (Endsley, 1990), an advanced bomber aircraft version (Endsley, 1989), and an air traffic control version (Endsley and Kiris, 1995). Further, many freeze probe methods based on the SAGAT approach have been developed for use in other domains. SALSA (Hauss and Eyferth, 2003), for example, was developed specifically for use in air traffic control, whereas the Situation Awareness Control Room Inventory method (SACRI; Hogg et al., 1995) was developed for use in nuclear power control rooms.

Approximate Training and Application Times

The training time for the SAGAT approach is low; however, if analysts require training in the SA requirements analysis procedure, then the training time incurred will increase significantly. The application time for the overall SAGAT procedure, including the conduct of an SA requirements analysis and the development of SAGAT queries, is high. The actual data collection process of administering queries and gathering responses requires relatively little time, although this is dependent upon the task under analysis.

Reliability and Validity

There is considerable validation evidence for the SAGAT approach presented in the literature. Jones and Kaber (2004) point out that numerous studies have been undertaken to assess the validity of the SAGAT, and the evidence suggests that the method is a valid metric of SA. Endsley (2000) reports that SAGAT has been shown to have a high degree of validity and reliability for measuring SA. Collier and Folleso (1995) also reported good reliability for SAGAT when measuring nuclear power plant operator SA. Fracker (1991), however, reported low reliability for SAGAT when measuring participant knowledge of aircraft location. Regarding validity, Endsley, Holder, Leibricht, Garland, Wampler, and Matthews (2000) reported a good level of sensitivity for SAGAT, but not for real-time probes (on-line queries with no freeze) and subjective SA measures. Endsley (1995a) also reported that SAGAT showed a degree of predictive validity when measuring pilot SA, with SAGAT scores indicative of pilot performance in a combat simulation.

Tools Needed

Typically, a high fidelity simulation of the task under analysis and computers with the ability to generate and score SAGAT queries are required. The simulation and computer used should possess the ability to randomly blank all operator displays and "window" displays, randomly administer relevant SA queries, and calculate participant SA scores.

Example

Based on the example soccer SA requirements analysis presented earlier, a series of SAGAT probes for a soccer game situation were developed. Level 1, 2, and 3 SA SAGAT probes for the task "make pass" are presented in Table 6.2.

Flowchart

(See Flowchart 6.2.)

Recommended Texts

Endsley, M. R. (1995). Measurement of situation awareness in dynamic systems. *Human Factors* 37(1):65–84.

Jones, D. G., and Kaber, D. B. (2004). Situation awareness measurement and the situation awareness global assessment technique. In *Handbook of Human Factors and ergonomics methods*, eds. N. Stanton, A. Hedge, H. Hendrick, K. Brookhuis, and E. Salas, 42.1–42.7. Boca Raton, FL: CRC Press.

SITUATION PRESENT ASSESSMENT METHOD

Background and Applications

The Situation Present Assessment Method (SPAM; Durso et al., 1998) is an on-line, real-time probe method that was developed for use in the assessment of air traffic controller SA. The idea behind real-time on-line probe methods is that they retain the objectivity of on-line freeze probe approaches but reduce the level of intrusion on task performance by not using task freezes. SPAM focuses on operator ability to locate information in the environment as an indicator of SA, rather than the recall of specific information regarding the current situation. The analyst probes the operator for SA using task-related on-line SA queries based on pertinent information in the environment (e.g., which of the two aircraft, A or B, has the highest altitude?) via telephone landline. Query response and query response time (for those responses that are correct) are taken as indicators of the operator's SA. Additionally, the time taken to answer the telephone acts as an indicator of mental workload.

Domain of Application

The SPAM method was developed for measuring air traffic controller SA. However, the principles behind the approach (assessing participant SA using real-time probes) could be applied in any domain.

TABLE 6.2
Example SAGAT Queries for "Make Pass" Task

Level 1 Queries
- Where are your team-mates?
- Where are the opposition players?
- Where are you?
- Where is the ball?
- What is the score?
- Who is requesting the ball?
- Where is the player(s) who wants the ball?
- Where is the free space on the pitch?
- What instructions have you just been given by your manager?
- What are the current weather conditions?
- Was the pass successful?

Level 2 Queries
- What passing opportunities are available at the present moment?
- Is the player who wants the ball being marked?
- Which players are currently being marked by opposition players?
- Are any of your own team-mates making a run into space?
- Which of the passes available is the most appropriate?
- What are your instructions/tactics for this sort of situation?
- What type of pass is required?
- What is the approximate distance of the pass?
- How much power is required for the pass?
- Which pass do you need to make to fulfill the instructions just given?

Level 3 Queries
- Will the chosen pass be successful?
- Which player will receive the ball?
- Where will the intended recipient of the ball be?
- What are the projected actions of player X?
- What is the opposition player X going to do next?
- Are there any opposition players who could potentially intercept the pass?

APPLICATION IN SPORT

SPAM can be applied in a sporting context for performance aid or sports technology evaluation in order to determine the effect that a new device has on SA. Further, comparisons of the SA levels achieved by novice and expert performers could also be made using SPAM type analyses.

PROCEDURE AND ADVICE

Step 1: Define Analysis Aims

First, the aims of the analysis should be clearly defined, since this affects the scenarios used and the SPAM queries administered. For example, the aims of the analysis may be to evaluate the impact that a new performance aid or technological device has on SA during task performance, or it may be to compare novice and expert performer SA during a particular task.

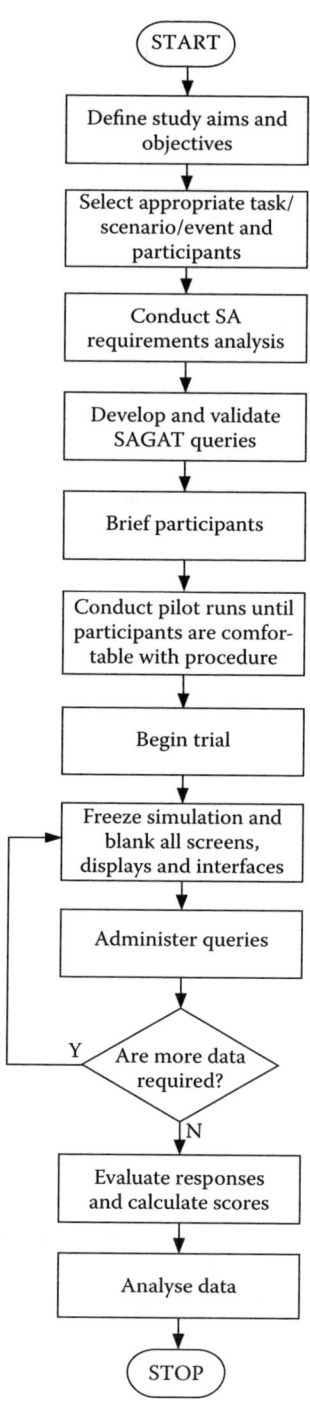

FLOWCHART 6.2 SAGAT flowchart.

Step 2: Define Task(s) or Scenario under Analysis

The next step involves clearly defining the task or scenario under analysis. It is recommended that the task be described clearly, including the different actors involved, the task goals, and the environment within which the task is to take place.

Step 3: Conduct SA Requirements Analysis and Generate Queries

Once the task or scenario under analysis is determined, an SA requirements analysis procedure should be undertaken and a series of appropriate queries should be developed based on the SA requirements analysis outputs. For SPAM analyses, the queries developed should be generic so that they can be applied throughout the task under analysis. Rather than concentrate on information regarding single aircraft (like the SAGAT method) SPAM queries normally ask for "gist type" information (Jeannott, Kelly, and Thompson, 2003).

Step 4: Brief Participants

Once appropriate participants have been recruited based on the analysis requirements, the data collection phase can begin. It is important first to brief the participants involved. This should include an introduction to the area of SA and a description and demonstration of the SPAM methodology. Participants should also be briefed on what the aims of the study are and what is required of them as participants.

Step 5: Conduct Pilot Runs

Before the data collection process begins, it is recommended that participants take part in a number of pilot runs of the SPAM data collection procedure. A number of small test scenarios should be used to iron out any problems with the data collection procedure, and the participants should be encouraged to ask any questions. The pilot runs should include multiple SPAM query administrations. Once the participant is familiar with the procedure and is comfortable with his or her role, the data collection process can begin.

Step 6: Undertake Task Performance

Once the participants fully understand the SPAM method and the data collection procedure, they are free to undertake the task(s) under analysis as normal. The task is normally performed using a simulation of the system and task under analysis; however, it may also be performed in the field (i.e., real-world task performance). Participants should be instructed to undertake the task as normal.

Step 7: Administer SPAM Queries

SPAM queries should be administered at random points during the task. This involves asking the participant a question or series of questions regarding the current situation. Once the analyst has asked the question, a stopwatch should be started in order to measure participant response time. The query answer, query response time, and time to answer the landline should be recorded for each query administered. Step 7 should be repeated until the required amount of data are collected.

Step 8: Calculate Participant SA/Workload Scores

Once the task is complete, the analyst(s) should calculate participant SA based upon the correct query responses and response times (only correct responses are taken into account). A measure of workload can also be derived from the landline response times recorded.

ADVANTAGES

1. SPAM is quick and easy to use, requiring minimal training.
2. There is no need for task freezes.

3. SPAM provides an objective measure of SA.
4. On-line administration removes the various problems associated with collecting SA data post-trial.

Disadvantages

1. Even without task freezes the method remains highly intrusive to task performance.
2. Various preparatory activities are required, including the conduct of SA requirements analysis and the generation of queries.
3. SPAM probes could potentially alert participants to aspects of the task that they are unaware of.
4. It is difficult to see how the method could be applied to real-world sporting tasks.
5. The total application time for the whole procedure (i.e., including SA requirements analysis and query development) is high.
6. There is only limited evidence of the method's application.
7. Often it is required that the SA queries are developed on-line during task performance, which places a great burden on the analyst involved.
8. SPAM is based on the three-level model of SA, which has various flaws (Salmon et al., 2008).
9. In order to use the approach one has to be able to determine what SA should be a priori.

Related Methods

SPAM is essentially the SAGAT approach but without the use of task freezes. An SA requirements analysis should be used when developing SPAM queries. Other real-time probe methods also exist, including the SASHA method (Jeannott, Kelly, and Thompson 2003), which is a modified version of the SPAM method.

Training and Application Times

Training time for the SPAM approach is low; however, if analysts require training in the SA requirements analysis procedure, then the training time incurred will increase significantly. The application time for the overall SPAM procedure, including the conduct of an SA requirements analysis and the development of queries, is high. The actual data collection process of administering queries and gathering responses requires relatively little time, although this is dependent upon the task under analysis.

Reliability and Validity

There are only limited data regarding the reliability and validity of the SPAM method available in the open literature. Jones and Endsley (2000) conducted a study to assess the validity of real-time probes as a measure of SA. In conclusion, it was reported that the real-time probe measure demonstrated a level of sensitivity to SA in two different scenarios and that the method was measuring participant SA, and not simply measuring participant response time.

Tools Needed

SPAM could be applied using pen and paper only; however, Durso et al. (1998) used a landline telephone located in close proximity to the participant's workstation in order to administer SPAM queries. A simulation of the task and system under analysis may also be required.

FLOWCHART

(See Flowchart 6.3.)

RECOMMENDED TEXT

Durso, F. T., Hackworth, C. A., Truitt, T., Crutchfield, J., and Manning, C. A. (1998). Situation awareness as a predictor of performance in en route air traffic controllers. *Air Traffic Quarterly* 6:1–20.

SITUATION AWARENESS RATING TECHNIQUE

BACKGROUND AND APPLICATIONS

The Situation Awareness Rating Technique (SART; Taylor, R. M., 1990) is a simplistic post-trial subjective rating method that uses the following 10 dimensions to measure operator SA: familiarity of the situation, focussing of attention, information quantity, information quality, instability of the situation, concentration of attention, complexity of the situation, variability of the situation, arousal, and spare mental capacity. SART is administered post trial and involves participants subjectively rating each dimension on a seven point rating scale (1 = Low, 7 = High) based on their performance of the task under analysis. The ratings are then combined in order to calculate a measure of participant SA. A quicker version of the SART approach also exists, known as the 3D SART. The 3D SART uses the 10 dimensions described above grouped into the following three dimensions:

1. *Demands on attentional resources*   A combination of complexity, variability, and instability of the situation
2. *Supply of attentional resources*   A combination of arousal, focussing of attention, spare mental capacity, and concentration of attention
3. *Understanding of the situation*   A combination of information quantity, information quality, and familiarity of the situation

DOMAIN OF APPLICATION

The SART approach was originally developed for use in the military aviation domain; however, SART has since been applied in various domains, and since the SART dimensions are generic it is feasible that it could be used in any domain to assess operator SA.

APPLICATION IN SPORT

The SART method could be applied in a sporting context in order to elicit post-trial, subjective ratings of SA. The method is administered post task and so SART could be used where it is not possible to use other on-line SA measures such as SAGAT and SPAM (i.e., during real-world performance of sports tasks).

PROCEDURE AND ADVICE

Step 1: Define Task(s) under Analysis

The first step in a SART analysis is to define the tasks that are to be subjected to analysis. The type of tasks analysed are dependent upon the focus of the analysis. For example, when assessing the effects on operator SA caused by a novel design or training programme, it is useful to analyse as representative a set of tasks as possible. To analyse a full set of tasks will often be too time consuming and labour intensive, and so it is pertinent to use a set of tasks that use all aspects of the system under analysis.

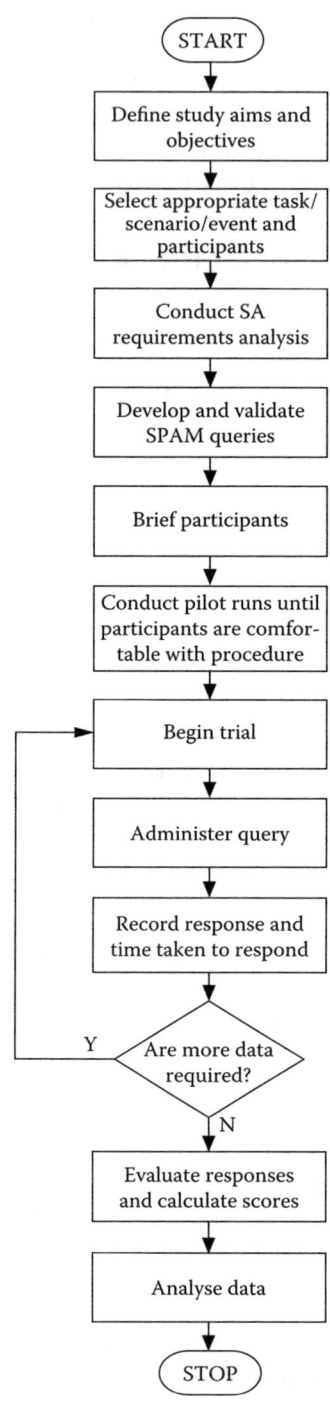

FLOWCHART 6.3 SPAM flowchart.

Step 2: Selection of Participants

Once the task(s) under analysis are clearly defined, it is useful to select the participants that are to be involved in the analysis. This may not always be necessary and it may suffice to select participants randomly on the day; however, if SA is being compared across rank or experience levels, then effort is required to select the appropriate participants.

Step 3: Brief Participants

Before the task(s) under analysis are performed, all of the participants involved should be briefed regarding the purpose of the study, the concept of SA, and the SART method. It may be useful at this stage to take the participants through an example SART analysis, so that they understand how the method works and what is required of them as participants.

Step 4: Conduct Pilot Run

It is recommended that participants take part in a pilot run of the SART data collection procedure. A number of small test scenarios should be used to iron out any problems with the data collection procedure, and the participants should be encouraged to ask any questions. Once the participant is familiar with the procedure and is comfortable with his or her role, the data collection process can begin.

Step 5: Performance of Task

The next step involves the performance of the task or scenario under analysis. For example, if the study is focussing on SA during fell racing, the race should now commence. Participants should be asked to perform the task as normal.

Step 6: Complete SART Questionnaires

Once the task is completed, participants should be given a SART questionnaire and asked to provide ratings for each dimension based on how they felt during task performance. The participants are permitted to ask questions in order to clarify the dimensions; however, the participants' ratings should not be influenced in any way by external sources. In order to reduce the correlation between SA ratings and performance, no performance feedback should be given until after the participant has completed the self-rating process.

Step 7: Calculate Participant SART Scores

The final step in the SART analysis involves calculating each participant's SA score. When using SART, participant SA is calculated using the following formula:

$$SA = U - (D - S)$$

Where
   U = summed understanding
   D = summed demand
   S = summed supply

Typically, for each participant an overall SART score is calculated along with total scores for the three dimensions: understanding, demand, and supply.

ADVANTAGES

1. SART is very quick and easy to apply and requires minimal training.
2. The data obtained are easily and quickly analysed.
3. SART provides a low-cost approach for assessing participant SA.

4. The SART dimensions are generic and so the method can be applied in any domain.
5. SART is non-intrusive to task performance.
6. SART is a widely used method and has been applied in a range of different domains.
7. SART provides a quantitative assessment of SA.

Disadvantages

1. SART suffers from a host of problems associated with collective subjective SA ratings, including a correlation between performance and SA ratings and questions regarding whether or not participants can accurately rate their own awareness (e.g., how can one be aware that they are not aware?).
2. SART suffers from a host of problems associated with collecting SA data post task, including memory degradation and poor recall, a correlation of SA ratings with performance, and also participants forgetting low SA portions of the task.
3. The SART dimensions are not representative of SA. Upon closer inspection the dimensions are more representative of workload than anything else.
4. SART has performed poorly in a number of SA methodology comparison studies (e.g., Endsley, Sollenberger, and Stein, 2000; Endsley, Selcon, Hardiman, and Croft, 1998; Salmon et al., 2009).
5. The SART dimensions were developed through interviews with RAF aircrew (Taylor, R. M., 1990).
6. The method is dated.

Related Methods

SART is a questionnaire-based methodology and is one of many subjective rating SA measurement approaches. Other subjective rating SA measurement methods include the Situation Awareness Rating Scale (SARS; Waag and Houck, 1994), the Crew Awareness Rating Scale (CARS; McGuinness and Foy, 2000), and the SA-SWORD method (Vidulich and Hughes, 1991).

Approximate Training and Application Times

The training and application times associated with the SART method are very low. As it is a self-rating questionnaire, there is very little training involved. In our experience, the SART questionnaire takes no longer than 5 minutes to complete, and it is possible to set up programmes that auto-calculate SART scores based on raw data entry.

Reliability and Validity

Along with SAGAT, SART is the most widely used and tested measure of SA (Endsley and Garland, 2000); however, SART has performed poorly in a number of validation and methods comparison studies (e.g., Endsley, Sollenberger, and Stein, 2000; Endsley et al., 1998; Salmon et al., 2009). In particular, SART has been found to be insensitive to display manipulations. Further, the construct validity of SART is limited, and many have raised concerns regarding the degree to which SART dimensions are actually representative of SA (e.g., Endsley, 1995a; Salmon et al., 2009; Selcon et al., 1991; Uhlarik and Comerford, 2002).

Tools Needed

SART is applied using pen and paper only; however, it can also be administered using Microsoft Excel, which can also be used to automate the SART score calculation process.

EXAMPLE

The SART rating scale is presented in Figure 6.5.

FLOWCHART

(See Flowchart 6.4.)

RECOMMENDED TEXT

Taylor, R. M. (1990). Situational Awareness Rating Technique (SART): The development of a tool for aircrew systems design. In *Situational awareness in aerospace operations (AGARD-CP-478)*, 3/1–3/17. Neuilly Sur Seine, France: NATO-AGARD.

SITUATION AWARENESS SUBJECTIVE WORKLOAD DOMINANCE

BACKGROUND AND APPLICATIONS

The Situation Awareness Subjective Workload Dominance method (SA-SWORD; Vidulich and Hughes, 1991) is an adaptation of the SWORD workload assessment method that is used to test the levels of SA afforded by two different displays or devices. SA-SWORD elicits subjective ratings of SA in terms of one task, or displays dominance over another based on the level of SA afforded. For example, Vidulich and Hughes (1991) used SA-SWORD to assess pilot SA when using two different display design concepts.

DOMAIN OF APPLICATION

The method was originally developed to test display design concepts within the military aviation domain; however, it is a generic approach and can be applied in any domain.

APPLICATION IN SPORT

SA-SWORD could potentially be used in a sporting context in order to test two devices, displays, or performance aids in terms of the level of SA that they afford.

PROCEDURE AND ADVICE

Step 1: Define the Aims of the Analysis

The first step involves clearly defining what the aims of the analysis are. When using the SA-SWORD method the aims may be to compare the levels of SA afforded by two different display design concepts or two different performance aids or devices. Alternatively, it may be to compare the different levels of SA achieved during performance of two different tasks.

Step 2: Define Task(s) under Analysis

The next step involves clearly defining the task(s) under analysis. Once this is done a task or scenario description should be created. Each task should be described individually in order to allow the creation of the SWORD rating sheet. It is recommended that HTA be used for this purpose.

Step 3: Create SWORD Rating Sheet

Once a task description (e.g., HTA) is developed, the SWORD rating sheet can be created. When using SA-SWORD, the analyst should define a set of comparison conditions. For example, when using SA-SWORD to compare to F-16 cockpit displays, the comparison conditions

SART

Instability of the situation
How changeable is the situation? Is the situation highly unstable and likely to change suddenly (high) or is it stable and straightforward (low)?

Complexity of the situation
How complicated is the situation? Is it complex with many interrelated components (high) or is it simple and straightforward (low)?

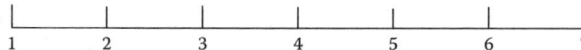

Variability of the situation
How many variables are changing within the situation? Are there a large number of factors varying (high) or are there very few variables changing (low)?

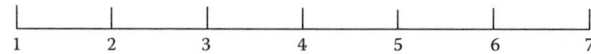

Arousal
How aroused are you in the situation? Are you alert and ready for activity (high) or do you have a low degree of alertness (low)?

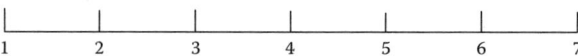

Concentration of attention
How much are you concentrating on the situation? Are you concentrating on many aspects of the situation (high) or focussed on only one (low)?

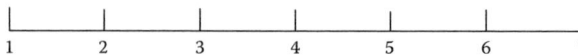

Division of attention
How much is your attention divided in the situation? Are you concentrating on many aspects of the situation (high) or focussed on only one (low)?

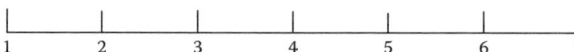

Spare mental capacity
How much mental capacity do you have to spare in the situation? Do you have sufficient to attend to many variables (high) or nothing to spare at all (low)?

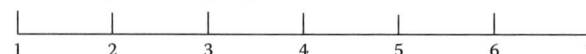

Information quantity
How much information have you gained about the situation? Have you received and understood a great deal of knowledge (high) or very little (low)?

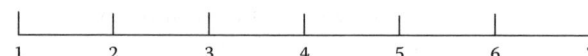

Familiarity with the situation
How familiar are you with the situation? Do you have a great deal of relevant experience (high) or is it a new situation (low)?

FIGURE 6.5 SART rating scale.

Situation Awareness Assessment Methods

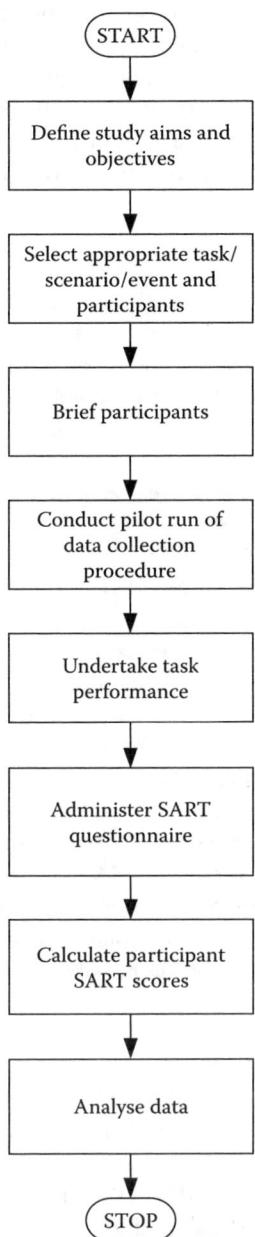

FLOWCHART 6.4 SART flowchart.

used were Fire Control Radar (FCR) display versus Horizontal Situational Format (HSF) display, flight segment (ingress and engagement), and threat level (low vs. high). To do this, the analyst should list all of the possible combinations of tasks or artefacts (e.g., A vs. B, A vs. C, B vs. C).

Step 4: SA and SA-SWORD Briefing

Once the trial and comparison conditions are defined, the participants should be briefed on the concept of SA, the SA-SWORD method, and the purposes of the study. It is critical that each participant has an identical, clear understanding of what SA actually is in order for the SA-SWORD method to provide reliable, valid results. Therefore, it is recommended that participants be given a group briefing, including an introduction to the concept of SA, a clear definition of SA, and an explanation of SA in terms of the operation of the system in question. It may also prove useful to define the SA requirements for the task under analysis. Once the participants clearly understand SA, an explanation of the SA-SWORD method should be provided. It may be useful here to demonstrate the completion of an example SA-SWORD questionnaire. Finally, the participants should then be briefed on the purpose of the study.

Step 5: Conduct Pilot Run
Next, a pilot run of the data collection process should be conducted. Participants should perform a small task and then complete an SA-SWORD rating sheet. The participants should be taken step-by-step through the SA-SWORD rating sheet, and be encouraged to ask any questions regarding any aspects of the data collection procedure that they are not sure about.

Step 6: Undertake Trial
SA-SWORD is administered post trial. Therefore, the task(s) under analysis should be performed next.

Step 7: Administer SA-SWORD Rating Sheet
Once task performance is complete, the SA-SWORD rating procedure can begin. This involves the administration of the SA-SWORD rating sheet immediately after task performance has ended. The SWORD rating sheet lists all possible SA paired comparisons of the task conducted in the scenario under analysis, e.g., display A versus display B, condition A versus condition B. The analyst has to rate the two variables (e.g., display A vs. display B) in terms of the level of SA that they provided during task performance. For example, if the participant feels that the two displays provided a similar level of SA, then they should mark the "Equal" point on the rating sheet. However, if the participant feels that display A provided a slightly higher level of SA than display B did, they would move towards task A on the sheet and mark the "Weak" point on the rating sheet. If the participant felt that display A imposed a much greater level of SA than display B, then they would move towards display A on the sheet and mark the "Absolute" point on the rating sheet. This allows the participant to provide a subjective rating of one display's dominance in terms of SA level afforded over the over. This procedure should continue until all of the possible combinations of SA variables in the task under analysis are rated.

Step 8: Constructing the Judgement Matrix
Once all ratings have been elicited, the SWORD judgement matrix should be conducted. Each cell in the matrix should represent the comparison of the variables in the row with the variable in the associated column. The analyst should fill each cell with the participant's dominance rating. For example, if a participant rated displays A and B as equal, a "1" is entered into the appropriate cell. If display A is rated as dominant, then the analyst simply counts from the "Equal" point to the marked point on the sheet, and enters the number in the appropriate cell. The rating for each variable (e.g., display) is calculated by determining the mean for each row of the matrix and then normalising the means (Vidulich, Ward, and Schueren, 1991).

Step 9: Matrix Consistency Evaluation
Once the SWORD matrix is complete, the consistency of the matrix can be evaluated by ensuring that there are transitive trends amongst the related judgements in the matrix.

ADVANTAGES
1. SA-SWORD is quick and easy to use and requires only minimal training.
2. The SA-SWORD method is generic can be applied in any domain.

3. SA-SWORD has a high level of face validity and user acceptance (Vidulich and Hughes, 1991).
4. The method is useful when comparing two different interface concepts and their effect upon operator SA.
5. Intrusiveness is reduced, as SA-SWORD is administered post trial.

Disadvantages

1. What is good SA for one operator may be poor SA for another, and a very clear definition of SA would need to be developed in order for the method to work. For example, each participant may have different ideas as to what SA actually is, and as a result, the data obtained would be incorrect. In a study testing the SA-SWORD method, it was reported that the participants had very different views on what SA actually was (Vidulich and Hughes, 1991).
2. The method does not provide a direct measure of SA. The analyst is merely given an assessment of the conditions in which SA is greatest.
3. The reporting of SA post trial has a number of problems associated with it, such as a correlation between SA rating and task performance, and participants forgetting low SA periods during task performance.
4. There is limited evidence of the use of the SA-SWORD method in the literature.

Related Methods

The SA-SWORD method is an adaptation of the SWORD workload assessment method. SA-SWORD appears to be unique in its use of paired comparisons to measure SA.

Approximate Training and Application Times

The SA-SWORD method is both easy to learn and apply and thus has a low training and application time associated with it.

Reliability and Validity

Administered in its current form, the SA-SWORD method suffers from a poor level of construct validity, i.e., the extent to which it is actually measuring SA. Vidulich and Hughes (1991) encountered this problem and found that half of the participants understood SA to represent the amount of information that they were attempting to track, while the other half understood SA to represent the amount of information that they may be missing. This problem could potentially be eradicated by incorporating an SA briefing session or a clear definition of what constitutes SA on the SA-SWORD rating sheet. In a study comparing two different cockpit displays, the SA-SWORD method demonstrated a strong sensitivity to display manipulation (Vidulich and Hughes, 1991). Vidulich and Hughes (1991) also calculated inter-rater reliability statistics for the SA-SWORD method, reporting a grand inter-rater correlation of 0.705. According to Vidulich and Hughes, this indicates that participant SA-SWORD ratings were reliably related to the conditions apparent during the trials.

Tools Needed

SA-SWORD can be administered using pen and paper. The system under analysis, or a simulation of the system under analysis, is also required for the task performance component of the data collection procedure.

Flowchart

(See Flowchart 6.5.)

Recommended Text

Vidulich, M. A., and Hughes, E. R. (1991). Testing a subjective metric of situation awareness. *Proceedings of the Human Factors Society 35th Annual Meeting* 1307–11.

PROPOSITIONAL NETWORKS

Background and Applications

The propositional network methodology is used for modelling DSA in collaborative systems. Moving away from the study of SA "in the heads" of individuals towards the study of the level of DSA held by the joint cognitive system brings with it a requirement for new methods of modelling and assessing SA. Existing individual-based measures such as SAGAT and SART are not applicable as they focus exclusively on awareness "in the head" of individual operators (Salmon et al., 2009). In response, Stanton et al. (2006, 2009) and Salmon et al. (2009) outline the propositional network methodology as a method for modelling a system's SA. Propositional networks use networks of linked information elements to depict the information underlying a system's awareness, the relationships between the different pieces of information, and also how each component of the system is using each piece of information.

When using propositional networks, DSA is represented as information elements (or concepts) and the relationships between them, which relates to the assumption that knowledge comprises concepts and the relationships between them (Shadbolt and Burton, 1995). Anderson (1983) first proposed the use of propositional networks to describe activation in memory. They are similar to semantic networks in that they contain linked nodes; however, they differ from semantic networks in two ways (Stanton et al., 2006): First, rather than being added to the network randomly, the words instead are added through the definition of propositions. A proposition in this sense represents a basic statement. Second, the links between the words are labelled in order to define the relationships between the propositions, i.e., elephant "has" tail, mouse "is" rodent. Following Crandall, Klein, and Hoffman (2006), a propositional network about propositional networks is presented in Figure 6.6.

Domain of Application

The approach was originally applied for modelling DSA in command and control scenarios in the military (e.g., Stanton et al., 2006; Salmon et al., 2009) and civilian (e.g., Salmon et al., 2008) domains; however, the method is generic and the approach can be applied in any domain. Propositional networks have since been applied in a range of domains, including naval warfare (Stanton et al., 2006), land warfare (Salmon et al., 2009), railway maintenance operations (Walker et al., 2006), energy distribution substation maintenance scenarios (Salmon et al., 2008), and military aviation airborne early warning systems (Stewart et al., 2008).

Application in Sport

Propositional networks can be used in a sporting context to model DSA during sports performance. The approach is likely to be particularly useful for modelling DSA and identifying the differences in DSA and DSA requirements between team members or between players (e.g., different positions), managers, coaching staff, and game officials (e.g., referees).

Situation Awareness Assessment Methods

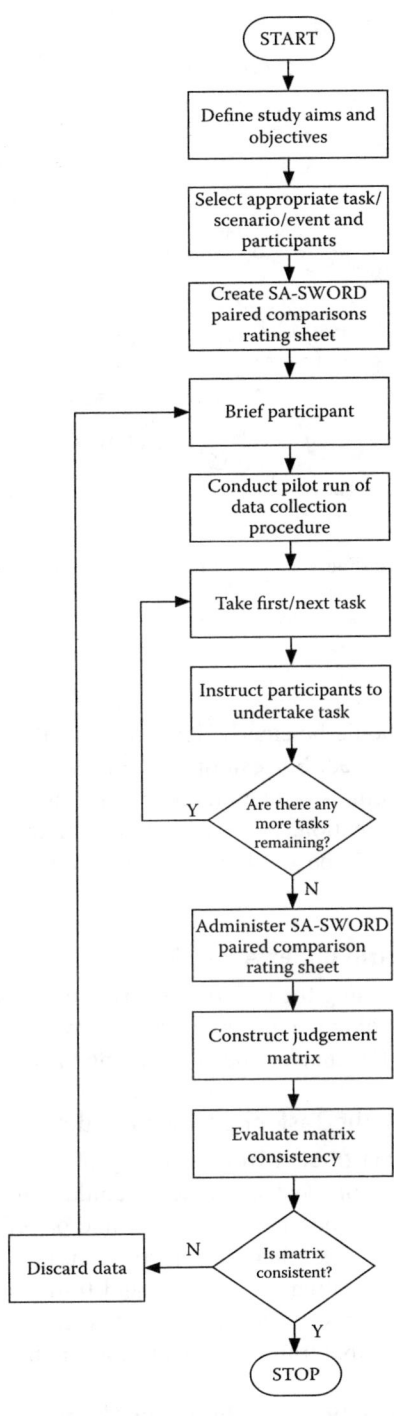

FLOWCHART 6.5 SA-SWORD flowchart.

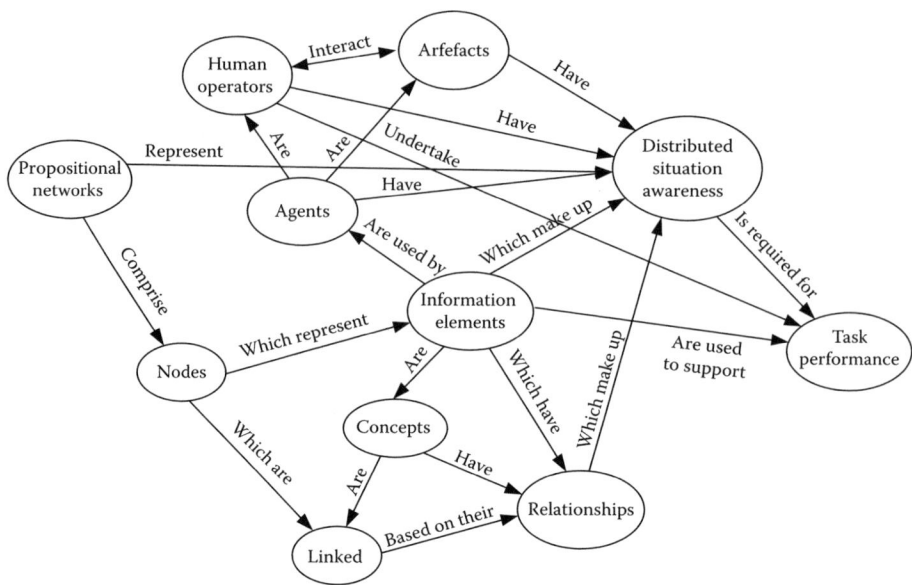

FIGURE 6.6 Propositional network diagram about propositional networks.

Procedure and Advice

Step 1: Define Analysis Aims

First, the aims of the analysis should be clearly defined, since this affects the scenarios used and the propositional networks developed. For example, the aims of the analysis may be to evaluate the DSA levels of different performers or team members during task performance; to evaluate the impact that a new training intervention, performance aid, or technological device has on DSA during task performance; or it may be to compare novice and expert performer DSA during a particular task.

Step 2: Define Task(s) or Scenario under Analysis

The next step involves clearly defining the task or scenario under analysis. It is recommended that the task is described clearly, including the different actors involved, the task goals, and the environment within which the task is to take place. The HTA method is useful for this purpose.

Step 3: Collect Data Regarding the Task or Scenario under Analysis

Propositional networks can be constructed from a variety of data sources, depending on whether DSA is being modelled (in terms of what it should or could comprise) or assessed (in terms of what it did comprise). These include observational study and/or verbal transcript data, CDM data, HTA data, or data derived from work-related artefacts such as Standard Operating Instructions (SOIs), user manuals, standard operating procedures, and training manuals. Data should be collected regarding the task based on opportunity, although it is recommended that, at a minimum, the task in question is observed and verbal transcript recordings are made.

Step 4: Define Concepts and Relationships between Them

It is normally useful to identify distinct task phases. This allows propositional networks to be developed for each phase, which is useful for depicting the dynamic and changing nature of DSA throughout a task or scenario. For example, a soccer game scenario might be divided temporally into game phases or into "attacking" and "defending" phases. In order to construct propositional networks, first the concepts need to be defined followed by the relationships between them. For the purposes of DSA assessments, the term "information elements" is used to refer to concepts. To identify the information elements related to the task under analysis, a simple content analysis is performed on the input data and keywords are extracted. These keywords represent the information elements, which are then linked based on their causal links during the activities in question (e.g., player "has" ball, defender "marks" striker, referee "enforces" rules, etc.). Links are represented by directional arrows and should be overlaid with the linking proposition. A simplistic example of the relationship between concepts is presented in Figure 6.7.

The output of this process is a network of linked information elements; the network contains all of the information that is used by the different actors and artefacts during task performance, and thus represents the system's awareness, or what the system "needed to know" in order to successfully undertake task performance.

Step 5: Define Information Element Usage

Information element usage is normally represented via shading of the different nodes within the network based on their usage by different actors during task performance. During this step the analyst identifies which information elements the different actors involved used during task performance. This can be done in a variety of ways, including by further analysing input data (e.g., verbal transcripts, HTA), and by holding discussions with those involved or relevant SMEs.

Step 6: Review and Refine Network

Constructing propositional networks is a highly iterative process that normally requires numerous reviews and reiterations. It is recommended that once a draft network is created, it be subject to at least three reviews. It is normally useful to involve domain SMEs or the participants who performed the task in this process. The review normally involves checking the information elements and the links between them and also the usage classification. Reiterations to the networks normally include the addition of new information elements and links, revision of existing information elements and links, and also modifying the information element usage based on SME opinion.

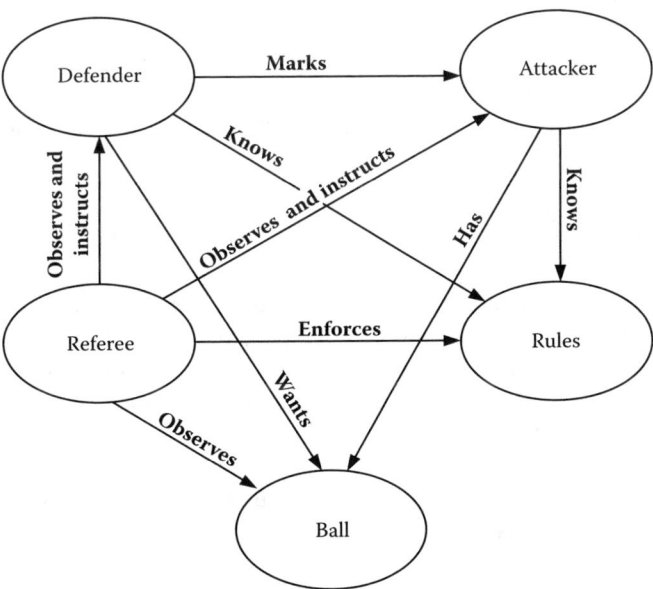

FIGURE 6.7 Example relationships between concepts.

Step 7: Analyse Networks Mathematically

Depending on the analysis aims and requirements, it may also be pertinent to analyse the propositional networks mathematically using social network statistics. For example, in the past we have used sociometric status and centrality calculations to identify the key information elements within propositional networks. Sociometric status provides a measure of how "busy" a node is relative to the total number of nodes present within the network under analysis (Houghton et al., 2006). In this case, sociometric status gives an indication of the relative prominence of information elements based on their links to other information elements in the network. Centrality is also a metric of the standing of a node within a network (Houghton et al., 2006), but here this standing is in terms of its "distance" from all other nodes in the network. A central node is one that is close to all other nodes in the network, and a message conveyed from that node to an arbitrarily selected other node in the network would, on average, arrive via the least number of relaying hops (Houghton et al., 2006). Key information elements are defined as those that have salience for each scenario phase; salience being defined as those information elements that act as hubs to other knowledge elements. Those information elements with a sociometric status value above the mean sociometric status value, and a centrality score above the mean centrality value, are identified as key information elements.

ADVANTAGES

1. Propositional networks depict the information elements underlying a system's DSA and the relationships between them.
2. In addition to modelling the system's awareness, propositional networks also depict the awareness of individuals and sub-teams working within the system.
3. The networks can be analysed mathematically in order to identify the key pieces of information underlying a system's awareness.
4. Unlike other SA measurement methods, propositional networks consider the mapping between the information elements underlying SA.
5. The propositional network procedure avoids some of the flaws typically associated with SA measurement methods, including intrusiveness, high levels of preparatory work (e.g., SA requirements analysis, development of probes), and the problems associated with collecting subjective and SA data or post trial.
6. The outputs can be used to inform training, system, device, and interface design and evaluation.
7. Propositional networks are easy to learn and use.
8. Software support is available via the WESTT software tool (see Houghton et al., 2008).

DISADVANTAGES

1. Constructing propositional networks for complex tasks can be highly time consuming and laborious.
2. It is difficult to present larger networks within articles, reports, and/or presentations.
3. No numerical value is assigned to the level of SA achieved by the system in question.
4. It is more of a modelling approach than a measure, although SA failures can be represented.
5. The initial data collection phase may involve a series of activities and often adds considerable time to the analysis.
6. Many find the departure from viewing SA in the heads of individual operators (i.e., what operators know) to viewing SA as a systemic property that resides in the interactions between actors and between actors and artefacts (i.e., what the system knows) to be a difficult one to take.
7. The reliability of the method is questionable, particularly when used by inexperienced analysts.

Related Methods

Propositional networks are similar to other network-based knowledge representation methods such as semantic networks (Eysenck and Keane, 1990) and concept maps (Crandall, Klein, and Hoffman, 2006). The data collection phase typically utilises a range of approaches, including observational study, CDM interviews, verbal protocol analysis, and HTA. The networks can also be analysed using metrics derived from social network analysis methods.

Approximate Training and Application Times

Providing the analysts involved have some understanding of DSA theory, the training time required for the propositional network method is low; our experiences suggest that around 1 to 2 hours of training is required. Following training, however, considerable practice is required before analysts become proficient in the method. The application time is typically high, although it can be low provided the task is simplistic and short.

Reliability and Validity

The content analysis procedure should ease some reliability concerns; however, the links between concepts are made on the basis of the analyst's subjective judgement and so the reliability of the method may be limited, particularly when being used by inexperienced analysts. The validity of the method is difficult to assess, although our experiences suggest that validity is high, particularly when appropriate SMEs are involved in the process.

Tools Needed

On a simple level, propositional networks can be conducted using pen and paper; however, video and audio recording devices are typically used during the data collection phase and a drawing package such as Microsoft Visio is used to construct the propositional networks. Houghton et al. (2008) describe the WESTT software tool, which contains a propositional network construction module that auto-builds propositional networks based on text data entry.

Example

Example propositional networks for a soccer game are presented below. The propositional network presented in Figure 6.8 represents a typical network for a soccer game. The propositional network presented in Figure 6.9 shows those information elements that are used by an attacking midfielder currently in possession of the ball and looking to build an attack. The propositional network presented in Figure 6.10 represents the opposition defender's awareness for the same (attacking midfielder in possession of ball) scenario.

Flowchart

(See Flowchart 6.6.)

Recommended Text

Salmon, P. M., Stanton, N. A., Walker, G. H., and Jenkins, D. P. (2009). *Distributed situation awareness: Advances in theory, modelling and application to teamwork.* Aldershot, UK: Ashgate.

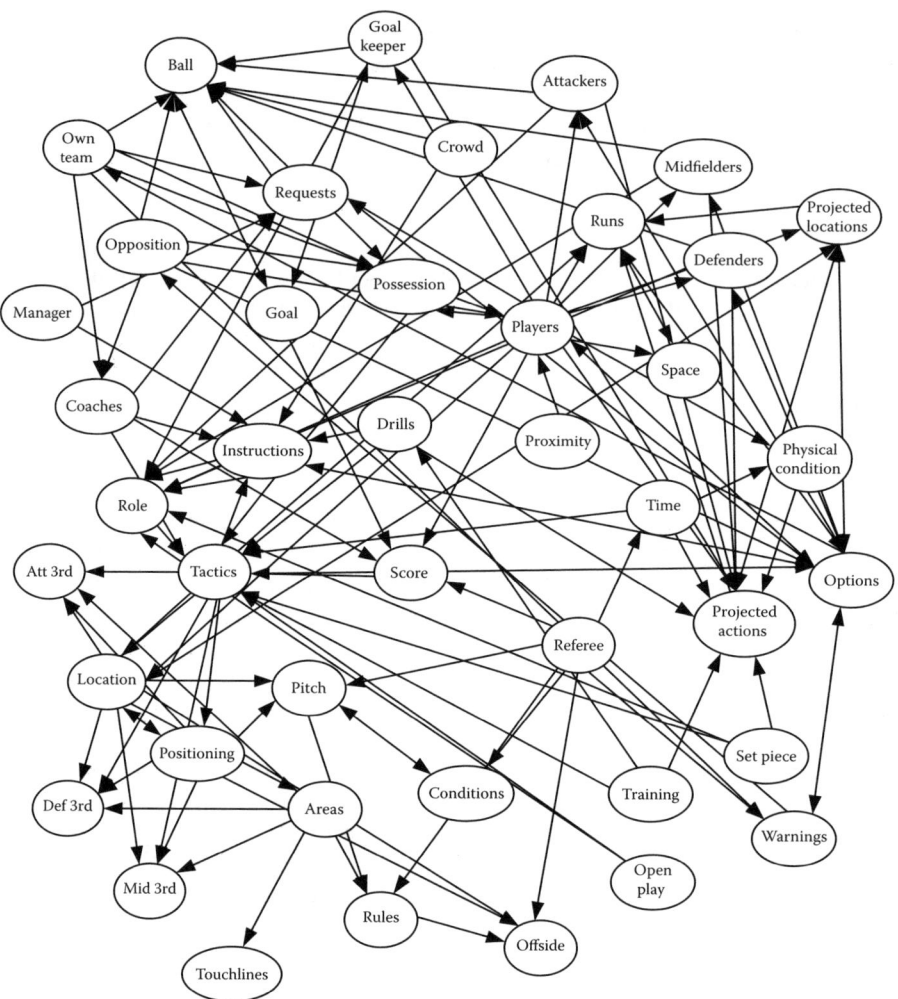

FIGURE 6.8 Soccer propositional network example.

Situation Awareness Assessment Methods

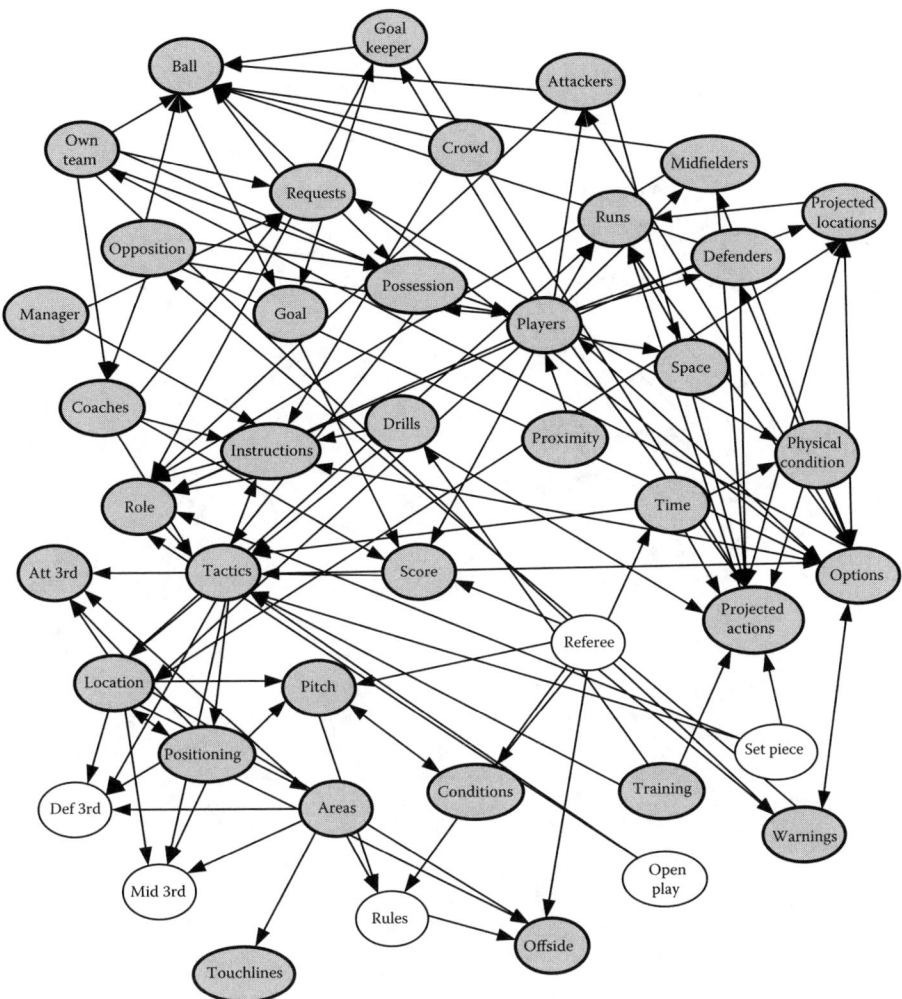

FIGURE 6.9 Attacking midfielder in possession of ball propositional network; the shaded information elements represent the attacking midfielder's awareness.

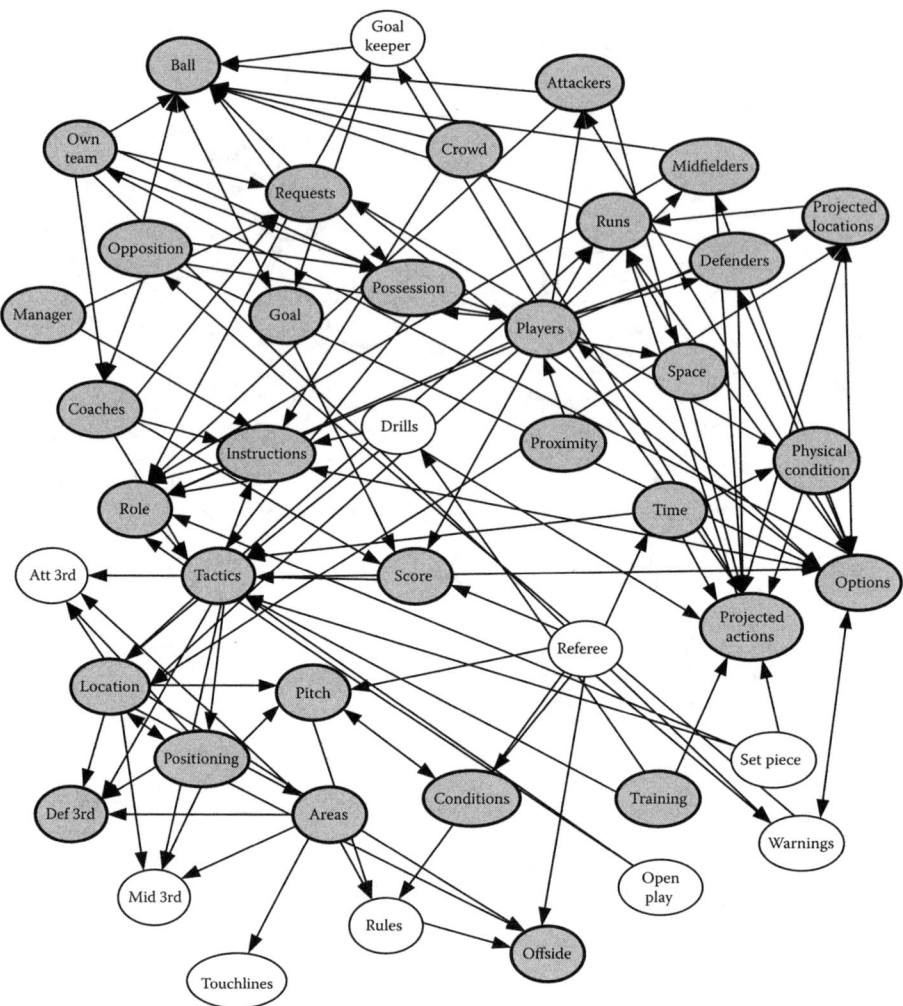

FIGURE 6.10 Defender without possession of the ball propositional network; the shaded information elements represent the defender's awareness.

Situation Awareness Assessment Methods

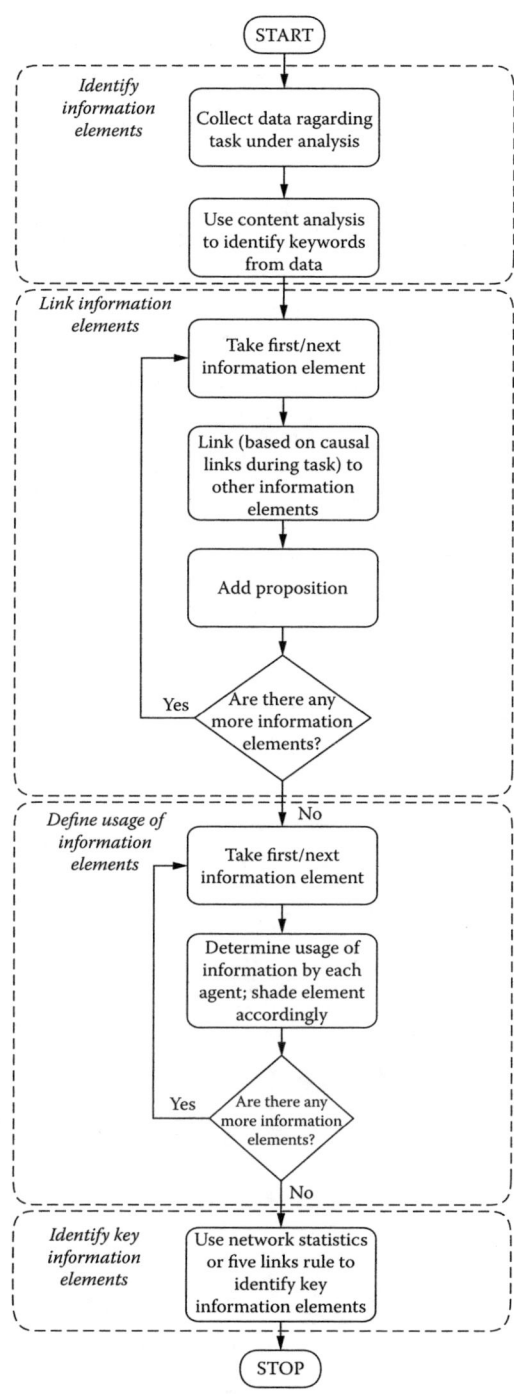

FLOWCHART 6.6 Propositional network flowchart.

7 Mental Workload Assessment Methods

INTRODUCTION

Workload represents the cost of accomplishing task requirements for the human element of systems (Hart and Wickens, 1990; cited in Gregoriades and Sutcliffe, 2007) and is a critical issue within Human Factors. For the purposes of this book, we will focus on the concept of mental workload, although there is a range of methods that Human Factors researchers also use to measure levels of physical workload (see Stanton et al., 2004). Inappropriate mental workload levels (both too high and too low), caused by factors such as poor system design, inappropriate procedures, and adverse environmental conditions, have a range of consequences, including fatigue, errors, monotony, mental saturation, reduced vigilance, and stress (Spath, Braun, and Hagenmeyer, 2007).

MENTAL WORKLOAD

Human operators possess a finite attentional capacity, and during task performance our attentional resources are allocated to component tasks. Workload is a function of the human operator's attentional capacity and the demand for resources imposed by the task, of the supply and demand of attentional resources (Tsang and Vidulich, 2006). The level of mental workload therefore represents the proportion of resources that are required to meet the task demands (Welford, 1978). Young and Stanton (2001) formally define mental workload as follows:

> [T]he level of attentional resources required to meet both objective and subjective performance criteria, which may be mediated by task demands, external support, and past experience.

According to Tsang and Vidulich (2006) there are two main determinants of workload: the exogenous task demand (as specified by factors such as task difficulty, priority, and contextual factors), and the endogenous supply of attentional resources available to support information processing activities. Mental workload is therefore a multidimensional construct that is characterised by the task (e.g., complexity, demands, etc.) and the individual involved (e.g., skill, experience, training, etc.) (Young and Stanton, 2001).

Sanders and McCormick (1993) also suggest that the concepts of stress and strain are relevant to human workload. Stress refers to "some undesirable condition, circumstance, task or other factor that impinges upon the individual" (Sanders and McCormick, 1993, p. 225) and sources of stress include heavy work, immobilisation, heat and cold, noise, sleep loss, danger, information overload, and boredom (Sanders and McCormick, 1993). Strain, on the other hand, refers to the effects of the stress on the individual and is measured by observing changes in elements such as blood oxygenation, electrical activity of the muscles, heart rate, body temperature, and work rate (Sanders and McCormick, 1993).

When operators are faced with excessive task demands and their attentional resources are exceeded they become *overloaded*. Mental overload therefore occurs when the demands of the task are so great that they are beyond the limited attentional capacity of the operator. Conversely, when an operator experiences excessively low task demands they may experience a state of mental *underload*. Both conditions can be detrimental to task performance (Wilson and Rajan, 1995), since operators become less likely to attend to potentially important sources of information (Lehto and Buck, 2007). There are various consequences associated with inappropriate levels of workload that can potentially lead to performance decrements. These include inattention, complacency, fatigue, monotony, mental saturation, mental satiation, reduced vigilance, and stress.

In an attempt to clarify the concept and the factors affecting it, Megaw (2005) presents a framework of interacting stressors on an individual (see Figure 7.1). According to Megaw's model, workload is affected by a variety of factors, including task factors (e.g., task demands, task constraints, etc.), environmental factors (e.g., workspace, interfaces, etc.), and organisational factors (e.g., staffing levels, team organisation, allocation of functions, etc.).

There are various other models of attention and workload presented in the literature. Kahneman (1973), for example, presented a capacity model of attention, which suggests that individuals possess a single pool of attentional resources that has a finite limit. The ability to perform concurrent tasks is therefore dependent upon the effective allocation of attention to each (Young and Stanton, 2002). More recently, multiple resource theory models have contended the notion of a single pool of attentional resources. For example, Wickens' (1992) multiple resource view of attentional resources argues that rather than simply having one single supply of resources, individuals instead have several different capacities with resource properties. This means that tasks will interfere more when resources are shared. Regardless of the theoretical standpoint, it is now generally agreed that individuals possess a finite pool of attentional resources and that these resources are allocated to tasks during task performance.

The importance of operator workload as a critical consideration in system design has been articulated by many. For example, Nachreiner (1995) points out that the level of workload imposed on operators is one of the key criteria for evaluating system design, since it is directly connected to the operator's capabilities and skills and also the impairing effects of work on operators. Most in the field agree that when considering operator workload in complex systems, designers should aim to maximise the match between task demands and human capacity (Young and Stanton, 2002). This is

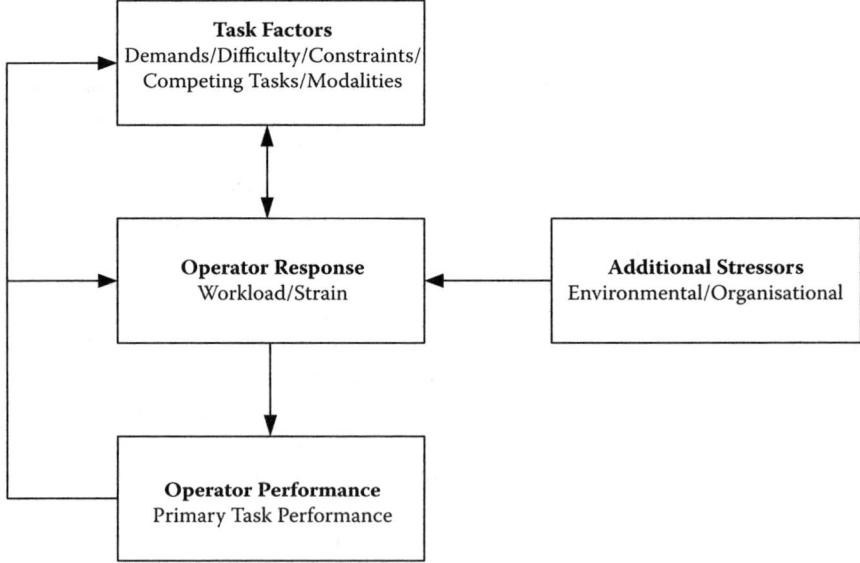

FIGURE 7.1 Framework of interacting stressors impacting workload. *Source*: Adapted from Megaw, T. (2005). The definition and measurement of mental workload. In *Evaluation of human work*, eds. J. R. Wilson and N. Corlett, 525–53. Boca Raton, FL: CRC Press.

known as workload optimisation. It is hypothesised that when operator workload falls below or rises above this optimal level, performance decrements will occur. System and procedure designers should therefore strive to optimise the level of workload experienced by process control operators. Workload optimisation involves achieving a balance between task demands and operator resources.

One issue that has dominated workload research in recent times is that of automation—the replacement of human operator sensing, planning, decision making, and manual activities with computer-based technology (Sharit, 2006). Originally conceived to reduce operator workload levels, automation is now increasingly being used in process control rooms and other complex safety critical systems. Despite its many well-reported benefits (e.g., reduced levels of operator physical and mental workload), many have identified effects of automation that adversely impact operator performance and system safety (e.g., Bainbridge, 1987; Lee, 2006; Young and Stanton, 2002, etc.). These include manual and cognitive skill degradation, operator underload, out-of-the loop performance decrements (Endsley and Kiris, 1995), and increased levels of workload due to system monitoring requirements. Lee (2006), for example, identifies the following automation pitfalls:

- Out-of-loop unfamiliarity
- Clumsy automation
- Automation-induced errors
- Inappropriate trust (misuse, disuses, and complacency)
- Behavioural adaptation
- Inadequate training and skill loss
- Job satisfaction and health
- Eutectic behaviour

Despite the impetus for most automated systems being a requirement to reduce operator workload, automation can potentially cause both decreases and increases in workload levels (Young and Stanton, 2002). It has been postulated, for example, that automation can lead to overload through a lack of feedback, increased vigilance demands (Hancock and Verwey, 1997), and increased decision options in certain situations (Hilburn, 1997). Various studies have also identified poorer performance in automated control conditions (e.g., Stanton, Young, and McCaulder, 1997). The general consensus is that great caution should be taken when using automation in an attempt to enhance performance efficiency and safety in complex systems. Lee (2006) articulates a series of strategies for designing effective automation that includes effective allocation of functions, the use of dynamic function allocation (adaptive and dynamic automation), matching automation to human performance characteristics, representation aiding and multimodal feedback, matching automation to user mental models, and the use of formal automation analysis methods.

One theory that has been posited in order to explain why automated systems may lead to performance decrements is the Malleable Attentional Resources Theory (MART; Young and Stanton, 2002). MART introduces the concept of malleable (as opposed to static) attentional resource pools and suggests that during low demand tasks an operator's resources shrink to accommodate the reduced level of demand. According to Young and Stanton, this reduced pool of attentional resources explains why operator performance suffers during low workload tasks. Further, the model suggests that operator performance suffers when the task demand increases quickly (e.g., emergency or non-routine scenarios) since the diminished resources cannot keep up.

WORKLOAD AND SPORT

Workload is of interest to sports scientists as it is one of the factors that influences exercise and sports performance. The physical level of workload experienced by athletes and sports performers has traditionally been a common research theme (e.g. Davidson and Trewartha, 2008; McClean, 1992; Reilly, 1986, 1997; Roberts, Trewartha, Higgitt, El-Abd, and Stokes, 2008; Wright and Peters,

2008), and is typically assessed using physiological measures (e.g., heart rate) and activity analysis (e.g., McClean, 1992; Roberts et al., 2008). The level of mental workload experienced by athletes and sports performers, however, has not received nearly as much attention. Mental workload levels, of course, are of interest. The difference in workload levels across different sports, different positions within sports teams, different events, different ability levels, and when using different tactics and sports technology, devices, and performance aids are all important lines of enquiry.

A limited number of studies focussing on mental workload in sports scenarios have been undertaken. For example, Helsen and Bultynck (2004) investigated the physical and perceptual-cognitive demands imposed on elite UEFA referees and assistant referees during the final round of the European Championships in 2000. Physical workload levels were measured using heart rate measurements, and the level of cognitive demand was assessed by counting (via notational analysis) the number of observable decisions made by referees during games. In conclusion, Helsen and Bultynck (2004) reported that, over the 31 games analysed, the mean number of observable decisions made by the referees was 137 (range 104–162). Helsen and Bultynck (2004) speculated that non-observable decisions (i.e., those decisions that the referee makes but do not interfere with play) would bring the total number of decisions made by elite referees to approximately 200, although they had no data to back this up (data of this type could be obtained using the subjective workload assessment measures described in this chapter). They reported that, when taken into consideration along with the actual playing time of around 51 minutes, elite referees are required to make three to four decisions every minute. Based on their study, Helsen and Bultynck (2004) recommended video training as a way of improving referee decision-making skills. It is notable that Helsen and Bultynck (2004) did not use a standard form of mental workload assessment method as we know them, and thus it seems that valid, structured workload assessment methods of the type used by Human Factors researchers in the safety critical domains do not seem to have yet crossed over to the discipline of Sports Science.

MENTAL WORKLOAD ASSESSMENT

There is a range of different mental workload assessment approaches available, including primary and secondary task performance measures, physiological measures, and subjective-rating methods. Typically, a combination of all four, known as a battery of workload assessment methods, is used during mental workload assessment efforts. A brief summary of each method type is presented below.

Primary task performance measures are used to assess an individual's ability to perform a particular task under varying levels of workload, based on the supposition that an individual's performance will diminish as their level of workload increases. For example, when assessing the level of workload imposed by driver support systems, Young and Stanton (2004) measured speed, lateral position, and headway as indicators of performance on a driving task. Poor performance on the primary task is taken to be indicative of an increased level of workload, while efficient task performance indicates that the level of workload is manageable. Wierwille and Eggemeier (1993) recommend that primary task performance measures should be included in any assessment of operator workload. The main advantages associated with the use of primary task measures are their reported sensitivity to variations in workload (Wierwille and Eggemeier, 1993) and their ease of use, since performance of the primary task is normally measured anyway. However, there are a number of disadvantages, including the ability of operators to perform efficiently under high levels of workload, due to factors such as experience and skill. Similarly, performance may also suffer during low workload parts of the task. It is recommended that great care be taken when interpreting the results obtained through primary task performance assessment of workload.

Secondary task performance measures are used to measure an individual's ability to perform a secondary task (in addition to the primary task), the hypothesis being that as the level of workload imposed by the primary task increases, performance on the secondary task will diminish, since the resources available for the secondary task are reduced. There is a range of secondary task performance measures available, including memory recall tasks, mental arithmetic tasks, reaction time measurement, and tracking tasks. The main disadvantages associated with secondary task performance methods are a reported lack of sensitivity to minor workload variations (Young and Stanton, 2004) and their intrusion on primary task performance. One way around this is the use of embedded secondary task measures, whereby the operator is required to perform a secondary task with the system under analysis. Since the secondary task is no longer external to that of operating the system, the level of intrusion is reduced. Researchers adopting a secondary task measurement approach to the assessment of workload are advised to adopt discrete stimuli, which occupy the same attentional resource pools as the primary task (Young and Stanton, 2004). For example, if the primary task is a driving one, then the secondary task should be a visual-spatial one involving manual response (Young and Stanton, 2004). This ensures that the method is actually measuring spare capacity and not an alternative resource pool.

Physiological measures of workload involve the measurement of physiological responses that may be affected by increased or decreased levels of workload. Examples of physiological measures include heart rate, heart rate variability, eye movement, and brain activity. The main advantage associated with the use of physiological measures of workload is that they do not intrude upon primary task performance and that they can be applied in the field, as opposed to simulated settings. There are a number of disadvantages, however, including the cost, physical obtrusiveness, and reliability of the technology used, and doubts regarding the construct validity and sensitivity of the methods.

Subjective-rating methods involve participants providing subjective ratings of their workload during or post task performance. Subjective rating methods can be unidimensional (assessing overall level of workload only) or multidimensional (assessing various dimensions associated with workload), and typically offer a quantitative rating of the level of workload imposed on an individual during task performance. Although the data obtained when using unidimensional methods are far simpler to analyse than the data obtained when using multidimensional methods, multidimensional approaches possess greater levels of diagnosticity. Subjective-rating methods are attractive due to their ease and speed of application, and also the low cost involved. They are also nonintrusive to primary task performance and can be applied in the field in "real-world" settings, rather than in simulated environments. That said, subjective workload assessment methods are mainly only used when there is an operational system available and therefore it is difficult to employ them during the design process, as the system under analysis may not actually exist, and simulation can be costly. There are also a host of problems associated with collecting subjective data post trial. Often, workload ratings correlate with performance on the task under analysis. Participants are also prone to forgetting certain parts of the task where variations in their workload may have occurred.

More recently, a number of analytical workload prediction models and methods have been developed (e.g., Gregoriades and Sutcliffe, 2007, Neerincx, 2003; Vidulich, Ward, and Schueren, 1991); however, due to their infancy we do not consider them in this chapter. For the purposes of this book, we focus on the following seven workload assessment methods: primary and secondary task performance measures, physiological measures, the NASA Task Load Index (NASA-TLX; Hart and Staveland, 1988), Subjective Workload Assessment Technique (SWAT; Reid and Nygren, 1988), Subjective Workload Dominance Technique (SWORD; Vidulich and Hughes, 1991), and instantaneous self-assessment (ISA) subjective rating methods.

Primary task performance measures involve the measurement of an operator's ability to perform the primary task under analysis. It is expected that operator performance of the task under analysis will diminish as workload increases. Secondary task performance measures of workload involve the measurement of an operator's ability to perform an additional secondary task, as well as the primary task involved in the task under analysis. Typical secondary task measures include memory recall tasks, mental arithmetic tasks, reaction time measurement, and tracking tasks. The use of secondary task performance measures is based upon the assumption that as operator workload increases, the ability to perform the secondary task will diminish due to a reduction in spare capacity, and so secondary task performance will suffer. Physiological measures involve the measurement of those physiological aspects that may be affected by increased or decreased levels of workload; for example, heart rate, heart rate variability, eye movement, and brain activity have all been used to provide a measure of operator workload. The NASA-TLX (Hart and Staveland, 1988) is a multidimensional subjective rating tool that is used to derive a workload rating based upon a weighted average of six workload sub-scale ratings (mental demand, physical demand, temporal demand, effort, performance, and frustration level). The SWAT method (Reid and Nygren, 1988) is a multidimensional tool that measures three dimensions of operator workload: time load, mental effort load, and stress load. After an initial weighting procedure, participants are asked to rate each dimension and an overall workload rating is calculated. The SWORD method (Vidulich and Hughes, 1991) uses paired comparison of tasks in order to provide a rating of workload for each individual task. Administered post trial, participants are required to rate one task's dominance over another in terms of workload imposed. The ISA method involves participants self-rating their workload throughout task performance (normally every 2 minutes) on a scale of 1 (low) to 5 (high). A summary of the mental workload assessment methods described is presented in Table 7.1.

PRIMARY AND SECONDARY TASK PERFORMANCE MEASURES

BACKGROUND AND APPLICATIONS

Primary task performance measures of workload involve assessing suitable aspects of participant performance during the task(s) under analysis, assuming that an increase in workload will facilitate a performance decrement of some sort. Secondary task performance measures typically involve participants performing an additional task in addition to that of primary task performance. Participants are required to maintain primary task performance and perform the secondary task as and when the primary task allows them to. The secondary task is designed to compete for the same resources as the primary task. Any differences in workload between primary tasks are then reflected in the performance of the secondary task. Examples of secondary tasks used in the past include tracking tasks, memory tasks, rotated figures tasks, and mental arithmetic tasks.

DOMAIN OF APPLICATION

Generic.

APPLICATION IN SPORT

Primary and secondary task performance measures could be used in a sporting context as part of a battery of workload assessment methods in order to assess sports performer mental workload. Mental workload (hereafter referred to as "workload"), in a sporting context, is likely to be of interest when new technology or performance aids are being tested. Alternatively, the workload levels experienced by performers of differing ability and experience levels are also likely to be of interest.

PROCEDURE AND ADVICE

Step 1: Define Primary Task under Analysis

The first step in an assessment of operator workload is to clearly define the task(s) under analysis. When assessing the workload associated with the use of a novel or existing system, procedure, or interface, it is recommended that the task(s) assessed are as representative of the system or device under analysis as possible.

Mental Workload Assessment Methods

Step 2: Define Primary Task Performance Measures
Once the task(s) under analysis is clearly defined and described, the analyst should next define those aspects of the task that can be used to measure participant performance. For example, in a driving task Young and Stanton (2004) used speed, lateral position, and headway as measures of primary task performance. The measures used are dependent upon the task in question and equipment that is used during the analysis.

Step 3: Design Secondary Task and Associated Performance Measures
Once the primary task performance measures are clearly defined, an appropriate secondary task measure should be selected. Young and Stanton (2004) recommend that great care be taken to ensure that the secondary task competes for the same attentional resources as the primary task. For example, Young and Stanton (2004) used a visual-spatial task that required a manual response as their secondary task when analysing driver workload.

Step 4: Test Primary and Secondary Tasks
Once the primary and secondary task performance measures are defined, they should be tested in order to ensure that they are sensitive to variations in task demand. The analyst should define a set of tests that are designed to ensure the validity of the primary and secondary task measures chosen. Trial runs of the measures using other researchers as participants are often used for this purpose.

Step 5: Brief Participants
Once the measurement procedure has been validated during step 4, the appropriate participants should be selected and then briefed regarding the area of workload, workload assessment, the purpose of the analysis, and the data collection procedure to be employed. Before data collection begins, participants should have a clear understanding of workload theory, and of the measurement methods being used. It may be useful at this stage to take the participants through an example workload assessment, so that they understand how the secondary task measure works and what is required of them as participants.

Step 6: Conduct Pilot Run
Once the participant(s) understand the data collection procedure, a small pilot run should be conducted to ensure that the process runs smoothly and efficiently. Participants should be instructed to perform a small task (separate from the task under analysis), and an associated secondary task. This acts as a pilot run of the data collection procedure and serves to highlight any potential problems. The participant(s) should be instructed to ask any questions regarding their role in the data collection procedure.

Step 7: Undertake Primary Task Performance
Once a pilot run of the data collection procedure has been successfully completed and the participants are comfortable with their role during the trial, the data collection procedure can begin. The participant should be instructed to perform the task under analysis, and to attend to the secondary task when they feel that they can. The task should run for a set amount of time, and the secondary task should run concurrently.

Step 8: Administer Subjective Workload Assessment Method
Typically, subjective workload assessment methods, such as the NASA-TLX (Hart and Staveland, 1988), are used in conjunction with primary and secondary task performance measures to assess participant workload. The chosen method should be administered immediately upon completion of the task under analysis, and participants should be instructed to rate the appropriate workload dimensions based upon the primary task that they have just completed.

TABLE 7.1
Mental Workload Assessment Methods Summary Table

Name	Domain	Application in Sport	Training Time	App. Time	Input Methods	Tools Needed	Main Advantages	Main Disadvantages	Outputs
Primary and secondary task performance measures	Generic	Measurement of workload during different tasks; Comparison of workload across performers of different levels (e.g., elite vs. novice); Measurement of workload imposed by different tactics, technology, etc.	Low	Low	Observational study	Simulator	1. Useful when measuring workload in tasks that are lengthy in duration (Young and Stanton, 2004); 2. Easy to use and the data gathered are easy to analyse; 3. Little training required	1. Difficult to apply during real-world tasks; 2. The sensitivity of primary and secondary task performance measures may be questionable; 3. May prove expensive, as a simulator is normally required	Indication of workload throughout task via primary and secondary task performance scores
Physiological measures	Generic	Measurement of workload during different tasks; Comparison of workload across performers of different levels (e.g., elite vs. novice); Measurement of workload imposed by different tactics, technology, etc.	High	Low	Observational study	Physiological measurement equipment, e.g., heart rate monitor	1. Various physiological measures have demonstrated a sensitivity to task demand variations; 2. Data are recorded continuously throughout task performance; 3. Equipment, such as heart rate monitors, is now cheap and easy to obtain	1. Data are easily confounded by extraneous interference; 2. The equipment can be physically obtrusive, temperamental, and difficult to operate; 3. Physiological data can be difficult to obtain and analyse, and analysts require a thorough understanding of physiological responses to workload	Indication of workload throughout task via physiological response data
NASA TLX (Hart and Staveland, 1988)	Generic	Measurement of workload during different tasks; Comparison of workload across performers of different levels (e.g., elite vs. novice); Measurement of workload imposed by different tactics, technology, etc.	Low	Low	Observational study	NASA TLX pro-forma; Pen and paper	1. Provides a low cost, quick, and simple-to-use method for assessing workload; 2. Is the most commonly used approach and has previously been used in a sporting context; 3. Software support is available and a number of studies have shown its superiority over other workload methods	1. Construct validity may be questionable (i.e., is it assessing mental workload or something else?); 2. Suffers from the problems associated with collecting subjective data post task performance; 3. The sub-scale weighting procedure is laborious and adds more time to the process	Subjective rating of workload

Mental Workload Assessment Methods

Method					Advantages	Disadvantages	
Subjective Workload Assessment Technique (SWAT; Reid and Nygren, 1988)	Generic	Measurement of workload during different tasks. Comparison of workload across performers of different levels (e.g., elite vs. novice). Measurement of workload imposed by different tactics, technology, etc.	Low	Low	Observational study	SWAT pro-forma. Pen and paper	1. Provides a low cost, quick, and simple-to-use method for assessing workload 2. The SWAT sub-scales are generic, so the method can be applied to any domain 3. Has been applied in a range of different domains and has significant validation evidence associated with it	1. Construct validity may be questionable (i.e., is it assessing mental workload or something else?) 2. Suffers from the problems associated with collecting subjective data post task performance 3. The scale development procedure is laborious and adds more time to the process	Subjective rating of workload
Subjective Workload Dominance (SWORD; Vidulich, 1989)	Generic	Comparison of workload levels imposed by different devices, tactics, positions, etc.	Low	Low	Observational study	Pen and paper	1. Quick and easy to use and requires only minimal training 2. Generic; can be applied in any domain 3. Useful when comparing two different devices and their effect upon operator workload	1. High workload for one participant may be low workload for another 2. Does not provide a direct measure of workload 3. There are various problems associated with collecting subjective data post trial	Rating of one task/device's workload dominance over another
Instantaneous Self-Assessment (ISA)	Generic	Measurement of workload during different tasks. Comparison of workload across performers of different levels (e.g., elite vs. novice). Measurement of workload imposed by different tactics, technology, etc.	Low	Low	Observational study	Audio recording device	1. Provides a simple, low cost, and quick means of analysing varying levels of workload throughout a particular task 2. Requires very little (if any) training 3. Output is immediately useful, providing a workload profile for the task under analysis	1. Intrusive to primary task performance 2. The data are subjective and may often be correlated with task performance 3. Participants are not very efficient at reporting mental events	Subjective rating of workload throughout a task (i.e., ratings elicited every 2 minutes)

Step 9: Analyse Data

Once the data collection procedure is completed, the data should be analysed appropriately. Young and Stanton (2004) used the frequency of correct responses on a secondary task to indicate the amount of spare capacity the participant had, i.e., the greater the correct responses on the primary task, the greater the participant's spare capacity was assumed to be.

ADVANTAGES

1. Primary task performance measures offer a direct index of performance.
2. Primary task performance measures are particularly effective when measuring workload in tasks that are lengthy in duration (Young and Stanton, 2004).
3. Primary task performance measures are also useful when measuring operator overload.
4. It requires no further effort on behalf of the analyst to set up and record, as primary task performance measures are normally taken anyway.
5. Secondary task performance measures are effective at discriminating between tasks when no difference was observed assessing performance alone.
6. Primary and secondary task performance measures are easy to use and the data gathered are easy to analyse.

DISADVANTAGES

1. The sensitivity of primary and secondary task performance measures may be questionable. For example, primary task performance measures alone may not distinguish between different levels of workload, particularly minimal ones. Operators may still achieve the same performance levels under different workload conditions.
2. When used in isolation, primary task performance measures have limited reliability.
3. Secondary task performance measures have been found to be only sensitive to gross changes in participant workload.
4. Secondary task performance measures are highly intrusive to primary task performance.
5. Great care is required during the design and selection of the secondary task to be used. The analyst must ensure that the secondary task competes for the same resources as the primary task.
6. Extra work and resources are required in developing the secondary task performance measure.
7. The methods need to be used together to be effective.
8. It is difficult to apply both approaches during real-world tasks undertaken in the field.
9. Using primary and secondary task performance measures may prove expensive, as simulators and computers are often required.

RELATED METHODS

Primary and secondary task performance measures are typically used in conjunction with physiological measures and subjective workload assessment methods in order to measure participant workload. A number of secondary task performance measurement methods exist, including task reaction times, tracking tasks, memory recall tasks, and mental arithmetic tasks. Physiological measures of workload include measuring participant heart rate, heart rate variability, blink rate, and brain activity. There are a number of subjective workload assessment methods available, including the NASA-TLX (Hart and Staveland, 1988), and the SWAT (Reid and Nygren, 1988).

Training and Application Times

The training and application times associated with both primary and secondary task performance measures of workload are typically low. However, when no applicable secondary task measure exists, substantial time may be required for the development of an appropriate secondary task measure. The application time is also dependent upon the duration of the task under analysis.

Reliability and Validity

It is not possible to comment on the reliability and validity of primary and secondary performance measures of workload, as they are developed specifically for the task and application under analysis. The reliability and validity of the methods used can be checked to an extent by using a battery of measures (primary task performance measures, secondary task performance measures, physiological measures, and subjective assessment methods). The validity of the secondary task measure can be assured by making sure that the secondary task competes for the same attentional resources as the primary task.

Tools Needed

The tools needed are dependent upon the nature of the analysis. For example, in the example described below a driving simulator and a PC were used. The secondary task is normally presented separately from the primary task via a desktop or laptop computer. A simulator or a PC is normally used to record participant performance on the primary and secondary tasks.

Example

Young and Stanton (2004) describe the measurement of workload in a driving simulator environment. Primary task performance measurement included recording data regarding speed, lateral position, and headway (distance from the vehicle in front). A secondary task was used to assess spare attentional capacity. The secondary task used was designed to compete for the same attentional resources as the primary task of driving the car. The secondary task was comprised of a rotated figures task whereby participants were randomly presented with a pair of stick figures (one upright, the other rotated through 0°, 90°, 180°, or 270°) holding one or two flags. The flags were made up of either squares or diamonds. Participants were required to make a judgement, via a push button device, as to whether the figures were the same or different, based upon the flags that they were holding. The participants were instructed to attend to the secondary task only when they felt that they had time to do so. Participant correct responses were measured, and it was assumed that the higher the frequency of correct responses, the greater participant spare capacity was assumed to be.

Flowchart

(See Flowchart 7.1.)

Recommended Text

Young, M. S., and Stanton, N. (2004). Mental workload. In *Handbook of Human Factors methods*, eds. N. A. Stanton, A. Hedge, K. Brookhuis, E. Salas, and H. Hendrick. London: Taylor & Francis.

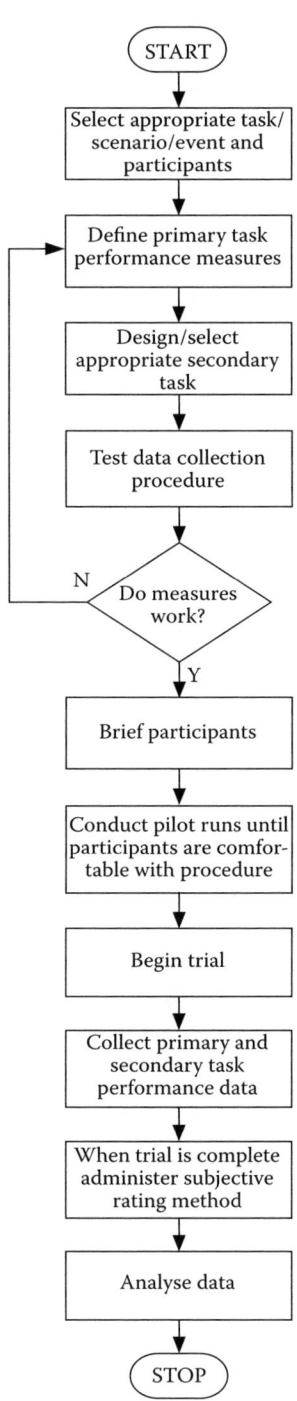

FLOWCHART 7.1 Primary and secondary task performance measures flowchart.

PHYSIOLOGICAL MEASURES

BACKGROUND AND APPLICATIONS

Physiological and psychophysiological measures have also been used extensively for assessing workload in many domains. Physiological methods are used to measure variations in participant physiological responses to the task under analysis. The indication of workload is based upon the assumption that as task demand increases, marked changes in physiological systems are apparent. In the past, heart rate, heart rate variability, endogenous blink rate, brain activity, electrodermal response, eye movements, papillary responses and event-related potentials have all been used to assess workload. Measuring heart rate, whereby increases in participant heart rate are taken to be indicative of increased workload, is one of the most common physiological measures of workload. Heart rate variability has also been used as a measure of workload, where a decrease in heart rate variability (heart rhythm) is taken to be indicative of an increased level workload (Salvendy, 1997). Endogenous eye blink rate has also been used in the assessment of operator workload. Increased visual demands have been shown to cause a decreased endogenous eye blink rate (Salvendy, 1997) and a relationship between blink rate and workload has been demonstrated in the flight environment (Wierwille and Eggemeier, 1993). It is assumed that a higher visual demand causes the operator to reduce his or her blink rate in order to achieve greater visual input. Measures of brain activity involve using EEG recordings to assess operator workload. According to Wierwille and Eggemeier (1993), measures of evoked potentials have demonstrated a capability of discriminating between levels of task demand.

DOMAIN OF APPLICATION

Generic.

APPLICATION IN SPORT

The use of physiological measures is popular within Sports Science research, although this is typically for assessing the levels of physical workload experienced by performers. For example, Helsen and Bultynck (2004) used heart rate measurements to investigate the physical workload levels experienced by UEFA referees and assistant referees during the final round of the Euro 2000 soccer championships; however, for assessing mental workload they instead used the number of decisions made during games. The use of physiological measures to assess mental workload has therefore not received significant attention within the Sports Science domain; however, the well-developed physiological measures that exist make it a potentially viable way of assessing mental workload levels.

PROCEDURE AND ADVICE

The following procedure offers advice on the measurement of heart rate as a physiological indicator of workload. When using other physiological methods, it is assumed that the procedure is the same, only with different equipment being used.

Step 1: Define Task under Analysis

The first step in an assessment of operator workload is to clearly define the task(s) under analysis. It is recommended that an HTA be conducted for the task(s) under analysis. When assessing the workload associated with the use of a novel or existing device or system, it is recommended that the task(s) assessed are as representative of the device or system under analysis as possible.

Step 2: Select the Appropriate Measuring Equipment

Once the task(s) under analysis is clearly defined and described, the analyst should select the appropriate measurement equipment. For example, when measuring workload in a driving task, Young and Stanton (2004) measured heart rate using a Polar Vantage NV Heart Rate Monitor. Polar heart rate monitors are relatively cheap to purchase and comprise a chest belt and a watch.

Step 3: Conduct Initial Testing of the Data Collection Procedure

It is recommended that various pilot runs of the data collection procedure are conducted, in order to test the measurement equipment used and the appropriateness of the data collected. Physiological measurement equipment can be temperamental and difficult to use and it may take some time for the analyst(s) to become proficient in its use.

Step 4: Brief Participants

Once the measurement procedure has been subjected to sufficient testing, the appropriate participants should be selected and briefed regarding the purpose of the study and the data collection procedure employed. Before the task(s) under analysis are performed, all of the participants involved should be briefed regarding the purpose of the study, workload, workload assessment, and the physiological measurements methods to be employed. It may be useful at this stage to take the participants through an example workload assessment, so that they understand how the chosen physiological measures work and what is required of them as participants.

Step 5: Fit Equipment

Next, the participant(s) should be fitted with the appropriate physiological measuring equipment. The heart rate monitor consists of a chest strap, which is placed around the participant's chest, and a watch, which the participant can wear on their wrist or the analyst can hold. The watch collects the data and is then connected to a computer post trial in order to download the data collected.

Step 6: Conduct Pilot Run

Once the participant(s) understand the data collection procedure, a small pilot run should be conducted to ensure that the process runs smoothly and efficiently. Participants should be instructed to perform a small task (separate from the task under analysis), and an associated secondary task while wearing the physiological measurement equipment. Upon completion of the task, the participant(s) should be instructed to complete the chosen subjective workload assessment method. This acts as a pilot run of the data collection procedure and serves to highlight any potential problems. The participant(s) should be permitted to ask any questions regarding their role in the data collection procedure.

Step 7: Begin Primary Task Performance

Once a pilot run of the data collection procedure has been successfully completed, and the participants fully understand their role during the trial, the data collection procedure can begin. The participant should be instructed to begin the task under analysis, and to attend to the secondary task (if one is being used) when they feel that they can. The task should run for a set amount of time, and the secondary task should run concurrently. The heart rate monitor continuously collects participant heart rate data throughout the task. Upon completion of the task, the heart rate monitor should be turned off and removed from the participant's chest.

Step 8: Administer Subjective Workload Assessment Method

Typically, subjective workload assessment methods, such as the NASA-TLX (Hart and Staveland, 1988), are used in conjunction with primary and secondary task performance measures and physiological measures to assess participant workload. The chosen method should be administered immediately after the task under analysis is completed, and participants should be instructed to rate the appropriate workload dimensions based upon the primary task that they have just completed.

Step 9: Download Data

The heart rate monitor data collection tool (typically a watch) can now be connected to a laptop computer in order to download the data collected. Normally, the associated software package can be used to analyse and represent the data in a number of different forms.

Step 10: Analyse Data

Once the data collection procedure is completed, the data should be analysed appropriately. It is typically assumed that an increase in workload causes an increase in operator heart rate. Heart rate variability has also been used as an indicator of operator workload, where decreased variability indicates an increase in workload (Salvendy, 1997).

Advantages

1. Various physiological measures have demonstrated a sensitivity to task demand variations.
2. When using physiological measures, data are recorded continuously throughout task performance.
3. Physiological measures are popular in the Sports Science domains and so analysts should be proficient in their use.
4. Physiological measurement equipment, such as heart rate monitors, is now cheap and easy to obtain.
5. Physiological measurement equipment typically comes with a software package that automates some of the data analysis and presentation.
6. Physiological measurements can often be taken in a real-world setting, removing the need for a simulation of the task.
7. Advances in technology have resulted in increased accuracy and sensitivity of the various physiological measurement tools.
8. Physiological measurement typically does not interfere with primary task performance.

Disadvantages

1. The data are easily confounded by extraneous interference (Young and Stanton, 2004).
2. Physical workload is likely to contaminate data.
3. The equipment used to measure physiological responses is typically physically obtrusive.
4. The equipment can be temperamental and difficult to operate.
5. Physiological data are difficult to obtain and analyse.
6. In order to use physiological methods effectively, the analyst(s) requires a thorough understanding of physiological responses to workload.
7. It may be difficult to use certain equipment in the field, e.g., brain and eye measurement equipment.

Related Methods

A number of different physiological measures have been used to assess operator workload, including heart rate, heart rate variability, and brain and eye activity. Physiological measures are typically used in conjunction with other workload assessment methods, such as primary and secondary task measures and subjective workload assessment methods. Primary task performance measures involve measuring certain aspects of participant performance on the task(s) under analysis. Secondary task performance measures involve measuring participant performance on an additional task, separate from the primary task under analysis. Subjective workload assessment methods are completed post trial by participants and involve participants rating specific dimensions of workload. There are a number of subjective workload assessment methods, including the NASA-TLX (Hart and Staveland, 1988), and the SWAT (Reid and Nygren, 1988).

Training and Application Times

The training time associated with physiological measurement methods is high. The equipment is often difficult to operate, and the data may be difficult to analyse and interpret correctly. The application time for physiological measurement methods is dependent upon the duration of the task under analysis. For lengthy, complex tasks, the application time may be high; however, the typical application time for a physiological measurement of workload is generally low.

Reliability and Validity

There is a considerable body of research that validates the use of physiological measures for assessing workload, and it is suggested that heart rate variability is probably the most promising approach (Young and Stanton, 2004). While a number of studies have reported the sensitivity of a number of physiological measures to variations in task demand, a number of studies have also demonstrated a lack of sensitivity to demand variations.

Tools Needed

When using physiological measurement methods, the relevant equipment is required, e.g., Polar heart rate monitor strap, watch, and software. A PC is also required to transfer the data from the measuring equipment to the software programme.

Flowchart

(See Flowchart 7.2.)

Recommended Text

Young, M. S., and Stanton, N. (2004). Mental workload. In *Handbook of Human Factors and ergonomics methods*, eds. N. A. Stanton, A. Hedge, K. Brookhuis, E. Salas, and H. Hendrick. Boca Raton, FL: CRC Press.

NASA TASK LOAD INDEX

Background and Applications

The NASA Task Load Index (NASA-TLX; Hart and Staveland, 1988) is a subjective workload assessment method that is used to elicit subjective ratings of workload upon completion of task performance. By far the most commonly used workload assessment method with applications in a range of domains, the NASA-TLX is a multidimensional approach that uses the following six sub-scales: mental demand, physical demand, temporal demand, effort, performance, and frustration level. Upon completion of the task under analysis, participants provide a rating of between 1 (low) and 20 (high) for each subscale and, based on a weighted average of the six sub-scales, an overall workload score is derived. Researchers using the TLX also often employ a paired comparisons procedure. This involves presenting 15 pair-wise combinations to the participants and asking them

Mental Workload Assessment Methods

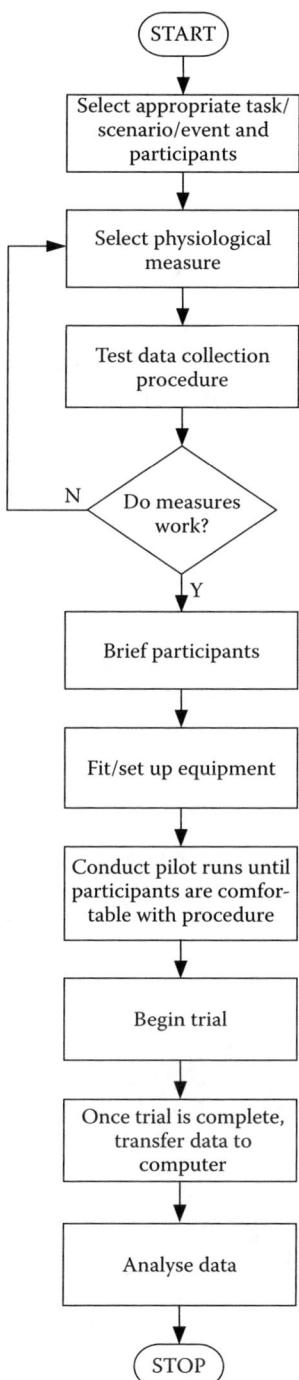

FLOWCHART 7.2 Physiologial measures flowchart.

to select the scale from each pair that has the most effect on the workload during the task under analysis. This procedure accounts for two potential sources of between-rater variability: differences in workload definition between the raters, and differences in the sources of workload between the tasks.

DOMAIN OF APPLICATION

The NASA-TLX uses generic sub-scales and can be applied in any domain.

APPLICATION IN SPORT

The NASA-TLX can be used to elicit subjective ratings of workload from sports performers. This may be to investigate differing workload levels of sports performers of different ability or experience levels (e.g., elite vs. amateur), or it may be to investigate differences in the level of workload experienced in different game situations and events, or brought about by the introduction of new technology, performance aids, or devices. The NASA-TLX has had some exposure in the Sports Science discipline. For example, McMorris, Delves, Sproule, Lauder, and Hale (2005) report the use of the TLX to elicit ratings of perceived workload during a cycling test in order to investigate how exercise at moderate and maximal intensities affects performance on a choice response time, whole body psychomotor task.

PROCEDURE AND ADVICE

Step 1: Define Analysis Aims

First, the aims of the analysis should be clearly defined. For example, the aims of the analysis may be to evaluate the impact that a new performance aid or technological device has on workload during task performance, or it may be to compare novice and expert performer workload during a particular task. Clearly defining the aims of the analysis is important as it affects the methods and tasks used during the analysis.

Step 2: Define Task(s) or Scenario under Analysis

The next step involves clearly defining the task or scenario under analysis. It is recommended that the task is described clearly, including the different actors involved, the task goals, and the environment within which the task is to take place. HTA may be useful for this purpose if there is sufficient time available.

Step 3: Select Participants

Once the task(s) under analysis is clearly defined and described, it may be useful to select the participants that are to be involved in the analysis. This may not always be necessary and it may suffice to select participants randomly on the day. However, if workload is being compared across rank or experience levels, then clearly effort is required to select the appropriate participants.

Step 4: Brief Participants

Before the task(s) under analysis is performed, all of the participants involved should be briefed regarding the purpose of the study, the area of workload and workload assessment, and the NASA-TLX method. It is recommended that participants be given a workshop on workload and workload assessment. It may also be useful at this stage to take the participants through an example NASA-TLX application, so that they understand how the method works, what it produces, and what is required of them as participants.

Step 5: Conduct Pilot Run

Before the data collection process begins, it is recommended that participants take part in a pilot run of the NASA-TLX data collection procedure. Pilot runs should be used to iron out any problems with the data collection procedure, and the participants should be encouraged to ask any question throughout this process. Once the participant is familiar with the procedure and is comfortable with his or her role, the data collection phase can begin.

Step 6: Performance of Task under Analysis

Next, the participant(s) should be instructed to perform the task under analysis. The NASA-TLX can be administered either during or post trial. However, it is recommended that it is administered post trial, as on-line administration is intrusive to primary task performance. If on-line administration is required, then the TLX should be administered at predefined points and responded to verbally.

Step 7: Conduct Weighting Procedure

If using the weighting procedure, then this should be undertaken directly after task performance is completed. The WEIGHT software presents 15 pair-wise comparisons of the six sub-scales (mental demand, physical demand, temporal demand, effort, performance, and frustration level) to the participant. Participants should be instructed to select, from each of the 15 pairs, the sub-scale from each pair that contributed the most to the workload of the task. The WEIGHT software then calculates the total number of times each sub-scale was selected by the participant. Each scale is then rated by the software based upon the number of times it is selected by the participant. This is done using a scale of 0 (not relevant) to 5 (more important than any other factor).

Step 8: Conduct NASA-TLX Rating Procedure

Upon completion of task performance (and the NASA-TLX weighting procedure if it is being applied), participants should be presented with the NASA-TLX pro-forma and asked to provide a rating of between 1 (low) and 20 (high) for each sub-scale based on the task undertaken. Ratings should be made purely on the basis of the participant's subjective judgement, and no assistance should be given during the ratings procedure.

Step 9: Calculate NASA-TLX Scores

The final step involves calculating NASA-TLX scores for each participant. This can be done via the NASA-TLX software or simply by hand. Normally, an overall workload score is calculated for each participant. This is calculated by multiplying each rating by the weight given to that sub-scale by the participant. The sum of the weighted ratings for each task is then divided by 15 (sum of weights). A workload score of between 0 and 100 is then derived for the task under analysis. NASA-TLX data are normally analysed across participants, tasks, and sub-scales. NASA-TLX data can also be presented simply as mean overall scores and mean ratings for each sub-scale.

ADVANTAGES

1. The NASA-TLX provides a low cost, quick, and simple method for assessing workload.
2. Its simplicity allows it to be easily understood by participants.
3. The results obtained can be compared across participants, trials, and sub-scales.
4. The NASA-TLX sub-scales are generic, so it can be applied to any domain.
5. It has previously been used in a sporting context (e.g., McMorris et al., 2005) and could potentially be applied in any sporting domain.
6. The NASA-TLX has been applied in a range of different domains and has significant validation evidence associated with it, e.g., Wierwille and Eggemeier (1993).
7. Software support is available via the NASA-TLX software tool (see http://humansystems.arc.nasa.gov/groups/TLX/computer.php).
8. It is by far the most commonly applied of all workload assessment methods.
9. A number of studies have shown its superiority over other approaches such as the SWAT method (Hart and Staveland, 1988; Nygren, 1991).
10. When administered post trial the approach is nonintrusive to primary task performance.

Disadvantages

1. When administered on-line, the TLX can be intrusive to primary task performance.
2. The level of construct validity may be questionable (i.e., is it assessing mental workload or something else?).
3. It suffers from the problems associated with collecting subjective data post task performance, such as poor recall of events (e.g., low workload parts of the task) and a correlation with performance (e.g., poor performance equals high workload).
4. The sub-scale weighting procedure is laborious and adds more time to the procedure.

Related Methods

The NASA-TLX is one of a number of multi-dimensional subjective workload assessment methods. Other multi-dimensional methods include the SWAT, Bedford scales, Defence Research Agency Workload Scales (DRAWS), and the Malvern Capacity Estimate (MACE) method. Often task analysis approaches, such as HTA, are used to describe the task so that analysts can determine which tasks should be involved in the data collection effort. Further, subjective workload assessment methods are typically used in conjunction with other forms of workload assessment, such as primary and secondary task performance measures and physiological measures. Finally, in order to weight the sub-scales, the TLX uses a pair-wise comparison weighting procedure.

Approximate Training and Application Times

The training time associated with the NASA-TLX method is low and the method is extremely quick to apply. Based on our experiences it should take participants no longer than 5 minutes to complete one NASA-TLX pro-forma for a task of any duration.

Reliability and Validity

A number of validation studies concerning the NASA-TLX method have been conducted (e.g., Hart and Staveland, 1988; Vidulich and Tsang, 1985, 1986). Vidulich and Tsang (1985, 1986) report that the NASA-TLX produced more consistent workload estimates for participants performing the same task than the SWAT (Reid and Nygren, 1988) did. Hart and Staveland (1988) report that the NASA-TLX workload scores suffer from substantially less between-rater variability than one-dimensional workload ratings did. Luximon and Goonetilleke (2001) also report that a number of studies have shown that the NASA-TLX is superior to SWAT in terms of sensitivity, particularly for low mental workloads. In a comparative study of the NASA-TLX, the Road NASA-TLX (RNASA-TLX), SWAT, and Modified Cooper Harper scale methods, Cha (2001) reported that the RNASA-TLX is the most sensitive and acceptable when used to assess driver mental workload during in-car navigation-based tasks.

Tools Needed

The NASA-TLX method is typically applied using pen and paper only; however, a software version of the tool also exists (see http://humansystems.arc.nasa.gov/groups/TLX/computer.php). It is also useful to get participants to enter ratings directly into a Microsoft Excel spreadsheet that then automatically calculates overall workload scores.

Mental Workload Assessment Methods

EXAMPLE

As part of a study focussing on workload and decision making during fell running, the NASA-TLX method was used to analyse fell runner workload during a recent local amateur fell race. Consenting runners were asked to complete two NASA-TLX pro-formas upon completion of the race, one for the ascent portion of the race (an 885-ft ascent to a local monument) and one for the descent portion of the race towards the finish. The NASA-TLX pro-forma used for the ascent portion of the race is presented in Figure 7.2. The results derived from this study are presented in the final chapter of this book.

FLOWCHART

(See Flowchart 7.3.)

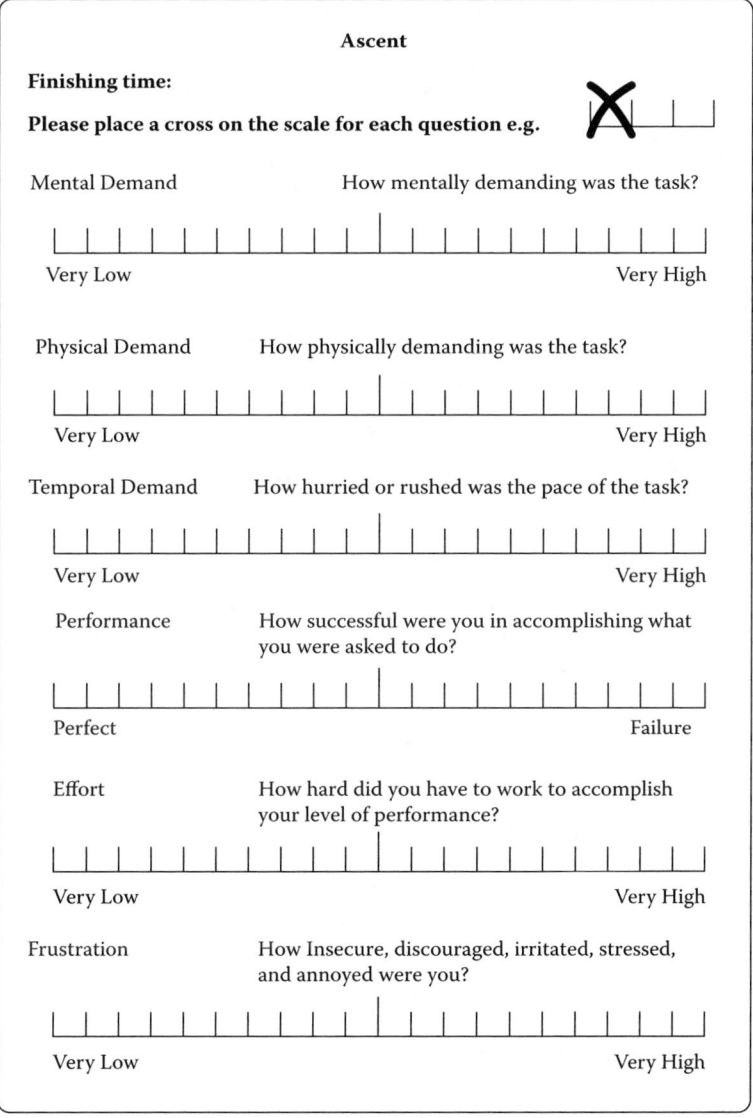

FIGURE 7.2 NASA-TLX pro-forma (Figure courtesy of NASA Ames Research Center).

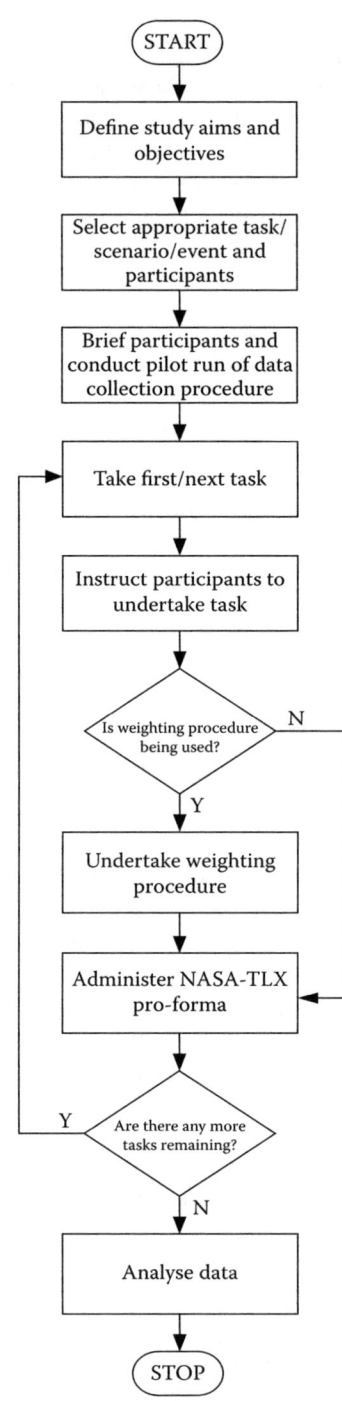

FLOWCHART 7.3 NASA-TLX flowchart.

RECOMMENDED TEXTS

Hart, S. G., and Staveland, L. E. (1988). Development of a multi-dimensional workload rating scale: Results of empirical and theoretical research. In *Human mental workload*, eds. P. A. Hancock and N. Meshkati. Amsterdam, The Netherlands: Elsevier.
Vidulich, M. A., and Tsang, P. S. (1986). Technique of subjective workload assessment: A comparison of SWAT and the NASA bipolar method. *Ergonomics* 29(11):1385–98.

SUBJECTIVE WORKLOAD ASSESSMENT TECHNIQUE

BACKGROUND AND APPLICATIONS

The Subjective Workload Assessment Technique (SWAT; Reid and Nygren, 1988) was developed by the US Air Force Armstrong Aerospace Medical Research Laboratory at the Wright Patterson Air Force Base, USA, to assess pilot workload in cockpit environments. SWAT is a multi-dimensional method that measures the following three dimensions of operator workload:

1. *Time load*   Refers to the time limit within which the task under analysis is performed, and also the extent to which multiple tasks must be performed concurrently
2. *Mental load*   Refers to the attentional, cognitive, or mental demands associated with the task under analysis.
3. *Stress load*   Refers to the level of stress imposed on the participant during performance of the task under analysis, and includes fatigue, confusion, risk, frustration, and anxiety

Upon completion of task performance, and after an initial weighting procedure, participants are asked to rate each dimension (time load, mental effort load, and stress load) on a scale of 1 to 3. A workload rating is then calculated for each dimension and an overall workload score between 1 and 100 is derived. The SWAT dimensions are presented in Table 7.2.

DOMAIN OF APPLICATION

Although originally developed for assessing fighter pilot workload, the SWAT dimensions are generic and can be applied in any domain.

APPLICATION IN SPORT

Much like the NASA-TLX, the SWAT method can be used to elicit subjective ratings of workload from sports performers. This may be to investigate differing workload levels of sports performers of different ability or experience levels, or it may be to investigate differences in the level of workload experienced in different game situations or events, or brought about by the introduction of new technology, performance aids, or devices.

PROCEDURE AND ADVICE

Step 1: Define Analysis Aims

First, the aims of the analysis should be clearly defined. For example, the aims of the analysis may be to evaluate the impact that a new performance aid or technological device has on workload during task performance, or it may be to compare novice and expert performer workload during a particular task.

TABLE 7.2
SWAT Dimensions

Time Load	Mental Effort Load	Stress Load
1. Often have spare time: interruptions or overlap among other activities occur infrequently or not at all	1. Very little conscious mental effort or concentration required: activity is almost automatic, requiring little or no attention	1. Little confusion, risk, frustration, or anxiety exists and can be easily accommodated
2. Occasionally have spare time: interruptions or overlap among activities occur frequently	2. Moderate conscious mental effort or concentration required: complexity of activity is moderately high due to uncertainty, unpredictability, or unfamiliarity; considerable attention is required	2. Moderate stress due to confusion, frustration, or anxiety noticeably adds to workload: significant compensation is required to maintain adequate performance
3. Almost never have spare time: interruptions or overlap among activities are very frequent, or occur all of the time	3. Extensive mental effort and concentration are necessary: very complex activity requiring total attention	3. High to very intense stress due to confusion, frustration, or anxiety: high to extreme determination and self-control required

Step 2: Define Task(s) or Scenario under Analysis

The next step involves clearly defining the task or scenario under analysis. It is recommended that the task be described clearly, including the different actors involved, the task goals, and the environment within which the task is to take place. HTA may be useful for this purpose if there is sufficient time available.

Step 3: Select Participants

Once the task(s) under analysis are clearly defined and described, it may be useful to select the participants that are to be involved in the analysis. This may not always be necessary, and it may suffice to select participants randomly on the day. However, if workload is being compared across rank or experience levels, then effort is required to select the appropriate participants.

Step 4: Brief Participants

Before the task(s) under analysis are performed, all of the participants involved should be briefed regarding the purpose of the study, the area of workload and workload assessment, and the SWAT method. It is recommended that participants be given a workshop on workload and workload assessment. It may also be useful at this stage to take the participants through an example SWAT application, so that they understand how the method works and what is required of them as participants.

Step 5: Undertake Scale Development

Once the participants understand how the SWAT method works, the SWAT scale development process can take place. This involves participants placing in rank order all possible 27 combinations of the three workload dimensions—time load, mental effort load, and stress load—according to their effect on workload. This "conjoint" measurement is used to develop an interval scale of workload rating, from 1 to 100.

Step 6: Conduct Pilot Run

Before the data collection process begins, it is recommended that the participants take part in a number of pilot runs of the SWAT data collection procedure. A number of small test scenarios should be used to iron out any problems with the data collection procedure, and the participants should be encouraged to ask any questions. Once the participant is familiar with the procedure and is comfortable with his or her role, the real data collection can begin.

Step 7: Performance of Task under Analysis

Next, the participant(s) should be instructed to perform the task under analysis. The SWAT can be administered either during or post trial. However, it is recommended that the SWAT is administered post trial, as on-line administration is intrusive to primary task performance. If on-line administration is required, then the SWAT should be administered and responded to verbally.

Step 8: Conduct SWAT Rating Procedure

Upon completion of task performance, participants should be presented with the SWAT pro-forma and asked to provide a rating between 1 (low) and 3 (high) for each SWAT dimension based on the task undertaken. Ratings should be made purely on the basis of the participant's subjective judgement and no assistance should be given in calculation of sub-scale ratings. Assistance may be given, however, if the participants do not understand any of the sub-scales (although participants should fully understand each sub-scale upon completion of the briefing during step 4).

Step 9: Calculate SWAT Scores

The final step involves calculating SWAT scores for each participant. Normally, an overall workload score is calculated for each participant. For the workload score, the analyst should take the scale value associated with the combination given by the participant. The scores are then translated into individual workload scores for each SWAT dimension. Finally, an overall workload score should be calculated.

ADVANTAGES

1. The SWAT method provides a low cost, quick, and simple-to-use method for assessing workload.
2. Its simplicity allows it to be easily understood by participants.
3. The results obtained can be compared across participants, trials, and dimensions.
4. The SWAT sub-scales are generic, so the method can be applied to any domain.
5. When administered post trial the approach is nonintrusive to primary task performance.
6. The PRO-SWAT (Kuperman,1985) variation allows the method to be used to predict operator workload.

DISADVANTAGES

1. When administered on-line, the SWAT method can be intrusive to primary task performance.
2. The level of construct validity may be questionable (i.e., is it assessing mental workload or something else?).
3. It suffers from the problems associated with collecting subjective data post task performance, such as poor recall of events (e.g., low workload parts of the task) and a correlation with performance (e.g., poor performance equals high workload).
4. The scale development procedure is laborious and adds more time to the procedure.
5. A number of validation studies report that the NASA-TLX is superior to SWAT in terms of sensitivity, particularly for low mental workload scenarios (Hart and Staveland, 1988; Nygren, 1991).
6. SWAT has been criticised for having a low level of sensitivity to workload variations (e.g., Luximon and Goonetilleke, 2001).

Related Methods

The SWAT is one of a number of multidimensional subjective workload assessment methods. Other multidimensional methods include the NASA-TLX, Bedford scales, DRAWS, and the MACE method. Often task analysis approaches, such as HTA, are used to describe the task so that analysts can determine which tasks should be involved in the data collection effort. Further, subjective workload assessment methods are typically used in conjunction with other forms of workload assessment, such as primary and secondary task performance measures and physiological measures. A predictive version of the SWAT, known as PRO-SWAT (Kuperman, 1985), also exists.

Approximate Training and Application Times

The training time associated with the SWAT method is minimal and the method is extremely quick to apply. Based on our experiences it should take participants no longer than 5 minutes to complete one SWAT pro-forma; however, in a study comparing the NASA-TLX, workload profile, and SWAT methods (Rubio et al., 2004), the SWAT incurred the longest application time of the three methods.

Reliability and Validity

A number of validation studies involving the SWAT method have been conducted (e.g., Hart and Staveland, 1988; Vidulich and Tsang, 1985, 1986). Vidulich and Tsang (1985, 1986) reported that NASA-TLX produced more consistent workload estimates for participants performing the same task than the SWAT approach did (Reid and Nygren, 1988). Luximon and Goonetilleke (2001) also reported that a number of studies have shown that the NASA-TLX is superior to SWAT in terms of sensitivity, particularly for low mental workloads (Hart and Staveland, 1988; Nygren, 1991).

Tools Needed

The SWAT method can be applied using pen and paper only; however, it is useful to get participants to enter ratings directly into a Microsoft Excel spreadsheet, which then automatically calculates overall workload scores.

Flowchart

(See Flowchart 7.4.)

Recommended Texts

Reid, G. B., and Nygren, T. E. (1988). The subjective workload assessment technique: A scaling procedure for measuring mental workload. In *Human mental workload*, eds. P. S. Hancock and N. Meshkati. Amsterdam, The Netherlands: Elsevier.

Vidulich, M. A., and Tsang, P. S. (1986) Technique of subjective workload assessment: A comparison of SWAT and the NASA bipolar method. *Ergonomics* 29(11):1385–98.

THE SUBJECTIVE WORKLOAD DOMINANCE METHOD

Background and Applications

The Subjective Workload Dominance method (SWORD; Vidulich, 1989) is a subjective workload assessment method that uses paired comparisons in order to elicit ratings of workload for different tasks. The SWORD method is administered post trial and requires participants to rate one task's dominance over another in terms of the level of workload imposed. The SWORD method has typically been used in assessments of the level of workload imposed by two or more different display or interface design concepts (e.g., Vidulich, Ward, and Schueren, 1991).

Mental Workload Assessment Methods

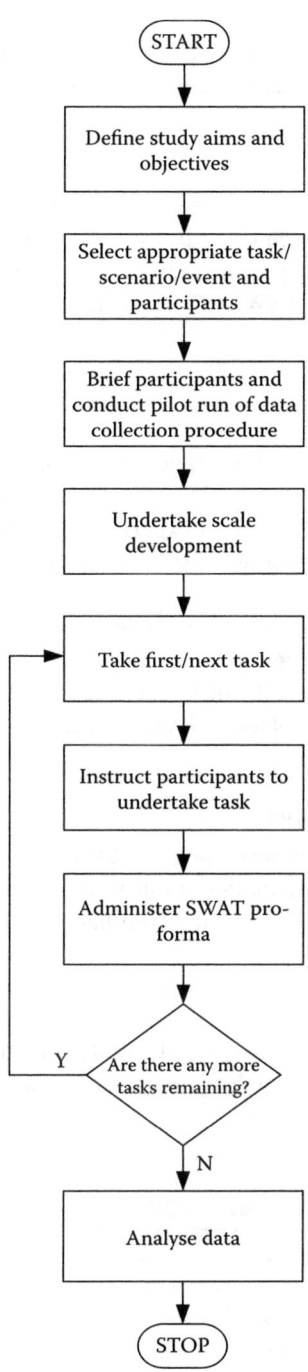

FLOWCHART 7.4 SWAT flowchart.

DOMAIN OF APPLICATION

The SWORD method was originally developed for the aviation domain to assess pilot workload; however, the approach is generic and can be applied in any domain.

APPLICATION IN SPORT

The SWORD method could potentially be used in a sporting context in order to assess the level of workload imposed on performers when using two different devices, displays, or performance aids. The level of workload imposed when playing in two different positions or roles within a team could also potentially be assessed using the SWORD method.

PROCEDURE AND ADVICE

Step 1: Define Analysis Aims

First, the aims of the analysis should be clearly defined. For example, the aim of the analysis may be to evaluate the levels of workload imposed by two different devices or procedures. Alternatively, it may be to compare the levels of workload experienced by novice and expert performers when performing two different tasks or using two different devices.

Step 2: Define Task(s) or Scenario under Analysis

The next step involves clearly defining the task or scenario under analysis. It is recommended that the task be described clearly, including the different actors involved, the task goals, and the environment within which the task is to take place. HTA may be useful for this purpose if there is sufficient time available.

Step 3: Create SWORD Rating Sheet

Once a task description (e.g., HTA) is developed, the SWORD rating sheet can be created. The analyst should list all of the possible combinations of tasks involved in the scenario under analysis (e.g., task A vs. B, A vs. C, B vs. C, etc.) and also the dominance rating scale. An example of a SWORD rating sheet is presented in Figure 7.3.

Step 4: Select Participants

Once the task(s) under analysis is clearly defined and described, it may be useful to select the participants that are to be involved in the analysis. This may not always be necessary and it may suffice to select participants randomly on the day.

Task	Absolute	Very Strong	Strong	Weak	EQUAL	Weak	Strong	Very Strong	Absolute	TASK
A										B
A										C
A										D
A										E
B										C
B										D
B										E
C										D
C										E
D										E

FIGURE 7.3 Example SWORD paired comparison rating sheet.

Step 5: Brief Participants

Before the task(s) under analysis is performed, all of the participants involved should be briefed regarding the purpose of the study, the area of workload and workload assessment, and the SWORD method. It is recommended that participants be given a workshop on workload and workload assessment. It may also be useful at this stage to take the participants through an example SWORD application, so that they understand how the method works and what is required of them as participants.

Step 6: Conduct Pilot Run

Before the data collection process begins, it is recommended that the participants take part in a number of pilot runs of the SWORD data collection procedure. A number of small test scenarios should be used to iron out any problems with the data collection procedure, and the participants should be encouraged to ask any questions. Once the participant is familiar with the procedure and is comfortable with his or her role, the real data collection can begin.

Step 7: Undertake Task(s) under Analysis

Next, the participant(s) should be instructed to perform the task under analysis. The SWORD method is administered post trial.

Step 8: Administration of SWORD Questionnaire

Once the tasks under analysis are completed, the SWORD data collection process begins. This involves the administration of the SWORD rating sheet. The participant should be presented with the SWORD rating sheet immediately after task performance has ended. The SWORD rating sheet lists all possible paired comparisons of the tasks conducted in the scenario under analysis. A 9-point rating scale is used. The analyst has to rate the two tasks (e.g., task A vs. B), in terms of the level of workload imposed, against each other. For example, if the participant feels that the two tasks imposed a similar level of workload, then they should mark the "Equal" point on the rating sheet. However, if the participant feels that task A imposed a slightly higher level of workload than task B did, they would move towards task A on the sheet and mark the "Weak" point on the rating sheet. If the participant felt that task A imposed a much greater level of workload than task B, then they would move towards task A on the sheet and mark the "Absolute" point on the rating sheet. This allows the participant to provide a subjective rating of one task's workload dominance over the over. This procedure should continue until all of the possible combinations of tasks in the scenario under analysis are assigned SWORD ratings. An example of a completed SWORD rating sheet is presented in Figure 7.4.

Step 9: Construct Judgement Matrix

Once all ratings have been elicited, the SWORD judgement matrix should be constructed. Each cell in the matrix should represent the comparison of the task in the row with the task in the associated column. The analyst should fill each cell with the participant's dominance rating. For example, if a participant rated tasks A and B as equal, a "1" is entered into the appropriate cell. If task A is rated as dominant, then the analyst simply counts from the "Equal" point to the marked point on the SWORD dominance rating sheet, and enters the number in the appropriate cell. The rating for each task is calculated by determining the mean for each row of the matrix and then normalising the means (Vidulich, Ward, and Schueren, 1991).

Step 10: Evaluate Matrix Consistency

Once the SWORD matrix is complete, the consistency of the matrix can be evaluated by ensuring that there are transitive trends amongst the related judgements in the matrix. For example, if task A is rated twice as hard as task B, and task B is rated three times as hard as task C, then task A should be rated as six times as hard as task C (Vidulich, Ward, and Schueren, 1991).

Task	Absolute	Very Strong	Strong	Weak	EQUAL	Weak	Strong	Very Strong	Absolute	TASK
A	✖									B
A	✖									C
A				✖						D
A				✖						E
B					✖					C
B			✖							D
B			✖							E
C							✖			D
C						✖				E
D					✖					E

FIGURE 7.4 Example of a completed SWORD paired comparison rating sheet.

Advantages

1. The SWORD method offers a simple, quick, and low cost approach to workload assessment.
2. SWORD is particularly useful when analysts wish to compare the levels of workload imposed by different tasks, devices, design concepts, or procedures.
3. SWORD is administered post trial and is non-intrusive to task performance.
4. SWORD has high face validity.
5. Studies have demonstrated SWORD's sensitivity to workload variations (e.g., Reid and Nygren, 1988).
6. It could be used in a sporting context to compare the levels of workload imposed by two different devices, performance aids, procedures, events, or design concepts.

Disadvantages

1. The SWORD method suffers from the problems associated with collecting subjective data post task performance, such as poor recall of events (e.g., low workload parts of the task) and a correlation with performance (e.g., poor performance equals high workload).
2. Only limited validation evidence is available in the literature.
3. The SWORD method has not been as widely applied as other subjective workload assessment methods, such as the SWAT and NASA-TLX.
4. The SWORD output does not offer a rating of participant workload as such, only a rating of which tasks or devices imposed greater workload.

Related Methods

SWORD is one of a number of subjective workload assessment methods. Others include the NASA-TLX, SWAT, Bedford scales, DRAWS, and the MACE method. Often task analysis approaches, such as HTA, are used so that analysts can determine which tasks should be involved in the analysis. Further, subjective workload assessment methods are typically used in conjunction with other forms of workload assessment, such as primary and secondary task performance measures. A predictive version of the SWORD, known as Pro-SWORD (Vidulich, Ward, and Schueren, 1991), also exists. In addition, an adaptation of the method known as SA-SWORD (Vidulich and Hughes, 1991) is used to measure differences in the levels of SA afforded by different tasks and devices.

Approximate Training and Application Times

The SWORD method requires only minimal training. The application time is dependent upon the tasks used, although completing SWORD ratings typically requires very little time.

Reliability and Validity

Vidulich, Ward, and Schueren (1991) tested SWORD for its accuracy in predicting the workload imposed upon F-16 pilots by a new Head-Up Display (HUD) attitude display system. Participants included F-16 pilots and college students and were divided into two groups. The first group (F-16 pilots experienced with the new HUD display) retrospectively rated the tasks using the traditional SWORD method, while the second group (F-16 pilots who had no experience of the new HUD display) used the Pro-SWORD variation to predict the workload associated with the HUD tasks. A third group (college students with no experience of the HUD) also used the Pro-SWORD method to predict the associated workload. In conclusion, it was reported that the pilot Pro-SWORD ratings correlated highly with the pilot SWORD (retrospective) ratings (Vidulich, Ward, and Schueren, 1991). Furthermore, the Pro-SWORD ratings correctly anticipated the recommendations made in an evaluation of the HUD system. Vidulich and Tsang (1986) also reported that the SWORD method was more reliable and sensitive than the NASA-TLX method.

Tools Needed

The SWORD method can be applied using pen and paper. The system or device under analysis is also required.

Flowchart

(See Flowchart 7.5.)

Recommended Texts

Vidulich, M. A. (1989). The use of judgement matrices in subjective workload assessment: The subjective Workload Dominance (SWORD) technique. In *Proceedings of the Human Factors Society 33rd Annual Meeting,* 1406–10). Santa Monica, CA: Human Factors Society.

Vidulich, M. A., Ward, G. F., and Schueren, J. (1991). Using Subjective Workload Dominance (SWORD) technique for Projective Workload Assessment. *Human Factors* 33(6):677–91.

INSTANTANEOUS SELF-ASSESSMENT METHOD

Background and Applications

The ISA workload method is a simplistic subjective workload assessment method that was developed for use in the assessment of air traffic controller workload (Kirwan et al., 1997). When using ISA, participants provide subjective ratings of their workload on-line during task performance (normally every 2 minutes) using a simplistic scale of 1 (low) to 5 (high). For example, Kirwan et al. (1997) used the ISA scale in Table 7.3 to assess air traffic controller workload.

Typically, ISA ratings are given either verbally or via a colour-coded keypad, which flashes when a workload rating is required. The frequency with which workload ratings are collected is particularly useful since it allows a workload profile throughout the task to be built, which allows the analyst to ascertain excessively high or low workload parts of the task under analysis.

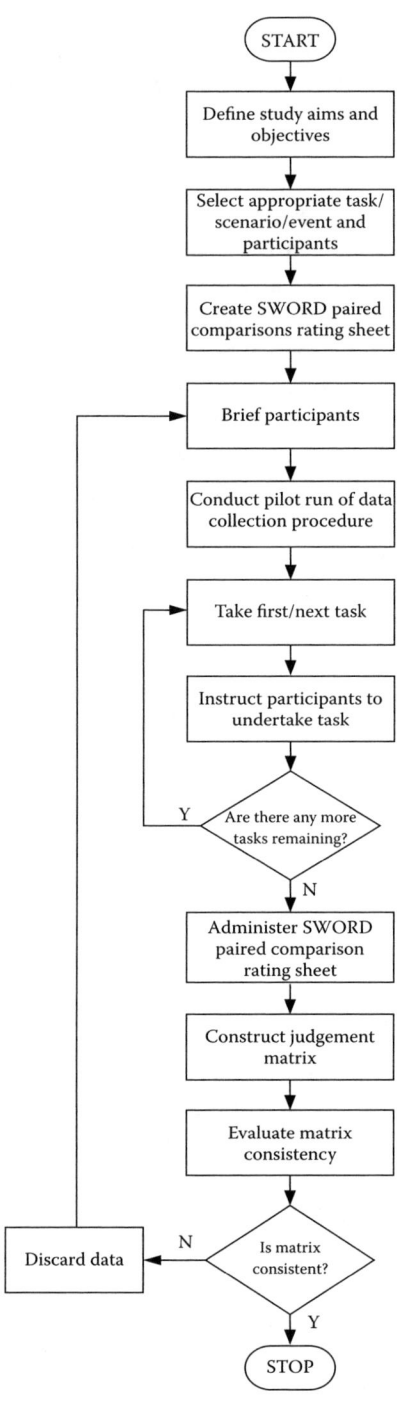

FLOWCHART 7.5 SWORD flowchart.

Mental Workload Assessment Methods

TABLE 7.3
ISA Scale

Level	Workload Heading	Spare Capacity	Description
5	Excessive	None	Behind on tasks; losing track of the full picture
4	High	Very little	Non-essential tasks suffering; could not work at this level very long
3	Comfortable busy pace	Some	All tasks well in hand; busy but stimulating pace; could keep going continuously at this level
2	Relaxed	Ample	More than enough time for all tasks; active on ATC task less than 50% of the time
1	Under utilised	Very much	Nothing to do; rather boring

Source: Kirwan, B., Evans, A., Donohoe, L., Kilner, A., Lamoureaux, T., Atkinson, T., and MacKendrick, H. (1997). Human Factors in the ATM System Design Life Cycle. FAA/Eurocontrol ATM R&D Seminar, Paris, France, June 16–20.

DOMAIN OF APPLICATION

Although the method was developed and originally applied in the air traffic control domain, the procedure and scale are generic, allowing it to be applied in any domain.

APPLICATION IN SPORT

ISA type approaches could potentially be used in sport to analyse workload throughout a particular task, event, or game. The method seems to be particularly suited to individual sports in which workload is likely to vary throughout performance, such as road and fell running and cycling.

PROCEDURE AND ADVICE

Step 1: Define Analysis Aims

First, the aims of the analysis should be clearly defined. For example, the aim of the analysis may be to evaluate the level of workload imposed on participants throughout a particular task or event, or it may be to compare the workload imposed by two different events, devices, or scenarios.

Step 2: Define Task(s) or Scenario under Analysis

The next step involves clearly defining the task or scenario under analysis. It is recommended that the task is described clearly, including the different actors involved, the task goals, and the environment within which the task is to take place. HTA may be useful for this purpose if there is sufficient time available.

Step 3: Select Participants

Once the task(s) under analysis is clearly defined and described, it may be useful to select the participants that are to be involved in the analysis. This may not always be necessary and it may suffice to select participants randomly on the day. However, if workload is being compared across rank or experience levels, then effort is required to select the appropriate participants.

Step 4: Brief Participants

Before the task(s) under analysis is performed, all of the participants involved should be briefed regarding the purpose of the study, the area of workload and workload assessment, and the ISA method. It is recommended that participants be given a workshop on workload and workload assessment. It may also be useful at this stage to take the participants through an example ISA application, so that they understand how the method works and what is required of them as participants.

Step 5: Conduct Pilot Run

Before the data collection process begins, it is recommended that the participants take part in a number of pilot runs of the ISA data collection procedure. A number of small test scenarios should be used to iron out any problems with the data collection procedure, and the participants should be encouraged to ask any questions. Once the participant is familiar with the procedure and is comfortable with his or her role, the real data collection can begin.

Step 6: Begin Task(s) under Analysis and Collect ISA Ratings

Next, the participant(s) should be instructed to perform the task under analysis. Workload ratings are typically requested verbally; however, in more sophisticated setups, ratings can be requested and provided via a flashing keypad device. The frequency and timing of workload ratings should be determined beforehand by the analyst, although every 2 minutes is a general rule of thumb. It is critical that the process of requesting and providing ratings is as unintrusive to task performance as is possible (i.e., for runners and cyclists verbally requesting and providing ratings is the most appropriate). Workload ratings should be elicited throughout the task until completion. The analyst should record each workload rating given.

Step 7: Construct Task Workload Profile

Once the task is complete and the workload ratings are collected, the analyst should construct a workload profile for the task under analysis. Typically, a graph is constructed, depicting the workload ratings over time throughout the task. This is useful for highlighting the high and low workload points of the task under analysis. An average workload rating for the task under analysis is also typically calculated.

ADVANTAGES

1. The ISA method provides a simple, low cost, and quick means of analysing varying levels of workload throughout a particular task.
2. It requires very little (if any) training.
3. The output is immediately useful, providing a workload profile for the task under analysis.
4. Little data analysis is required.
5. The method is generic and can be applied in any domain.
6. It is useful when other methods are too intrusive, such as physiological measures.
7. It requires very little in the way of resources.
8. While the method is intrusive to the primary task, it is probably the least intrusive of all on-line workload assessment methods.
9. ISA of workload is normally sensitive to varying task demands.

DISADVANTAGES

1. ISA is intrusive to primary task performance.
2. The data are subjective and may often be correlated with task performance.
3. There is only limited validation evidence associated with the method.
4. Participants are not very efficient at reporting mental events.

RELATED METHODS

ISA is a subjective workload assessment method of which there are many, such as the NASA-TLX, SWORD, MACE, DRAWS, and the Bedford scales. To ensure comprehensiveness, ISA is often used in conjunction with other subjective methods, such as the NASA-TLX.

Mental Workload Assessment Methods

TRAINING AND APPLICATION TIMES

The ISA method requires very little (if any) training. The application time is dependent upon the task under analysis, but is normally low since the data are collected on-line during task performance.

RELIABILITY AND VALIDITY

No data regarding the reliability and validity of the method are available in the literature.

TOOLS NEEDED

ISA is normally applied using pen and paper; however, for sporting applications where ratings are requested and made verbally, an audio recording device is required. In addition, a timing device of some sort (i.e., stopwatch or watch) is required to ensure that workload ratings are requested at the correct time.

EXAMPLE

The ISA method was used to evaluate runner physical and mental workload during a training run. The runner in question provided physical and mental workload ratings on a scale of 1 (low) to 5 (high) over the course of a 6-mile training run. Workload ratings were provided every 2 minutes. A heart rate monitor was also used to measure the runner's heart rate throughout the run. The graph presented in Figure 7.5 depicts the runner's ISA physical and mental workload ratings and heart rate throughout the 6-mile run.

FLOWCHART

(See Flowchart 7.6.)

RECOMMENDED TEXT

Kirwan, B., Evans, A., Donohoe, L., Kilner, A., Lamoureux, T., Atkinson, T., and MacKendrick, H. (1997). Human Factors in the ATM System Design Life Cycle. FAA/Eurocontrol ATM R&D Seminar, Paris, France, June 16–20.

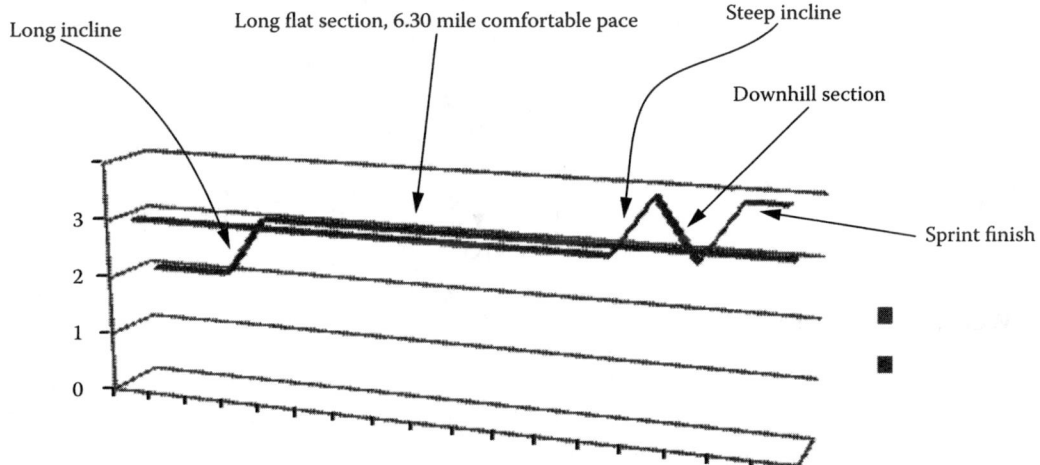

FIGURE 7.5 ISA ratings and heart rate throughout 6-mile run.

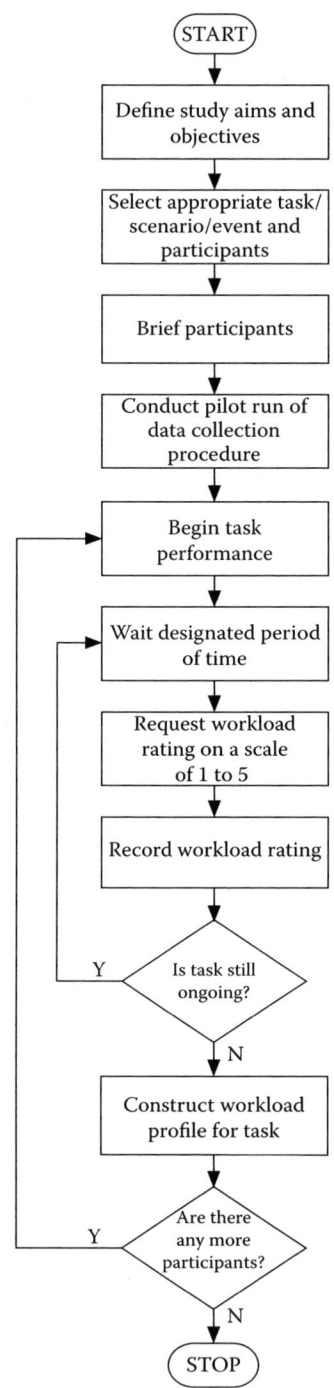

FLOWCHART 7.6 ISA flowchart.

8 Teamwork Assessment Methods

INTRODUCTION

Team processes and performance have been a common line of enquiry for sports scientists for many years (Fiore and Salas, 2006; Pedersen and Cooke, 2006). To date, various aspects of teamwork have been discussed in a sporting context, including team performance assessment (e.g., Jones, James, and Mellalieu, 2008), team cognition (e.g., Reimer, Park, and Hinsz, 2006), team coordination (e.g., Ferarro, Sforza, Dugnani, Michielon, and Mauro, 1999), and team attitudes such as collective efficacy (e.g., Chow and Feltz, 2008) and team cohesion (e.g., Carron, Bray, and Eys, 2002). Similarly, teamwork has received considerable attention from the Human Factors community, and a number of teamwork issues have been investigated across a wide range of domains. Examples include coordination (e.g., Fiore, Salas, Cuevas, and Bowers, 2003), communication (e.g., Svensson and Andersson, 2006), shared mental models (e.g., Fox, Code, and Langfield-Smith, 2000), team situation awareness (e.g., Salmon et al., 2009), distributed cognition (e.g., Hutchins, 1995), team errors (e.g., Wilson et al., 2007), team training (e.g., Salas, 2004), leadership (e.g., Bell and Kozlowski, 2002), trust (e.g., Salas, Burke, and Samman, 2001), and team attitudes, such as team cohesion, collective efficacy, and collective orientation (e.g., Fiore et al., 2003; Salas, Burke, and Samman, 2001; Wilson et al., 2007).

The potential for cross-disciplinary interaction between Human Factors researchers and sports scientists is perhaps greatest in the area of teamwork, due to the high presence of teams in the domains in which both work, and also the amount of knowledge that both disciplines have amassed over the years. Indeed, in attempt to promote interaction between team cognition researchers and sports psychologists, a recent special issue of the *International Journal of Sports and Exercise Psychology* (Fiore and Salas, 2006) focussed on the discussion of key teamwork concepts by leading team cognition researchers in a sporting context. Goodwin (2008) also points out there is great potential for gaining further insight into teamwork in the military through the study of sports teams. In particular, Goodwin (2008) outlines team dynamics, shared cognition, and team development as areas in which opportunities for learning more exist. Disappointingly, despite the ubiquitous presence of teams within sport and the Human Factors domains, the various teamwork assessment methods used by Human Factors researchers have not yet been applied in a sporting context.

TEAMWORK

From a Human Factors perspective at least, the use of teams has increased significantly over the past three decades (Savoie, cited in Salas, 2004). This is primarily due to two factors. First, the increasing complexity of work and work procedures, and second, because appropriately trained and constructed teams potentially offer a number of advantages over and above the use of individual operators. These include the ability to better perform more difficult and complex tasks, greater productivity and improved decision making (Orasanu and Fischer, 1997), more efficient performance under stress (Salas and Cannon-Bowers, 2000), and a reduction in the number of errors made (Wiener, Kanki, and Helmreich, 1993; cited in Salas and Cannon-Bowers, 2000).

A team is characterised as consisting of two or more people, dealing with multiple information sources and working to accomplish a common goal of some sort. Salas, Sims, and Burke (2005) define a team as "two or more individuals with specified roles interacting adaptively, interdependently, and dynamically toward a common and valued goal" (p. 561). Teams also have a range of distinct characteristics that distinguish them from small groups; Salas (2004), for example, suggests that characteristics of teams include meaningful task interdependency, coordination among team members, specialised member roles and responsibilities, and intensive communication.

Collaborative work comprises two forms of activity: teamwork and taskwork. Teamwork refers to those instances where individuals interact or coordinate behaviour in order to achieve tasks that are important to the team's goals (i.e., behavioural, attitudinal, and cognitive responses coordinated with fellow team members), while taskwork (i.e., task-oriented skills) describes those instances where team members are performing individual tasks separate from their team counterparts, i.e., those tasks that do not require interdependent interaction with other team members (Salas, Cooke, and Rosen, 2008). Teamwork is formally defined by Wilson et al. (2007) as "a multidimensional, dynamic construct that refers to a set of interrelated cognitions, behaviours and attitudes that occur as team members perform a task that results in a coordinated and synchronised collective action" (p. 5). According to Glickman et al. (1987; cited in Burke, 2004), team tasks require a combination of taskwork and teamwork skills in order to be completed effectively.

There have been many attempts to postulate models of teamwork (e.g., Fleishman and Zaccaro, 1992; Helmreich and Foushee, 1993; McIntyre and Dickinson, 1992; Salas, Sims, and Burke, 2005; Zsambok, Klein, Kyne, and Klinger, 1993, etc.), far more than there is room to include here (a recent review by Salas, Sims, and Burke [2005] identified over 130 models). Most of the models presented in the academic literature attempt to define the different teamwork processes involved and/or the different attributes that teams possess. A summary of the more prominent teamwork models is presented in Table 8.1.

For example, Salas, Sims, and Burke (2005) outlined the "big five" model of teamwork, arguing that the five most important teamwork processes are leadership, mutual performance monitoring, back up behaviour, adaptability, and team orientation. Salas, Sims, and Burke (2005) suggested that these factors would improve performance in any team, regardless of type, so long as the following three supporting mechanisms were also present within the team: shared mental models, closed loop communication, and mutual trust. One pertinent line of enquiry is the comparison of expert sports teams and teams performing within the safety critical domains (e.g., the military, emergency services). The extent to which teamwork models developed in both areas apply across both disciplines is of interest, as are the similarities, and differences, between the processes used by teams performing in both areas.

Team competencies are also heavily discussed within the teamwork literature. Salas and Cannon-Bowers (2000) define team competencies as resources that team members draw from in order to function; they refer to what team members need to know, how they need to behave, and what attitudes they need to hold (Salas, Cannon-Bowers, and Smith-Jentsch, 2001; cited in Salas, 2004). Knowledge-based competencies refer to what team members "think" during teamwork performance; they refer to understanding facts, concepts, relations, and underlying foundations of information that a team member must have to perform a task (Salas and Cannon-Bowers, 2000). Salas and Cannon-Bowers (2000) cite cue-strategy associations, team-mate characteristics, shared mental models, and task sequencing as examples of knowledge-based competencies. Skill-based competencies refer to the things that team members "do" during teamwork performance and, according to Salas and Cannon-Bowers (2000), are the necessary behavioural sequences and procedures required during task performance. Examples of skill-based competencies include adaptability, situational awareness, communication, and decision-making (Salas and Cannon-Bowers, 2000). Attitude-based competencies refer to what team members "feel" during teamwork performance and are those affective components that are required during task performance (Salas and Cannon-Bowers, 2000). Examples of attitude-based competencies include motivation, mutual trust, shared vision, teamwork efficacy, and collective orientation.

TABLE 8.1
Teamwork Models

Teamwork Model	Teamwork Behaviours/Processes
Normative Model of Group Effectiveness (Hackman, 1987)	Group design
	Group synergy
	Organizational variables
	Team strategies
	Effort by team
	Team knowledge and abilities
Big 5 Teamwork Model (Salas, Sims, and Burke, 2005)	Team leadership
	Team orientation
	Mutual performance monitoring
	Back up behaviour
	Adaptability
	Shared mental models
	Closed loop communication
	Mutual trust
Sociotechnical Systems Theory Perspective (Pasmore, Francis, Haldeman, and Shani, 1982)	Task interdependencies
	Task requirements
	Organisational context
	Team design
	Shared mental models
	Team effort
	Task strategies
	Team knowledge
	Team skills
	Common goals
Team Performance Model (McIntyre and Dickinson, 1992)	Orientation
	Leadership
	Communication
	Feedback
	Back up behaviour
	Monitoring Coordination
Teamwork Model (Flieshman and Zaccaro, 1992)	External conditions
	Member resources
	Team characteristics
	Task characteristics
	Individual task performance
	Team performance function

TEAMWORK ASSESSMENT METHODS

The complex, multi-dimensional nature of team-based activity has led to a range of different assessment methods being developed and applied by Human Factors researchers. These include team performance assessment frameworks (e.g., Stanton et al., 2005), Team Task Analysis (TTA; e.g., Burke, 2004), team cognitive task analysis (e.g., Klien, 2000), team communications analysis (e.g., Driskell and Mullen, 2004), team situation awareness assessment (e.g., Salmon et al., 2009), team workload assessment (e.g., Bowers and Jentsch, 2004), team coordination assessment (e.g., Burke, 2004), team behavioural assessment (e.g., Baker, 2004), and team training requirements analysis methods (e.g., Swezey, Owens, Bergondy, and Salas, 2000). For the purposes of this book we focus on the following four teamwork assessment methods: Social Network Analysis (SNA; Driskell and Mullen, 2004), Team Task Analysis (TTA; Burke, 2004), Coordination Demands Analysis (CDA;

Burke, 2004), and the Event Analysis of Systemic Teamwork framework (EAST; Stanton et al., 2005). In addition, some of the methods already described in earlier chapters, such as propositional networks (team situation awareness), HTA (team task analysis), and the CDM (team cognitive task analysis) can be applied in a teamwork context. A brief description of each method focussed on in this chapter is given below.

TTA (Burke, 2004) is used to describe the different knowledge, skills, and attitudes required by different team members. TTA outputs are typically used in the development of team-training interventions, for the evaluation of team performance, and also to identify operational and teamwork skills required within teams (Burke, 2004). SNA (Driskell and Mullen, 2004) is used to analyse and represent the communications or associations between actors within a network or team. SNA diagrams depict the communications or associations that occurred between team members during task performance, and network statistics are used to analyse the networks mathematically in order to identify key actors or communication "hubs." Coordination Demands Analysis (CDA; Burke, 2004) is used to evaluate the level of coordination between team members during task performance. Tasks are described using HTA, and a teamwork taxonomy is used to rate coordination levels on teamwork tasks. Finally, EAST (Stanton et al., 2005) provides an integrated framework of Human Factors methods for analysing team activities. EAST uses HTA, CDA, communications usage diagrams, SNA, the CDM, and propositional networks to analyse teamwork from a task, social, and knowledge network perspective. A summary of the teamwork assessment methods described is presented in Table 8.2.

SOCIAL NETWORK ANALYSIS

BACKGROUND AND APPLICATIONS

Social Network Analysis (SNA; Driskell and Mullen, 2004) is used to understand network structures via description, visualization, and statistical modelling (Van Duijn and Vermunt, 2006). The approach is increasingly being used across domains to analyse the relationships between individuals, teams, and technology. For example, recent applications have been undertaken within the domains of emergency services (e.g., Houghton et al., 2006), the military (e.g., Dekker, A. H., 2002), the Internet (e.g., Adamic, Buyukkokten, and Adar, 2003), terrorism (e.g., Skillicorn, 2004), and railway maintenance (e.g., Walker et al., 2006). The approach involves collecting data, typically via observation, interview, or questionnaire (Van Duijn and Vermunt, 2006), regarding the relationship (e.g., communications, transactions, etc.) between the entities in the group or network under analysis. These data are then used to construct a social network diagram that depicts the connections between the entities in a way in which the relationships between agents and the structure of the network can be easily ascertained (Houghton et al., 2006). Statistical modelling is then used to analyse the network mathematically in order to quantify aspects of interest, such as key communication "hubs."

APPLICATION IN SPORT

There is great potential for applying SNA in a sporting context, not only to analyse communications between sports performers (e.g., team members, officiators), but also other associations, such as passing connections and marking. According to Wasserman and Faust (1994), the approach can be used in any domain where the connections between entities are important; indeed, SNA has previously been applied to the analysis of sports team performance. Reifman (2006), for example, used SNA to analyse the passing patterns of U.S. college basketball teams, and Gould and Gatrell (1980) used the approach to analyse network structures during the 1977 soccer FA Cup Final.

Teamwork Assessment Methods

PROCEDURE AND ADVICE

Step 1: Define Analysis Aims

First, the aims of the analysis should be clearly defined. For example, the aims of the analysis may be to evaluate the communications or passing associations between players during different game scenarios (e.g., defensive vs. attacking), or it may be to identify the key player in terms of passes made and received during the entire game. The analysis aims should be clearly defined so that appropriate scenarios are used and relevant data are collected. Further, the aims of the analysis will dictate which network statistics are used to analyse the social networks produced.

Step 2: Define Task(s) or Scenario under Analysis

The next step involves clearly defining the task or scenario under analysis. It is recommended that the task be described clearly, including the different actors involved, the task goals, and the environment within which the task is to take place. HTA may be useful for this purpose if there is sufficient time available. It is important to clearly define the task as it may be useful to produce social networks for different task phases, such as the first half and second half, or attacking and defensive phases of a particular game.

Step 3: Collect Data

The next step involves collecting the data that are to be used to construct the social networks. This typically involves observing or recording the task under analysis and recording the links of interest that occur during task performance. Typically, the direction (i.e., from actor A to actor B), frequency, type, and content of associations are recorded. It is recommended that the scenario be recorded for data validation purposes.

Step 4: Validate Data Collected

More often than not it is useful to record the task in question and revisit the data to check for errors or missing data. This involves observing the task again and checking each communication originally recorded. Often all communications/associations cannot be recorded on-line during the observation, and so this step is critical in ensuring that the data are accurate. It may also be pertinent for reliability purposes to get another analyst to analyse the data in order to compute reliability statistics.

Step 5: Construct Agent Association Matrix

Once the data are checked and validated, the data analysis phase can begin. The first step involves the construction of an agent association matrix. This involves constructing a simple matrix and entering the frequency of communications between each of the actors involved. For example purposes, a simple association matrix, showing the passing associations between a five-a-side soccer team, is presented in Table 8.3.

Step 6: Construct Social Network Diagram

Next, a social network diagram should be constructed. The social network diagram depicts each actor in the network and the associations that occurred between them during the scenario under analysis. Within the social network diagram, associations between actors are represented by directional arrows linking the actors involved. The frequency of associations is represented numerically and via the thickness of the arrows. For example purposes, a social network diagram for the five-a-side soccer team associations described in Table 8.3 is presented in Figure 8.1.

TABLE 8.2
Teamwork Assessment Methods Summary Table

Name	Domain	Application in Sport	Training Time	App. Time	Input Methods	Tools Needed	Main Advantages	Main Disadvantages	Outputs
Social Network Analysis (SNA; Driskell & Mullen, 2004)	Generic	Analysis of associations (e.g. communications, passing) between team members during task performance. Performance evaluation. Coaching/training design. Tactics development and selection	Low	High	Observational study	Pen and paper. Agna SNA software. Microsoft Visio	1. Has previously been applied in a sporting context to analyse the passing associations between team members in soccer and basketball. 2. Social network diagrams provide a powerful way of representing the associations between players. 3. Highly suited to performance evaluation in a sporting context	1. Data collection procedure can be hugely time consuming. 2. For large, complex networks the data analysis procedure is highly time consuming. 3. For complex collaborative tasks in which a large number of associations occur, SNA outputs can become complex and unwieldy	Graphical representation of associations between players in a network (i.e. team members). Statistical analysis of associations. Classification of network structure
Team Task Analysis (TTA; Burke, 2004)	Generic	Coaching/training design. Tactics development and selection	Med	High	Observational study. Hierarchial task analysis. Coordination demands analysis	Pen and paper	1. Goes further than other task analysis methods by specifying the knowledge, skills, and abilities (KSAs) required to complete each task step. 2. Exhaustive, considering team tasks, the levels of coordination required, and the KSAs required. 3. The output is useful for training programme design	1. Highly time consuming to apply. 2. A high level of access to SMEs is required throughout. 3. A high level of training may be required and the reliability of the method may be questionable due to lack of a rigid procedure	Description of team and individual tasks. Rating of coordination levels required. Identification of the KSAs required during task performance

Method	Type	Purpose	Training time	Application time	Related methods	Tools needed	Advantages	Disadvantages	Output
Coordination demands analysis	Generic	Performance evaluation; Coaching/training design	Low	High	Observational study; Hierarchical task analysis	Pen and paper; Microsoft Excel	1. Highly useful output that offers insight into the use of teamwork behaviours and also a rating of coordination between team members 2. Can potentially be used to analyse the coordination levels exhibited by sports teams 3. Coordination levels can be compared across scenarios, different teams, and also different domains	1. Can be highly time consuming and laborious 2. Intra-analyst and inter-analyst reliability can be a problem 3. High levels of access to SMEs is required	Rating of coordination levels observed or required during teamwork tasks
Event Analysis of Systemic Teamwork (EAST; Stanton et al., 2005)	Generic	Performance evaluation; Task analysis; Analysis of associations (e.g. communications, passing) between team members during task performance; Coaching/training design	High	High	Observational study	Pen and paper; WESTT software tool; HTA software tool	1. Highly comprehensive; activities are analysed from various perspectives 2. The analysis produced provides compelling views of teamwork activities 3. A number of Human Factors concepts are evaluated, including distributed situation awareness, decision making, teamwork, and communications	1. When undertaken in full, the EAST framework is a highly time consuming approach 2. The use of various methods ensures that the framework incurs a high training time 3. A high level of access to the domain, task and SMEs	Description of tasks performed; Rating of coordination levels; Description and analysis of communications between team members; Description of SA during task performance

TABLE 8.3
Social Network Agent Association Matrix Example

Player	Goalkeeper	Defender A	Defender B	Midfielder	Attacker A
Goalkeeper	—	13	13	7	0
Defender A	3	—	10	13	9
Defender B	0	5	—	2	0
Midfielder	0	0	0	—	7
Attacker	0	1	0	4	—

Step 7: Analyse Network Mathematically

Finally, the network should be analysed using appropriate social network analysis metrics. Various metrics exist and the ones used are dependent on the analysis aims. In the past, we have found sociometric status, centrality, and network density useful. Sociometric status provides a measure of how "busy" a node is relative to the total number of nodes present within the network under analysis (Houghton et al., 2006). Thus, sociometric status gives an indication of the relative prominence of actors based on their links to other actors in the network. Centrality is also a metric of the standing of a node within a network (Houghton et al., 2006), but here this standing is in terms of its "distance" from all other nodes in the network. A central node is one that is close to all other nodes in the network and a message conveyed from that node to an arbitrarily selected other node in the network would, on average, arrive via the least number of relaying hops (Houghton et al., 2006). Network density provides an indication of how dense, in terms of associations between actors, a particular network is. Various software tools are available for analysing social networks, such as Agna and WESTT (Houghton et al., 2008).

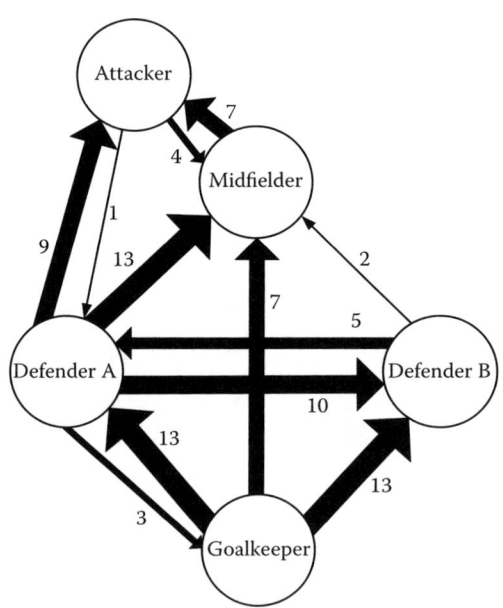

FIGURE 8.1 Example social network diagram for five-a-side soccer team.

Advantages

1. There is great potential for applying SNA within sport to analyse a range of associations, including communications, passing, marking, and interactions with technology.
2. SNA has previously been applied in a sporting context to analyse network structure (Gould and Gatrell, 1980) and the passing associations between team members in basketball (Reifman, 2006).
3. Social network diagrams provide a powerful, and easily interpretable, means of representing the associations between players.
4. SNA can be used to identify the importance of players within a team, based on a range of associations.
5. Highly suited to performance evaluation in a sporting context.
6. Networks can be classified according to their structure. This is particularly useful when analysing networks across different teams and domains.
7. SNA has been used in a range of domains for a number of different purposes.
8. SNA is simple to learn and apply.
9. Various free software programmes are available for analysing social networks (e.g., Agna).
10. SNA is generic and can be applied in any domain in which associations between actors exist.

Disadvantages

1. The data collection procedure for SNA can be hugely time consuming.
2. For large, complex networks the data analysis procedure is also highly time consuming.
3. For complex collaborative tasks in which a large number of associations occur, SNA outputs can become complex and unwieldy.
4. Some knowledge of network statistics is required to understand the analysis outputs.
5. It can be very difficult to collect SNA data for large teams.
6. Without the provision of software support, analysing the networks mathematically is a difficult and laborious procedure.

Related Methods

SNA uses observational study as its primary means of data collection, although questionnaires can also be used whereby actors indicate who in the network they had associations with during the task under analysis. The method itself is similar to link analysis, which describes the links between a user and device interface during task performance.

Approximate Training and Application Times

The SNA method requires only minimal training, although some knowledge of network statistics is required to interpret the results correctly. The application time is dependent upon the task and network involved. For short tasks involving small networks with only minimal associations between actors, the application time is low, particularly if an SNA software package is used for data analysis. For tasks of a long duration involving large, complex networks, the application time is likely to be high, due to lengthy data collection and analysis processes. The application time is reduced dramatically via the use of software support such as the Agna SNA programme, which automates the social network diagram construction and data analysis processes.

Reliability and Validity

Provided the data collected are validated as described above, the reliability of the SNA method is high. The method also has a high level of validity.

Tools Needed

At a simplistic level, SNA can be conducted using pen and paper only; however, it is recommended that video and/or audio recording devices are used to record the task under analysis for data checking and validation purposes, and that software support, such as Agna, is used for the data analysis part of SNA.

Example

The following example is taken from an analysis of the passing patterns of an international soccer team during the 2006 FIFA World Cup in Germany (Salmon et al., "Network analysis," 2009). Passing has often been used as a means of analysing the performance of soccer teams (e.g., Hughes and Franks, 2005; Scoulding, James, and Taylor, 2004). Typically, passing assessment involves analysing the total number of passes made (e.g., Scoulding, James, and Taylor, 2004); analysing passing sequences (e.g., Hughes and Franks, 2005); or analysing the percentage of passes successfully completed. We argued that the typical focus on passes linked to goals, although worthwhile when looking at the role of passes in goals scored, tells us relatively little about the overall passing performance of soccer teams in terms of each player's passing contribution. With this in mind, we investigated the use of social network analysis to analyse the passing performance of soccer teams where the relationships between the players are defined by the passes made between them. We used video recordings of an international soccer team's games to undertake SNAs of their passing performance. Networks were developed for different match phases (e.g., first and second half), for different areas of the pitch (e.g., defensive third, middle third, attacking third), and for passes that linked different areas of the pitch (e.g., defensive to middle, defensive to attacking, defensive to middle, middle to attacking third, etc.). Sociometric status was used to analyse the networks developed. In this case, the sociometric status of each player represents how prominent he was in terms of passes successfully made and received during the course of the game.

An example social network diagram, taken from one of the games analysed, is presented in Figure 8.2. Additionally, social network diagrams for each area of the pitch (e.g., defensive third, middle third, attacking third) and each sub-team are presented in Figure 8.3 and Figure 8.4. The sociometric status statistics for this game are presented in Table 8.4.

The analysis for the game presented revealed that player F (a central midfielder) was the key agent in terms of sociometric status, which indicates that he was the busiest player in terms of successful passes made and received overall throughout the 90 minutes. With regard to the different areas of the pitch, players C and D (both centre halfs) were the key agents within the defensive third; players F and H (both central midfielders) were the key agents within the middle third; and player I (central midfielder) was the key agent in the attacking third. In terms of passes between the areas, there were no key agents linking the defensive third to the middle third; player A (the goalkeeper) was the key agent in passing between the defensive third and the attacking third; and player F was the key agent in terms of passes made from the middle third to the attacking third of the pitch.

During the game in question the team analysed used a 4-5-1 formation with player K deployed as a lone striker. The analysis findings suggest that player K's overall contribution to the "network" during the game was only minimal. This is surprising since, as a lone striker, one would expect player K's role to primarily involve receiving the ball in the attacking third and then bringing other players into the game in the same area; effective performance of a lone striking role should arguably lead to the lone striker being highly connected within the network, and produce a high sociometric

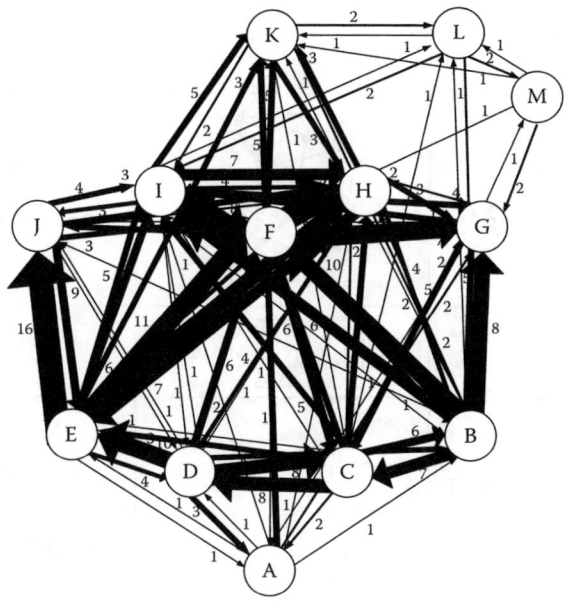

FIGURE 8.2 Social network diagram for 2006 FIFA World Cup Final game.

status score, both overall and in the attacking third area. In fact, player K scored poorly overall and also, even more surprisingly, in the attacking third in terms of sociometric status, an area in which one would expect a lone striker to make the most telling contributions. As would be expected, player K did achieve key agent status in relation to the passes made and received between the middle third and the attacking third, which is indicative of the lone striker's role of receiving the ball in advanced areas and bringing other players into the game; however, since he was not a key agent in the attacking third alone, the findings suggest that either the midfielders were not getting forward to support player K in the attacking third, that on receiving the ball he found himself to be isolated, or rather simply that player K's collection and distribution in the attacking third was poor. Further investigation via the social network diagrams reveals that player K made only five successful passes to only two of the five midfielders, which corroborates the notion that his "connectedness" in the attacking third was limited. These findings suggest that the lone striker formation was unsuccessful on this

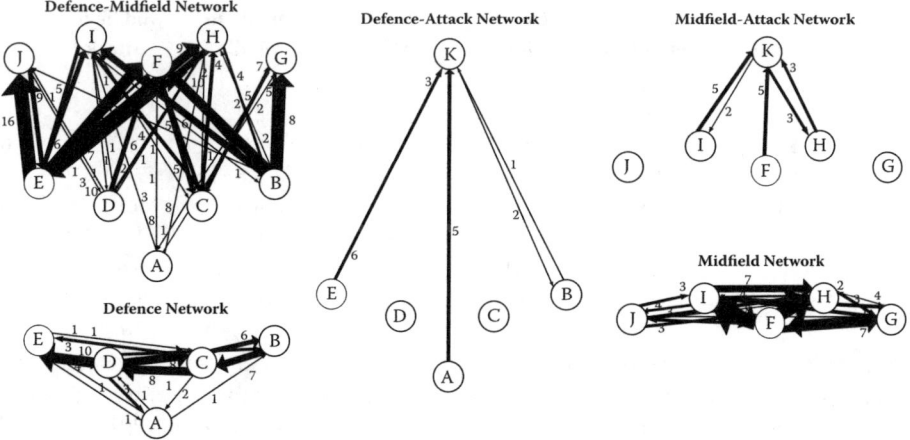

FIGURE 8.3 Social network diagrams for sub-teams and passing links between them.

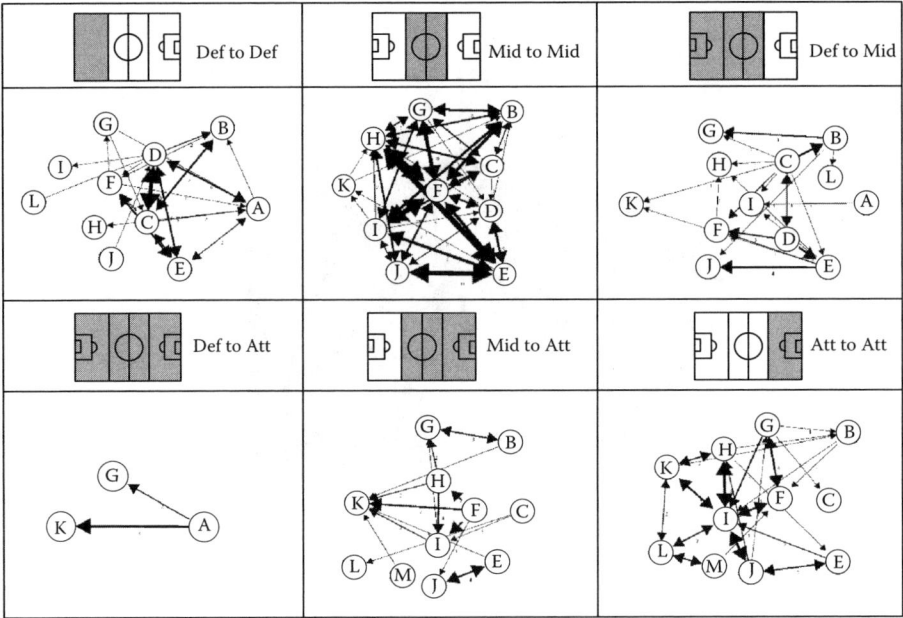

FIGURE 8.4 Social network diagrams for different pitch areas.

occasion. Further, the "agility" of the midfield in supporting player K during attacking moves was limited.

It was concluded that SNA can be used to meaningfully analyse passing performance in sports teams. We found that the outputs of such analyses can be used to make informed judgements on the passing performance of the different players involved in terms of the passing relationships between players, positions, and in different areas of the pitch. This includes identifying their overall contribution to the passing performance of the team, determining how well they passed and received the ball in different areas of the pitch, and also identifying the passing relationships between the different players involved. In addition to merely offering a means of objective performance analysis,

TABLE 8.4
Sociometric Status Statistics

Player	Overall	Def to Def	Def to Middle	Def to Attack	Middle to Middle	Middle to Attack	Attack to Attack
A	1.66	1.2	0.1	3	—	—	—
B	5.33	0.9	0.8	—	2.9	0.4	0.5
C	5.58	2.0	1.3	—	2.1	0.2	0.1
D	4.58	2.2	0.9	—	1.8	—	—
E	4.58	1.3	1.3	—	4.9	0.5	0.6
F	9.75	0.9	0.7	—	6.4	1.2	1
G	4.66	0.3	0.4	1	2.7	0.6	1
H	6.58	0.1	0.3	—	5.1	1	1.3
I	7	0.1	0.3	—	4.2	1.1	2.8
J	3.83	0.1	0.5	—	2.6	0.5	1.1
K	2.83	—	0.2	2	0.4	1.1	1
L	1.08	0.1	0.1	—	—	0.1	0.8
C	0.66	—	—	—	—	0.1	0.8

it is our contention that such findings can be used, either in isolation or together with other data, to inform tactics and team selection, and to analyse the key players and passing performance of opposing teams. For example, a description of the opposition in terms of key passing players could be used to inform tactical development and selection. Information on key passing players and locations on the pitch could be used to determine who should be tightly marked and where on the pitch most attention should be given. Additionally, the graphical networks provided are useful in themselves in that they can be quickly interpreted in terms of the passing relationships between players, positions, and passing in different areas of the pitch. On-line (i.e., during game) construction of such networks would provide coaches with a powerful and easily interpretable depiction of passing performance during games that could inform dynamic tactical; positional, and player changes.

Flowchart

(See Flowchart 8.1.)

Recommended Texts

Driskell, J. E., and Mullen, B. (2004). Social network analysis. In *Handbook of Human Factors and ergonomics methods*, eds. N. A. Stanton, A. Hedge, K, Brookhuis, E. Salas, and H. Hendrick, 58.1–58.6. Boca Raton, FL: CRC Press.

Wasserman, S., and Faust, K. (1994). *Social network analysis: Methods and applications.* Cambridge: Cambridge University Press.

TEAM TASK ANALYSIS

Background and Applications

Team Task Analysis (TTA) is a general class of method used to describe and analyse tasks performed by teams with a view to identifying the knowledge, skills, and attitudes (or team competencies) required for effective task performance (Baker, Salas, and Cannon-Bowers, 1998). TTA is typically used to inform the design and development of team training interventions, such as Crew Resource Management (CRM) training, the design of teams and teamwork procedures, and also for team performance evaluation. Although a set methodological procedure for TTA does not exist, Burke (2004) attempted to integrate the existing TTA literature into a set of guidelines for conducting a TTA.

Domain of Application

The TTA method is generic and can be applied in any domain in which teamwork is a feature. Burke (2004) points out that the TTA procedure has not yet been widely adopted by organisations, with the exception of the U.S. military and aviation communities.

Application in Sport

The TTA analysis method can be applied in a sporting context and is likely to be most useful for identifying the team competencies (i.e., knowledge, skills, and attitudes) required for different team sports, the outputs of which can be used to inform the design of coaching and training interventions.

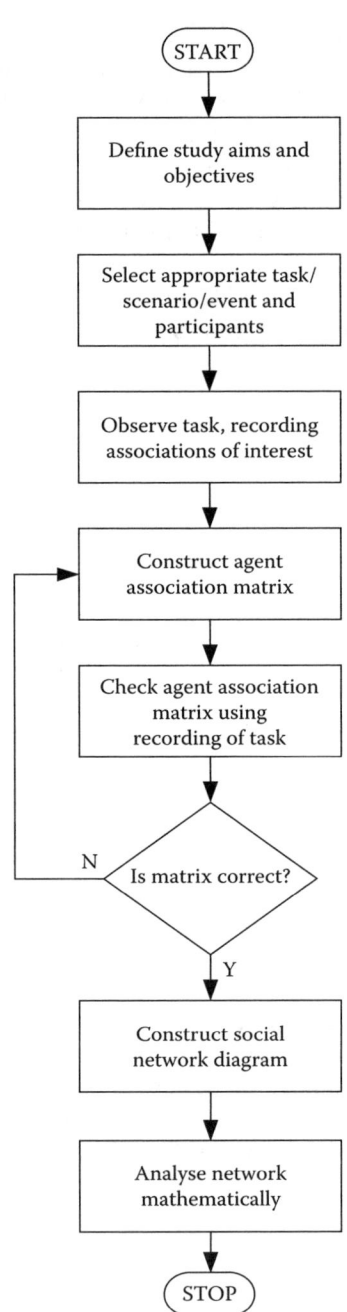

FLOWCHART 8.1 Social network analysis flowchart.

Procedure and Advice (Adapted from Burke, 2004)

Step 1: Conduct Requirements Analysis

First, a requirements analysis should be conducted. This involves clearly defining the task scenario to be analysed, including describing all duties involved and also conditions under which the task is to be performed. Burke (2004) also suggests that when conducting the requirements analysis, the methods of data collection to be used during the TTA should be determined. Typical TTA data collection methods include observational study, interviews, questionnaires, and surveys. The requirements analysis also involves identifying the participants that will be involved in the data collection process, including occupation and number.

Step 2: Define Task(s) or Scenario under Analysis

Next, the tasks involved in the scenario under analysis should be defined and described clearly. Burke (2004) recommends that interviews with SMEs, observational study, and source documents should be used to identify the full set of tasks. Once each individual task step is identified, a task statement should be written (for component tasks), including the following information:

- Task name
- Task goals
- What the individual has to do to perform the task
- How the individual performs the task
- Which devices, controls, and interfaces are involved in the task
- Why the task is required

Step 3: Identify Teamwork Taxonomy

Once all of the tasks involved in the scenario under analysis have been identified and described fully, an appropriate teamwork taxonomy should be selected for use in the analysis (Burke, 2004). A teamwork taxonomy is presented in Table 8.5.

Step 4: Conduct Coordination Demands Analysis

Once an appropriate teamwork taxonomy is selected, a Coordination Demands Analysis should be conducted. The CDA involves classifying the tasks under analysis into teamwork and taskwork activities, and then rating each teamwork task step for the level of coordination required between

TABLE 8.5
Teamwork Taxonomy

Coordination Dimension	Definition
Communication	Includes sending, receiving, and acknowledging information among crew members
Situational awareness (SA)	Refers to identifying the source and nature of problems, maintaining an accurate perception of the aircraft's location relative to the external environment, and detecting situations that require action
Decision making (DM)	Includes identifying possible solutions to problems, evaluating the consequences of each alternative, selecting the best alternative, and gathering information needed prior to arriving at a decision
Mission analysis (MA)	Includes monitoring, allocating, and coordinating the resources of the crew and aircraft; prioritizing tasks; setting goals and developing plans to accomplish the goals; creating contingency plans
Leadership	Refers to directing activities of others, monitoring and assessing the performance of crew members, motivating members, and communicating mission requirements
Adaptability	Refers to the ability to alter one's course of action as necessary, maintain constructive behaviour under pressure, and adapt to internal or external changes
Assertiveness	Refers to the willingness to make decisions, demonstrating initiative, and maintaining one's position until convinced otherwise by facts
Total coordination	Refers to the overall need for interaction and coordination among crew members

Source: Burke, S. C. (2004). Team task analysis. In *Handbook of Human Factors and ergonomics methods*, eds. N. A. Stanton, A. Hedge, K. Brookhuis, E. Salas, and H. Hendrick, 56.1–56.8. Boca Raton, FL, CRC Press.

team members for each behaviour identified in the teamwork taxonomy. For example, if the teamwork task step requires a high level of distributed situation awareness across the team in order for it to be performed successfully, a rating of 3 (high) is given to the situation awareness dimension. Conversely, if the level of communication between team members required is only minimal, then a rating of 1 (low) should be given for the communication dimension.

Step 5: Determine Relevant Taskwork and Teamwork Tasks

The next step in the TTA procedure involves determining the relevance of each of the component tasks involved in the scenario under analysis, including both teamwork and taskwork tasks. Burke (2004) recommends that a Likert scale questionnaire be used for this step and that the following task factors should be rated:

- Importance to train
- Task frequency
- Task difficulty
- Difficulty of learning
- Importance to job

It is recommended that the task indices used should be developed based upon the overall aims and objectives of the TTA.

Step 6: Translation of Tasks into Knowledge, Skills, and Attitudes

Next, the knowledge, skills, and attitudes (KSAs) for each of the relevant task steps should be determined. This is normally done in conjunction with SMEs for the task or system under analysis.

Step 7: Link KSAs to Team Tasks

The final step of a TTA is to link the KSAs identified in step 6 to the team tasks identified. According to Burke (2004), this is most often achieved through the use of surveys completed by SMEs.

ADVANTAGES

1. TTA goes farther than other task analysis methods by specifying the knowledge, skills, and attitudes required to complete each task step.
2. The approach is exhaustive, considering team tasks, the levels of coordination required, and the KSAs required on behalf of team members.
3. The output of TTA is useful for training programme design.
4. The TTA output states which of the component tasks involved are team based and which tasks are performed individually.

DISADVANTAGES

1. TTA is a highly time consuming method to apply.
2. SMEs are normally required throughout the procedure.
3. A high level of training may be required.
4. There is no rigid procedure for the TTA method. As a result, the reliability of the method may be questionable.
5. Great skill is required on behalf of the analyst in order to elicit the required information throughout the TTA procedure.

Related Methods

There are a number of different approaches to team task analysis, such as, Communications Usage Diagram (CUD), and SNA. TTA also utilises a number of Human Factors data collection methods, such as observational study, interviews, questionnaires, and surveys.

Approximate Training and Application Times

The training time associated with the TTA method is likely to be high, particularly for analysts with no experience of teamwork concepts such as team KSAs. Due to the exhaustive nature of the method, the application time is also typically high, including the data collection procedure, the CDA phase, and the KSA identification phases, all of which are time consuming even when conducted in isolation.

Tools Needed

The tools required for conducting a TTA are dependent upon the methodologies employed. TTA can be conducted using pen and paper, and a visual or audio recording device. A PC with a word processing package such as Microsoft Word is normally used to transcribe the data.

Example

For example purposes, an extract of a TTA of rugby union team tasks is presented. The extract focuses on identifying the KSAs required for the scrum in rugby union. A scrum is used in rugby union to restart play when the ball has either been knocked on, gone forward, has not emerged from a ruck or maul, or when there has been an accidental offside (BBC sport, 2008). Nine players from either side are involved in the scrum, including the hooker, loose-head prop, tight-head prop, two second rows, blind-side flanker, open-side flanker, and number 8, who form the scrum, and the scrum half, who feeds the ball into the scrum and/or retrieves the ball from the scrum. A standard scrum formation is presented in Figure 8.5.

For the task definition part of the analysis, an HTA was conducted. The HTA is presented in Figure 8.6. Based on the HTA description, a CDA was undertaken using the teamwork taxonomy presented in Table 8.5. The CDA results are presented in Table 8.6. Following the CDA, the competencies (i.e., KSAs) required for each task were identified. Knowledge-based competencies refer to what team members "think" during teamwork performance. Skill-based competencies refer to the things that team members "do" during teamwork performance and, according to Salas and Cannon-Bowers (2000), are the necessary behavioural sequences and procedures required during task performance. Attitude-based competencies refer to what team members "feel" during teamwork performance and are those affective components that are required during task performance (Salas and Cannon-Bowers, 2000). The KSAs identified for the scrum task are presented in Table 8.7.

Recommended Text

Burke, S. C. (2004). Team task analysis. In *Handbook of Human Factors and ergonomics methods*, eds. N. A. Stanton, A. Hedge, K. Brookhuis, E. Salas, and H. Hendrick, 56.1–56.8). Boca Raton, FL: CRC Press.

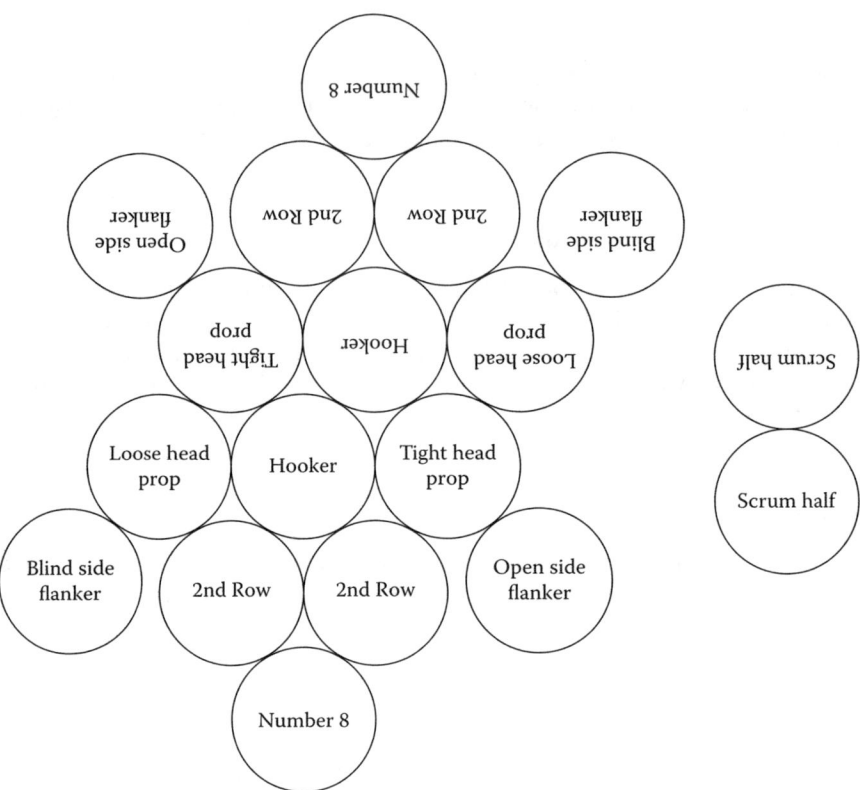

FIGURE 8.5 Standard scrum formation.

COORDINATION DEMANDS ANALYSIS

BACKGROUND AND APPLICATIONS

Coordination Demands Analysis (CDA) is used to rate the level of coordination between team members during collaborative activity. The CDA method focuses on the following teamwork behaviours (adapted from Burke, 2004):

- *Communication* Includes sending, receiving, and acknowledging information among team members.
- *Situation awareness (SA)* Involves identifying the source and nature of problems, maintaining an accurate perception of one's location relative to the external environment, and detecting situations that require action.
- *Decision making (DM)* Involves identifying possible solutions to problems, evaluating the consequences of each alternative, selecting the best alternative, and gathering information needed prior to arriving at a decision.
- *Mission analysis (MA)* Involves monitoring, allocating, and coordinating the resources of the crew, prioritising tasks, setting goals and developing plans to accomplish the goals, and creating contingency plans.
- *Leadership* Involves directing the activities of others, monitoring and assessing the performance of team members, motivating team members, and communicating mission requirements.
- *Adaptability* Refers to the ability to alter one's course of action as necessary, maintain constructive behaviour under pressure, and adapt to internal or external changes.

Teamwork Assessment Methods 261

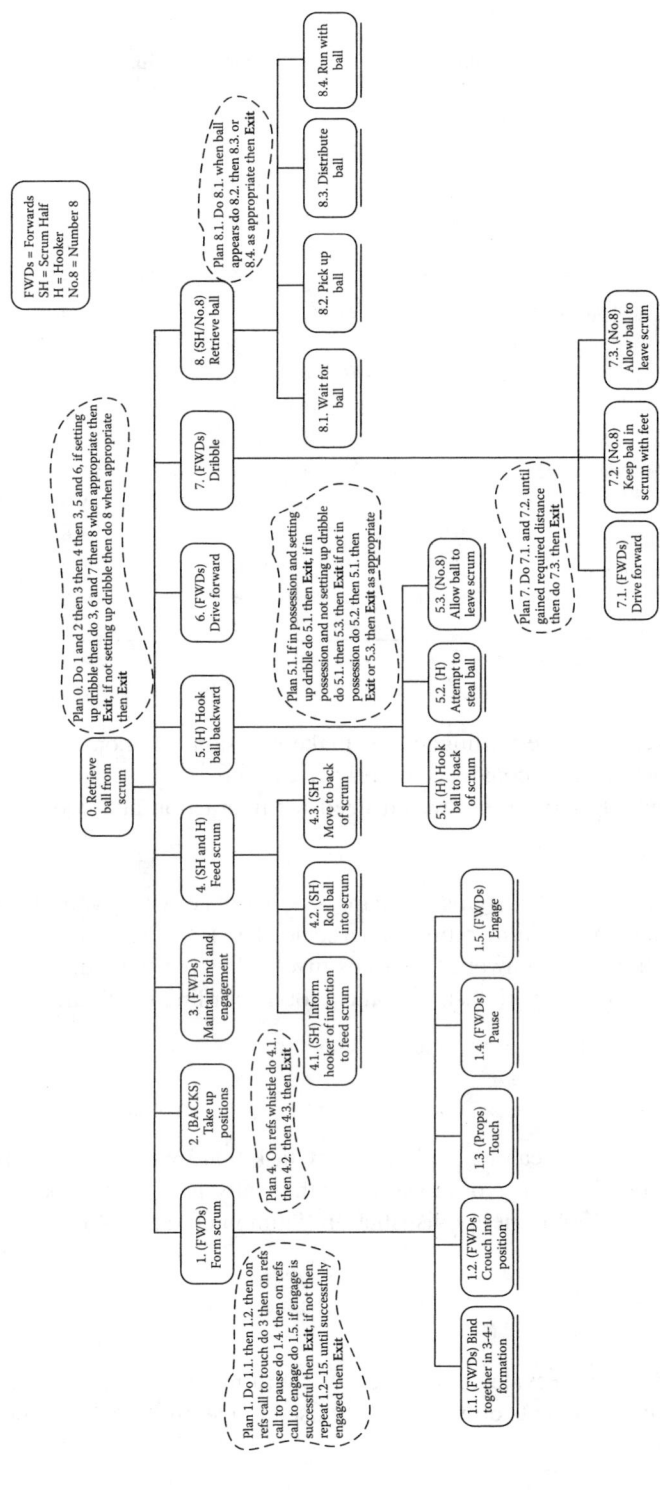

FIGURE 8.6 Retrieve ball from scrum HTA.

TABLE 8.6
CDA Results for Scrum Task

Task	Comms.	SA	DM	MA	Lship.	Adapt.	Assert.	Total Coord.
1.1. (FWDs) Bind together in 3-4-1 formation	2	3	1	1	3	3	3	3
1.2. (FWDs) Crouch into position	3	3	1	1	2	2	2	3
1.3. (Props) Touch	3	3	1	1	1	1	1	1
1.4. (FWDs) Pause	3	3	1	1	2	1	1	3
1.5. (FWDs) Engage	3	3	2	2	3	3	3	3
2.(BACKS) Take up positions	2	3	2	3	2	2	2	3
3. (FWDs) Maintain bind and engagement	3	3	1	1	3	3	3	3
4.1. Roll ball into scrum	3	3	2	2	1	1	1	3
5.1. Hook ball to back of scrum	3	3	3	3	3	2	3	3
5.2. Attempt to steal ball	3	3	3	3	3	2	3	3
5.3. Allow ball to leave scrum	3	3	3	3	3	3	3	3
6. (FWDs) Drive forward	3	3	3	3	3	3	3	3
7.1. (FWDs) Drive forward	3	3	3	3	3	3	3	3
7.2. Keep ball in the scrum with feet	3	3	3	3	3	3	3	3
7.3. (No. 8) Allow ball to leave scrum	3	3	3	3	3	3	3	3
8.3. (SH) Distribute ball	3	3	3	3	2	3	3	3

- *Assertiveness*   Refers to the willingness to make decisions, demonstrate initiative, and maintain one's position until convinced otherwise by facts.
- *Total coordination*   Refers to the overall need for interaction and coordination among team members

The CDA procedure involves identifying the teamwork-based activity involved in the task or scenario under analysis and then providing ratings, on a scale of 1 (low) to 3 (high), for each behaviour described above for each of the teamwork task steps involved. From the individual ratings, a total coordination figure for each teamwork task step and a total coordination figure for the overall task is derived.

DOMAIN OF APPLICATION

The CDA method is generic and can be applied to any task that involves teamwork or collaboration. CDA has been applied in a range of domains for the analysis of teamwork activity, including the military (Stewart et al., 2008), energy distribution (Salmon et al., 2008), and rail maintenance (Walker et al., 2006).

APPLICATION IN SPORT

The CDA method can be used to evaluate the levels of coordination between sports team members during task performance, the outputs of which can be used to identify areas in which further training/coaching is required.

TABLE 8.7
Scrum Task KSA Analysis

Task	Knowledge	Skills	Attitudes
1.1. (FWDs) Bind together in 3-4-1 formation	Knowledge of scrum formation and body positions Knowledge of scrum rules and regulations Knowledge of scrum positions/bindings Compatible mental models of scrum forming process	Ability to take up appropriate body positioning Ability to bind together in scrum formation Communication Situation awareness	Mutual trust Motivation Collective orientation Team cohesion Teamwork
1.2. (FWDs) Crouch into position	Knowledge of scrum formation and body positions Knowledge of scrum rules and regulations Knowledge of scrum positions/bindings Compatible mental models of scrum forming process	Ability to take up appropriate body positioning Ability to crouch together in scrum formation and maintain bind Communication Situation awareness	Mutual trust Motivation Collective orientation Team cohesion Teamwork
1.3. (Props) Touch	Knowledge of scrum formation and body positions Knowledge of scrum rules and regulations Knowledge of scrum positions/bindings Compatible mental models of scrum forming process	Ability to touch opposition scrum while in scrum formation Ability to maintain bind Ability to maintain appropriate body position Communication Situation awareness	Mutual trust Motivation Collective orientation Team cohesion Teamwork Adherence to rules
1.4. (FWDs) Pause	Knowledge of scrum formation and body positions Knowledge of scrum rules and regulations Knowledge of scrum positions/bindings Compatible mental models of scrum forming process	Ability to pause while in scrum formation Ability to maintain bind Ability to maintain appropriate body position Communication Situation awareness	Mutual trust Motivation Collective orientation Team cohesion Teamwork Adherence to rules
1.5. (FWDs) Engage	Knowledge of scrum formation and body positions Knowledge of scrum rules and regulations Knowledge of scrum positions/bindings Compatible mental models of scrum forming process	Ability to pause while in scrum formation Ability to maintain bind Ability to maintain appropriate body position Communication Situation awareness	Mutual trust Motivation Collective orientation Team cohesion Teamwork Adherence to rules
2. (BACKS) Take up positions	Knowledge of backs positions Knowledge of scrum tactics	Ability to take up appropriate position for tactics/situation Communication Situation awareness Decision making	Motivation Collective orientation Team cohesion Teamwork

TABLE 8.7 (CONTINUED)
Scrum Task KSA Analysis

Task	Knowledge	Skills	Attitudes
3. (FWDs) Maintain bind and engagement	Knowledge of scrum formation and body positions Knowledge of scrum rules and regulations Knowledge of scrum positions/bindings Compatible mental models of scrum forming process Knowledge of scrum tactics	Ability to maintain bind, body position, and engagement Communication Situation awareness	Mutual trust Motivation Collective orientation Team cohesion Teamwork Adherence to rules
4.1. Roll ball into scrum	Knowledge of scrum tactics Compatible mental models of play being used	Ability to roll ball into scrum Communication Situation awareness Decision making	Mutual trust Motivation Collective orientation Team cohesion Teamwork Adherence to rules
5.1. Hook ball to back of scrum	Knowledge of scrum tactics Compatible mental models of play being used Knowledge of scrum rules and regulations Knowledge of scrum positions/bindings	Ability to hook ball to back of scrum Decision making Situation awareness Communication	Motivation Teamwork Adherence to rules
5.2. Attempt to steal ball	Knowledge of scrum tactics Compatible mental models of play being used Knowledge of scrum rules and regulations Knowledge of scrum positions/bindings	Ability to steal ball in scrum Decision making Situation awareness Communication	Motivation Teamwork Adherence to rules
5.3. Allow ball to leave scrum	Knowledge of scrum tactics Compatible mental models of play being used Knowledge of scrum rules and regulations Knowledge of scrum positions/bindings	Ability to negotiate ball's exit from scrum Decision making Situation awareness Communication	Mutual trust Motivation Collective orientation Team cohesion Teamwork Adherence to rules
6. (FWDs) Drive forward	Knowledge of scrum tactics Compatible mental models for driving forward in scrum formation Knowledge of scrum rules and regulations Knowledge of scrum positions/bindings	Ability to drive forward while maintaining bind and scrum formation Ability to maintain bind Ability to maintain appropriate body position Communications Situation awareness Decision making	Mutual trust Motivation Collective orientation Team cohesion Teamwork Adherence to rules

TABLE 8.7 (CONTINUED)
Scrum Task KSA Analysis

Task	Knowledge	Skills	Attitudes
7.1. (FWDs) Drive forward	Knowledge of scrum tactics Compatible mental models for driving forward in scrum formation Knowledge of scrum rules and regulations Knowledge of scrum positions/bindings	Ability to drive forward while maintaining bind and scrum formation Ability to maintain bind Ability to maintain appropriate body position Communications Situation awareness Decision making	Mutual trust Motivation Collective orientation Team cohesion Teamwork Adherence to rules
7.2. Keep ball in the scrum with feet	Knowledge of scrum tactics Knowledge of scrum rules and regulations	Ability to locate and bring ball under control in scrum Ability to dribble ball in moving scrum Communications Situation awareness Decision making	Mutual trust Motivation Collective orientation Team cohesion Teamwork Adherence to rules
7.3. (No. 8) Allow ball to leave scrum	Knowledge of scrum tactics Compatible mental models of play being used Knowledge of scrum rules and regulations Knowledge of scrum positions/bindings	Ability to negotiate ball's exit from scrum Decision making Situation awareness Communication	Mutual trust Motivation Collective orientation Team cohesion Teamwork Adherence to rules
8.3. (SH) Distribute ball	Knowledge of tactics Knowledge of own and opposition backs positions Knowledge of position on pitch Compatible mental models of scrum retrieval tactics	Ability to pick up ball Ability to select appropriate distribution Ability to kick ball Ability to pass ball Decision making Situation awareness Communication	Mutual trust Motivation Collective orientation Team cohesion Teamwork Adherence to rules

PROCEDURE AND ADVICE

Step 1: Define Analysis Aims

First, the aims of the analysis should be clearly defined. For example, the aims of the analysis may be to evaluate coordination levels between team members during a particular game situation or during a range of game situations. The analysis aims should be clearly defined so that appropriate scenarios are used and relevant data are collected.

Step 2: Define Task(s) under Analysis

Next, the task(s) under analysis should be clearly defined. This is dependent upon analysis aims. It is recommended that if team coordination in a particular type of sport is under investigation, then a set of scenarios that are representative of all aspects of team performance in the sport in question should be used.

Step 3: Select Appropriate Teamwork Taxonomy

Once the task(s) under analysis is defined, an appropriate teamwork taxonomy should be selected. Again, this is dependent upon the aims of the analysis. However, it is recommended that the taxonomy used covers all aspects of teamwork, such as the generic CDA teamwork taxonomy described above.

Step 4: Data Collection

The next step involves collecting the data that will be used to inform the CDA. Typically, observational study of the task or scenario under analysis is used as the primary data source for a CDA, although other procedures can also be used, such as task simulations, interviews, and questionnaires. It is recommended that specific data regarding the task under analysis should be collected during this process, including information regarding each task step, each team member's roles, and all communications made. Particular attention is given to the teamwork activity involved in the task under analysis. Further, it is recommended that video and audio recording equipment be used to record any observations or interviews conducted during this process.

Step 5: Conduct an HTA for the Task under Analysis

Once sufficient data regarding the task under analysis have been collected, an HTA should be conducted. The purpose of the HTA is to decompose the task into sub-goals so that coordination levels during each component task step can be analysed.

Step 6: Taskwork/Teamwork Classification

Only those task steps that involve teamwork are rated for the level of coordination between team members. The next step of the CDA procedure therefore involves identifying the teamwork and taskwork task steps involved in the scenario under analysis. Teamwork refers to those instances where individuals interact or coordinate behaviour in order to achieve tasks that are important to the team's goals (i.e., behavioural, attitudinal, and cognitive responses coordinated with fellow team members), while taskwork (i.e., task-oriented skills) describes those instances where team members are performing individual tasks separate from their team counterparts, i.e., those tasks that do not require interdependent interaction with other team members (Salas, Cooke, and Rosen, 2008).

Step 7: Construct CDA Rating Sheet

Next, a CDA rating sheet should be created. The CDA rating sheet includes a column containing each bottom level teamwork operation from the HTA and the teamwork behaviours running across the top of the table.

Step 8: Rate Coordination Levels

The next step involves rating each teamwork behaviour based on the task performance observed or SME judgement. It is important to involve SMEs in this process; however, if they are not available at this stage they can validate ratings at a later stage. The rating procedure involves taking each task step and rating the extent to which the teamwork behaviour was involved or is required during performance of the task under analysis. Normally a rating scale of 1 (low) to 3 (high) is used.

Step 9: Validate Ratings

It is important to validate the ratings made, either through the use of SMEs who observed the task, or through those team members who were involved in the task.

Step 10: Calculate Summary Statistics

Once all of the teamwork task steps have been rated, the final step involves calculating appropriate summary statistics. Typically, the following are calculated:

- Mean overall coordination value for the entire task
- Mean coordination values for each teamwork dimension
- Mean overall coordination value for each task step

Advantages

1. CDA provides a useful output that offers insight into the use of teamwork behaviours and also a rating of coordination between actors in a particular network or team.
2. The output allows judgements to be made on those team tasks in which coordination is a problem.
3. CDA permits the identification of tasks that require high levels of coordination, which can be used to inform procedure/system/training programme redesign.
4. It can potentially be used to analyse the coordination levels exhibited by sports teams and to identify areas where further training/coaching is required.
5. Coordination levels can be compared across scenarios, different teams, and also different domains.
6. The teamwork taxonomy presented by Burke (2004) covers all aspects of team performance and coordination and is generic, allowing the method to be applied in any domain.
7. CDA is simple to apply and incurs minimal cost.
8. It requires only minimal training.

Disadvantages

1. The method can be highly time consuming, including observation of the task under analysis, the development of an HTA, and the ratings procedure.
2. The rating procedure is time consuming and laborious.
3. High levels of access to SMEs are required for the method to provide worthwhile results.
4. Intra-analyst and inter-analyst reliability may be poor.

Related Methods

CDA uses various data collection procedures, such as observational study, interviews, and questionnaires. HTA is also used to describe the task under analysis, the output of which provides the task steps to be rated. CDA is normally conducted as part of a wider TTA effort (Burke, 2004); for example, CDA has been used as part of the EAST framework (Stanton et al., 2005) for evaluating team performance (e.g., Salmon et al., 2008; Stanton et al., 2006; Walker et al., 2006).

Approximate Training and Application Times

Provided analysts are already well versed in HTA, the training time for the CDA method is minimal, requiring only that the SMEs used understand each of the behaviours specified in the teamwork taxonomy and also the rating procedure. The application time is high, involving observation of the task under analysis, the construction of an HTA, decomposition of the task into teamwork and taskwork behaviours, and the lengthy ratings procedure. The duration of the ratings procedure is dependent upon the task under analysis; however, from our own experiences with the method, a period of between 2 and 4 hours is normally required for the ratings procedure. In worst case scenarios, the overall process of observing the task, constructing the HTA, identifying teamwork task steps, rating each task step, and then validating ratings can take up to 2 weeks.

Reliability and Validity

Both the intra-analyst and inter-analyst reliability of the method may be questionable, and this may be dependent upon the type of rating scale used, e.g., it is estimated that reliability may be low when using a scale of 1–10, while it may be improved using a scale of 1–3 (e.g., low, medium, and high).

Tools Needed

During the data collection phase, video and audio recording equipment are required in order to make a recording of the task or scenario under analysis. Once the data collection phase is complete, the CDA method can be conducted using pen and paper; however, Microsoft Excel is useful as it can be programmed to auto-calculate CDA summary statistics.

Example

CDA was used to rate the levels of coordination between team members during the scrum in rugby union. The CDA output is presented in Table 8.6 in the TTA example section.

Flowchart

(See Flowchart 8.2.)

Recommended Text

Burke, S. C. (2004). Team task analysis. In *Handbook of Human Factors and ergonomics methods*, eds. N. A. Stanton, A. Hedge, K. Brookhuis, E. Salas, and H. Hendrick, 56.1–56.8). Boca Raton, FL: CRC Press.

EVENT ANALYSIS OF SYSTEMIC TEAMWORK

Background and Applications

The Event Analysis of Systemic Teamwork framework (EAST; Stanton et al., 2005) provides an integrated suite of methods for analysing teamwork activities. To date the approach has been used by the authors to analyse teamwork in a number of different domains, including land warfare (Stanton et al., in press), multinational warfare (Salmon et al., 2006), airborne early warning and control (Stewart et al., 2008), naval warfare (Stanton et al., 2006), air traffic control (Walker et al., in press), railway maintenance (Walker et al., 2006), energy distribution (Salmon et al., 2008), and emergency service (Houghton et al., 2006).

Underpinning the approach is the notion that distributed teamwork can be meaningfully described via a "network of networks" approach; to this end EAST is used to analyse teamwork from three different but interlinked perspectives—the task, social, and knowledge networks—that underlie teamwork activity. Task networks represent a summary of the goals and subsequent tasks being performed within a system. Social networks analyse the organisation of the team and the communications taking place between the actors working in the team, and knowledge networks describe the information and knowledge (distributed situation awareness) that the actors use and share in order to perform the teamwork activities in question. This so-called "network of networks" approach to understanding collaborative endeavour is represented in Figure 8.7.

EAST uses a framework of Human Factors methods to produce these networks; HTA (Annett et al., 1971) is typically used to construct task networks, SNA (Driskell and Mullen, 2004) is used

Teamwork Assessment Methods

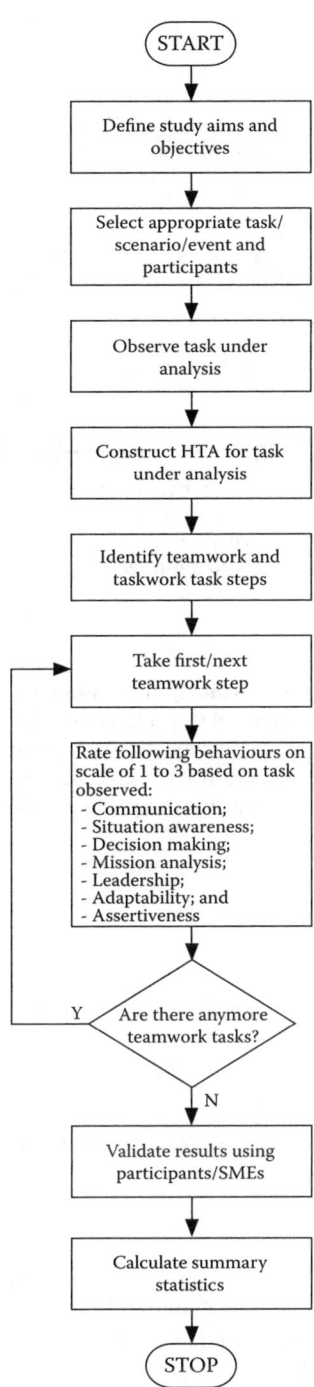

FLOWCHART 8.2 Coordination demands analysis flowchart.

to construct and analyse the social networks involved, and propositional networks (Salmon et al., 2009) are used to construct and analyse knowledge networks. In addition, OSDs, CDA, the CDM, and Communications Usage Diagrams (Watts and Monk, 2000) are used to evaluate team cognition and decision making, coordination, and the technology used for communications.

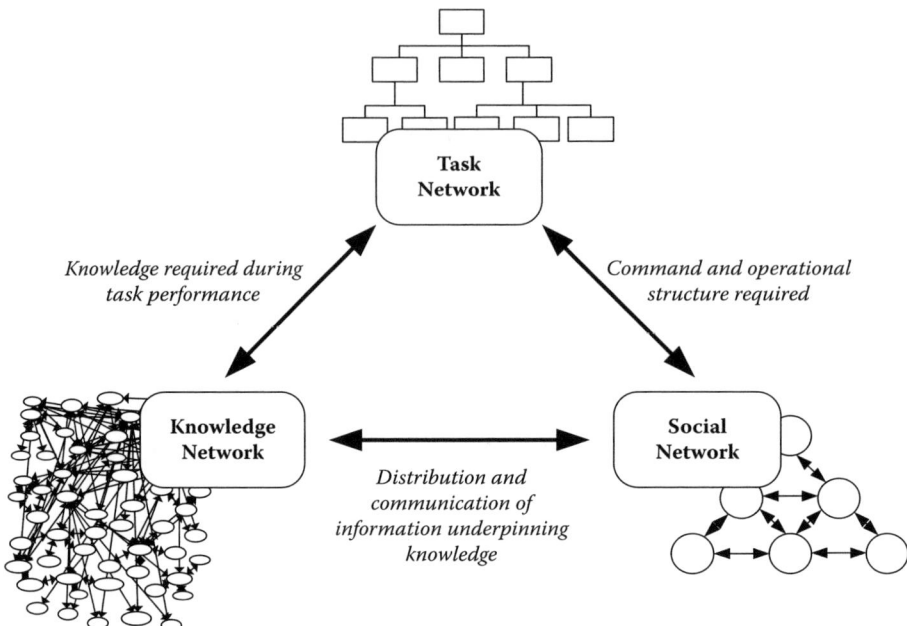

FIGURE 8.7 Network of networks approach to analysing distributed teamwork; figure shows example representations of each network, including hierarchical task analysis (task network), social network analysis (social network), and propositional network (knowledge network) representations. *Source*: Adapted from Houghton, R. J., Baber, C., Cowton, M., Stanton, M. A., and Walker, G. H. (2008). WESTT (Workload, Error, Situational Awareness, Time and Teamwork): An analytical prototyping system for command and control. *Cognition Technology and Work* 10(3):199–207.

Domain of Application

EAST is a generic approach that was developed specifically for the analysis of teamwork.

Application in Sport

EAST can be applied in a sporting context to evaluate sports team performance.

Procedure and Advice

Step 1: Define Analysis Aims

First, the aims of the analysis should be clearly defined. The analysis aims should be clearly defined so that appropriate scenarios are used and relevant data are collected. In addition, not all components of the EAST framework may be required, so it is important to clearly define the aims of the analysis to ensure that the appropriate EAST methods are applied.

Step 2: Define Task(s) under Analysis

Next, the task(s) or scenario(s) under analysis should be clearly defined. This is dependent upon the aims of the analysis and may include a range of tasks or one task in particular. It is normally standard practice to develop an HTA for the task under analysis if sufficient data and SME access are available. This is useful later on in the analysis and is also enlightening, allowing analysts to gain an understanding of the task before the observation and analysis begins.

Step 3: Conduct Observational Study of the Task or Scenario under Analysis

The observation step is the most important part of the EAST procedure. Typically, a number of analyst(s) are used in scenario observation. All activity involved in the scenario under analysis should be recorded along an incident timeline, including a description of the activity undertaken, the agents involved, any communications made between agents, and the technology involved. Additional notes should be made where required, including the purpose of the activity observed; any tools, documents, or instructions used to support activity; the outcomes of activities; any errors made; and also any information that the agent involved feels is relevant. It is also useful to video record the task and record verbal transcripts of all communications, if possible.

Step 4: Conduct CDM Interviews

Once the task under analysis is complete, each "key" actor (e.g., scenario commander, actors performing critical tasks) involved should be subjected to a CDM interview. This involves dividing the scenario into key incident phases and then interviewing the actor involved in each key incident phase using a set of predefined CDM probes (e.g., O'Hare et al., 2000; see Chapter 4 for CDM method description).

Step 5: Transcribe Data

Once all of the data are collected, they should be transcribed in order to make them compatible with the EAST analysis phase. The transcript should describe the scenario over a timeline, including descriptions of activity, the actors involved, any communications made, and the technology used. In order to ensure the validity of the data, the scenario transcript should be reviewed by one of the SMEs involved.

Step 6: Reiterate HTA

The data transcription process allows the analyst(s) to gain a deeper and more accurate understanding of the scenario under analysis. It also allows any discrepancies between the initial HTA scenario description and the actual activity observed to be resolved. Typically, teamwork activity does not run entirely according to protocol, and certain tasks may have been performed during the scenario that were not described in the initial HTA description. The analyst should compare the scenario transcript to the initial HTA, and add any changes as required.

Step 7: Conduct Coordination Demands Analysis

The CDA method involves extracting teamwork tasks from the HTA and rating them against the associated CDA taxonomy. Each teamwork task is rated against each CDA behaviour on a scale of 1 (low) to 3 (high). Total coordination for each teamwork step can be derived by calculating the mean across the CDA behaviours. The mean total coordination figure for the scenario under analysis should also be calculated.

Step 8: Construct Communications Usage Diagram

A Communications Usage Diagram (CUD; Watts and Monk 2000) is used to describe the communications between teams of actors dispersed across different geographical locations. A CUD output describes how and why communications between actors occur, which technology is involved in the communication, and the advantages and disadvantages associated with the technology used. A CUD analysis is typically based upon observational data of the task or scenario under analysis, although talk-through analysis and interview data can also be used (Watts and Monk, 2000).

Step 9: Conduct Social Network Analysis

SNA is used to analyse the relationships between the actors involved in the scenario under analysis. It is normally useful to conduct a series of SNAs representing different phases of the task under analysis (using the task phases defined during the CDM part of the analysis). It is recommended that the Agna SNA software package be used for the SNA phase of the EAST methodology.

Step 10: Construct Operation Sequence Diagram

The OSD represents the activity observed during the scenario under analysis. The analyst should construct the OSD using the scenario transcript and the associated HTA as inputs. Once the initial OSD is completed, the analyst should then add the results of the CDA to each teamwork task step.

Step 11: Construct Propositional Networks

The final step of the EAST analysis involves constructing propositional networks for each scenario phase identified during the CDM interviews. The WESTT software package should be used to construct the propositional networks. Following construction, information usage should be defined for each actor involved via shading of the information elements within the propositional networks.

Step 12: Validate Analysis Outputs

Once the EAST analysis is complete, it is pertinent to validate the outputs using appropriate SMEs and recordings of the scenario under analysis. Any problems identified should be corrected at this point.

Advantages

1. The analysis produced is extremely comprehensive, and activities are analysed from various perspectives.
2. The framework approach allows methods to be chosen based on analysis requirements.
3. EAST has been applied in a wide range of different domains.
4. The network of networks analysis produced provides compelling views of teamwork activities.
5. The approach is generic and can be used to evaluate teamwork in any domain.
6. A number of Human Factors concepts are evaluated, including distributed situation awareness, cognition, decision making, teamwork, and communications.
7. It has great potential to be applied in a sporting context to analyse team performance.
8. It uses structured and valid Human Factors methods and has a sound theoretical underpinning.

Disadvantages

1. When undertaken in full, the EAST framework is a highly time-consuming approach.
2. The use of various methods ensures that the framework incurs a high training time.
3. In order to conduct an EAST analysis properly, a high level of access to the domain, task, and SMEs is required.
4. Some parts of the analysis can become overly time consuming and laborious to complete.
5. Some of the outputs can be large, unwieldy, and difficult to present in reports, papers, and presentations.

Related Methods

The EAST framework comprises a range of different Human Factors methods, including observational study, HTA (Annett et al., 1971), CDA (Burke, 2004), SNA (Driskell and Mullen, 2004), the CDM (Klein, Calderwood, and McGregor, 1989), OSDs (Stanton et al., 2005), CUD (Watts and Monk, 2000), and propositional networks (Salmon et al., 2009).

Approximate Training and Application Times

Due to the number of different methods involved, the training time associated with the EAST framework is high. Similarly, application time is typically high, although this is dependent upon the task under analysis and the component methods used.

Reliability and Validity

Due to the number of different methods involved, the reliability and validity of the EAST method is difficult to assess. Some of the methods have high reliability and validity, such as SNA and CUD, whereas others may suffer from low levels of reliability and validity, such as the CDM and propositional networks approaches.

Tools Needed

Normally, video and audio recording devices are used to record the activities under analysis. The WESTT software package (Houghton et al., 2008) supports the development of most EAST outputs. Various HTA software packages exist for supporting analysts in the construction of HTAs for the task under analysis. A drawing software package such as Microsoft Visio is also typically used for the representational methods such as OSDs and CUD.

Example

An example EAST style analysis is presented in the final chapter.

Flowchart

(See Flowchart 8.3.)

Recommended Texts

Stanton, N. A., Salmon, P. M., Walker, G., Baber, C., and Jenkins, D. P. (2005). *Human Factors methods: A practical guide for engineering and design*. Aldershot, UK: Ashgate.

Walker. G. H., Gibson, H., Stanton, N. A., Baber, C., Salmon, P. M., and Green, D. (2006). Event analysis of systemic teamwork (EAST): A novel integration of ergonomics methods to analyse C4i activity. *Ergonomics* 49(12-13):1345–69.

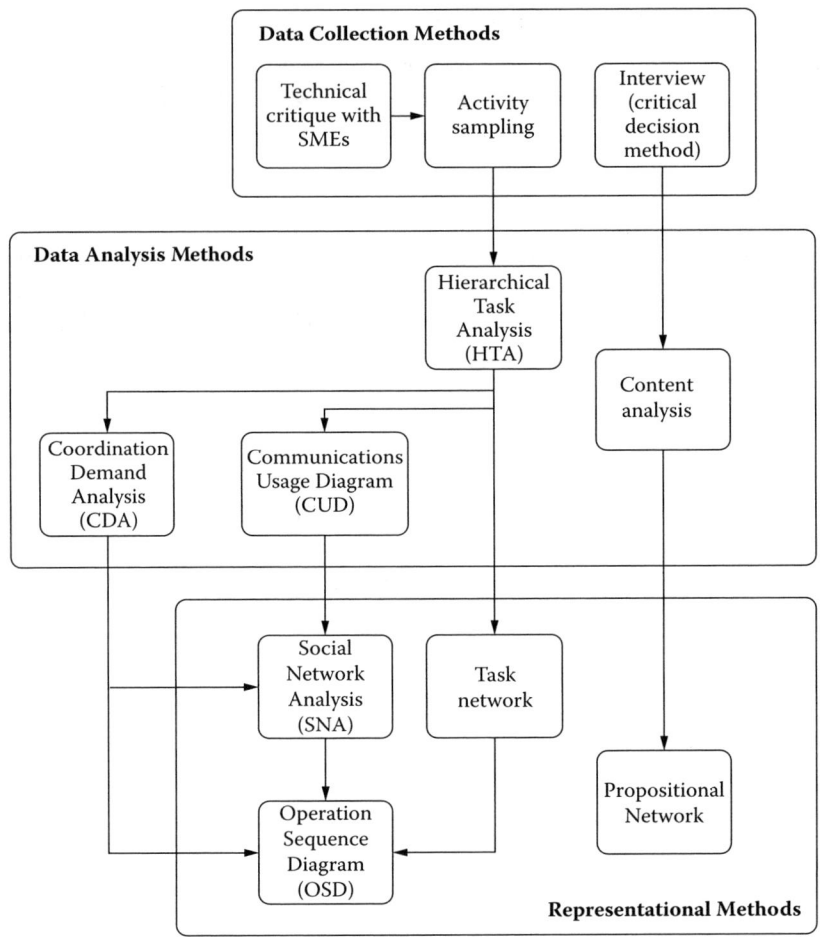

FLOWCHART 8.3   Event analysis of systemic teamwork flowchart.

9 Interface Evaluation

INTRODUCTION

Most of the methods described up to this point have focussed on the human elements of performance in sociotechnical systems (e.g., physical and cognitive tasks, situation awareness, human error, teamwork); however, methods for evaluating and designing products and interfaces also form a significant portion of the many Human Factors methods available. This chapter focuses on those methods that Human Factors researchers use to evaluate and design products and interfaces. Labelled here as *interface evaluation* methods, but also referred to elsewhere as usability evaluation methods, these methods represent a general class of methods that can be used to assess the usability or interface of a product or device. Normally based on models of human performance, interface evaluation methods allow various aspects of a device's interface to be assessed, including interface layout, usability (ease of use, effectiveness, efficiency, and attitude), colour coding, user satisfaction, and error. Interface evaluation methods are essentially designed to improve product design by understanding or predicting user interaction with those devices (Stanton and Young, 1999).

INTERFACE EVALUATION METHODS

A number of different types of interface evaluation method are available, including usability assessment, error analysis, interface layout analysis, and general interface assessment methods. This book focuses on the following five popular and simple forms of interface evaluation method: checklists, heuristic evaluation, link analysis, layout analysis, and interface surveys. Checklists, such as Ravden and Johnson's (1989) usability checklist, are used to assess the usability of a particular device or interface. When using checklists, the analyst performs a range of tasks with the device and then checks the device or interface against a predefined set of criteria in order to evaluate its usability. Heuristics evaluation is one of the simplest interface evaluation approaches available, and involves simply obtaining subjective opinions from analysts based upon their interactions with a particular device or product. Link analysis is used to evaluate an interface in terms of the nature, frequency, and importance of links between elements of the interface in question. The interface is then redesigned based upon these links, with the most often linked elements of the interface relocated to increase their proximity to one another. Layout analysis is also used to evaluate and redesign the layout of device interfaces, and involves arranging interface components into functional groupings, and then reorganising these groups based on importance of use, sequence of use, and frequency of use. The layout analysis output offers a redesign based upon the user's model of the task. Interface surveys (Kirwan and Ainsworth, 1992) are a group of surveys that are used to assess the interface under analysis in terms of controls and displays used, and their layout, labelling, and ease of use. Each survey is completed after a user trial and conclusions regarding the usability and design of the interface are made. A summary of the interface evaluation methods described in this chapter is presented in Table 9.1.

To maintain consistency, each example analysis presented in this chapter focuses on the same sports product, the Garmin 305 Forerunner device. The 305 Forerunner device is a GPS-based training and performance aid for runners that consists of a wrist unit and heart rate monitor strap. The wrist unit uses GPS technology to present time, pace, distance ran, heart rate, and positional information to runners. Other key features include an alert system, whereby the device can be

TABLE 9.1
Interface Evaluation Methods Summary Table

Name	Domain	Application in Sport	Training Time	App. Time	Input Methods	Tools Needed	Main Advantages	Main Disadvantages	Outputs
Checklists	Generic	Evaluation of sports technology, devices, performance aids, etc. Evaluation of sports technology, device, and performance aid design concepts	Low	Low	User trial Walkthrough Hierarchical task analysis	Pen and paper Device under analysis	1. Quick and simple to apply, require minimal training and incur only a minimal cost 2. Offer an immediately useful output 3. Checklists are based upon established knowledge about human performance (Stanton & Young, 1999)	1. Ignore context 2. Data are totally subjective—what may represent poor design for one rater may represent a high level usability for another 3. Poor reliability and consistency	Subjective ratings of device usability
Heuristic analysis	Generic	Evaluation of sports technology, devices, performance aids, etc. Evaluation of sports technology, device, and performance aid design concepts	Low	Low	User trial Walkthrough Hierarchical task analysis	Pen and paper Device under analysis	1. Quick and simple to apply, requires minimal training, and incurs only a minimal cost 2. Offers an immediately useful output and can be used throughout design process 3. Can be used to look at a range of factors, including usability, errors, interface design, and workload	1. Suffers from poor reliability, validity, and comprehensiveness 2. Devices designed on end-user recommendations only are typically flawed 3. Entirely subjective. What is usable to one analyst may be completely unusable to another	Subjective opinions of device/interface/design concept usability
Link analysis	Generic	Performance evaluation Coaching/training design Evaluation of sports technology, devices, performance aids, etc. Evaluation of sports technology, device, and performance aid design concepts	Low	High	Observational study Hierarchical task analysis	Pen and paper Device under analysis	1. Link analysis diagram provides a powerful means of representing the associations present in a particular system 2. Simple and quick to apply and requires only minimal training 3. The output is immediately useful and has a range of potential uses in a sporting context, including as an interface/device evaluation approach, a coaching aid, and/or a performance evaluation tool	1. The data collection phase can be time consuming 2. For large, complex tasks involving many actors and devices, the analysis may be difficult to conduct and the outputs may become large and unwieldy 3. Only considers the basic physical relationships	Graphical representation of the links present in a particular system

Method	Type	Domain	Training time	Application time	Related methods	Tools needed	Advantages	Disadvantages	Outputs
Layout analysis	Generic	Evaluation of sports technology, device, and performance aid design concepts	Low	Low	User trial Walkthrough Hierarchical task analysis	Pen and paper Device under analysis	1. Offers a quick, easy to use, and low cost approach to interface evaluation and design 2. The output is immediately useful and is easily communicated to designers 3. Low resource usage and requires very little training	1. Poor reliability and validity since the redesign of the interface components is still based to a high degree on the analyst's subjective judgement 2. If an initial HTA is required, application time can rise significantly 3. Can be difficult and time consuming for large, complex systems, and the outputs may be complex and unwieldy	Redesign of product/ device interface based on user's model of the task (importance, sequence, and frequency of use of interface elements)
Interface surveys	Generic	Evaluation of sports technology, device, and performance aid design concepts	Low	High	User trial Walkthrough Hierarchical task analysis	Pen and paper Device under analysis	1. Potentially exhaustive 2. The surveys are easy to apply and require very little training 3. The output offers a useful analysis of the interface under analysis, highlighting instances of bad design and problems arising from the man-machine interaction	1. Can be highly time consuming 2. An operational system is required for most of the methods 3. Reliability is questionable	Analysis of product/ device interface or interface design concept

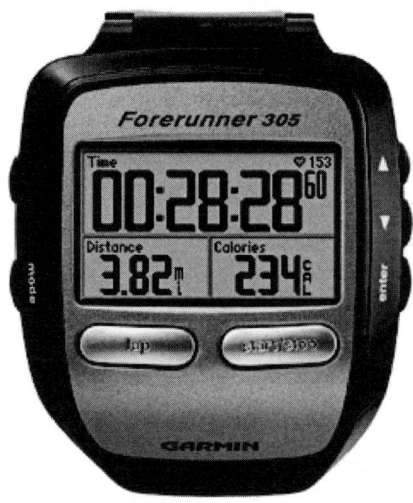

FIGURE 9.1 Garmin 305 Forerunner wrist unit. *Source*: www8.garmin.com, with permission.

programmed to alert the runner if he or she exceeds or fails to exceed a preset pace; a virtual partner mode, whereby data can be entered into the device enabling runners to race a virtual partner; and a navigation and waypoint dropping feature. The Garmin 305 Forerunner wrist unit is presented in Figure 9.1.

CHECKLISTS

BACKGROUND AND APPLICATIONS

Checklists offer a quick, simple, and low cost approach to interface evaluation and usability assessment. Checklist approaches involve analysts or participants evaluating an interface or device against a checklist containing a predefined set of assessment criteria. A number of Human Factors checklists exist, such as Ravden and Johnson's (1989) Human Computer Interaction (HCI) usability checklist, and Woodson, Tillman, and Tillman's (1992) checklists. Performing a checklist analysis is a matter of simply inspecting the device against each point on the chosen checklist and providing an appropriate rating (i.e., good or bad) and any associated comments. Due to their simplicity, checklist-style evaluations can be conducted throughout the design life cycle, and their generic nature allows them to be applied in any domain to any device.

For example purposes we will use Ravden and Johnson's (1989) usability checklist to demonstrate how checklist methods are applied. Their checklist comprises 10 sections of questions designed to assess the overall usability of a particular device or system. The 10 sections are briefly outlined below:

1. *Visual clarity* Refers to the clarity with which the system displays information. According to Ravden and Johnson (1989), information displayed should be clear, well organised, unambiguous, and easy to read.
2. *Consistency* Refers to the consistency of the interface in terms of how it looks, the ways in which it presents information, and the ways in which users perform tasks. According to Ravden and Johnson (1989), the way the system looks and works should be consistent at all times.
3. *Compatibility* Refers to the system's compatibility with other related systems. According to Ravden and Johnson (1989), the way the system looks and works should be compatible with user conventions and expectations.

Interface Evaluation

4. *Informative feedback*   Refers to the level, clarity, and appropriateness of the feedback provided by the system. According to Ravden and Johnson (1989), users should be given clear, informative feedback on where they are in the system, what actions they have taken, whether these actions have been successful, and what actions should be taken next.
5. *Explicitness*   Refers to the clarity with which the system transmits its functionality, structure, and capability. According to Ravden and Johnson (1989), the way the system works and is structured should be clear to the user.
6. *Appropriate functionality*   Refers to the level of appropriateness of the system's functionality in relation to the activities that it is used for. According to Ravden and Johnson (1989), the system should meet the needs and requirements of users when carrying out tasks.
7. *Flexibility and control*   Refers to the flexibility of the system and the level of control that the user has over the system. According to Ravden and Johnson (1989), the interface should be sufficiently flexible in structure, in the way information is presented, and in terms of what the user can do, to suit the needs and requirements of all users and to allow them to feel in control of the system.
8. *Error prevention and correction*   Refers to the extent to which the system prevents user errors from either being made or impacting task performance. According to Ravden and Johnson (1989), systems should minimise the possibility of user error and possess facilities for detecting and handling the errors that do occur. Further, they suggest that users should be able to check inputs and correct errors or potential error situations before inputs are processed.
9. *User guide and support*   Refers to the level of guidance and support that the system provides to its end users. According to Ravden and Johnson (1989), systems should be informative, easy to use, and provide relevant guidance and support, both on the system itself (e.g., help function) and in document form (e.g., user guide), to help the user understand and use the system.
10. *System usability*   Refers to the overall usability of the system in question.

In addition to the sections described above, an 11th section allows participants to record any additional comments that they have regarding the device or system under analysis.

Domain of Application

The concept of checklists first emerged in the context of HCI; however, most checklists are generic and can be applied in any domain.

Application in Sport

Checklist-style analyses can be used within a sporting context to evaluate sports technology, devices, and performance aids. Checklists can be used throughout the design process to evaluate design concepts or for the evaluation of operational designs.

Procedure and Advice

Step 1: Define Analysis Aims

First, the aims of the analysis should be clearly defined. For example, the aims of the analysis may be to evaluate the usability of a particular device or to assess the error potential associated with a particular device. The analysis aims should be clearly defined so that an appropriate checklist is selected and relevant data are collected.

Step 2: Select Appropriate Checklist

Next, an appropriate checklist should be selected based on the aims of the analysis defined during step 1. Various checklists are available (e.g., Ravden and Johnson, 1989; Woodson, Tillman, and Tillman, 1992). The checklist used may be an existing one or an adaptation of an existing one, or the analyst may choose to develop a completely new checklist. One of the endearing features of checklists is that they are flexible and can be adapted or modified according to the demands of the analysis; Stanton and Young (1999), for example, used a portion of Ravden and Johnson's (1989) HCI checklist for the usability evaluation of in-car entertainment systems.

Step 3: Identify Set of Representative Tasks for the Device in Question

Once the aims of the analysis are clearly defined, a set of representative tasks for the device or product under analysis should be identified. It is recommended that checklist analyses be based upon analysts performing an exhaustive set of tasks with the device in question. The tasks defined should then be placed in a task list. It is normally useful to conduct an HTA for this purpose, based on the operation of the device in question. The HTA output then acts as a task list for the checklist analysis.

Step 4: Brief Participants

Before the checklist analysis can proceed, all participants should be briefed regarding the purpose of the study, the area of focus (e.g., usability, error, interface evaluation), and the checklist method. It may also be useful at this stage to take the participants through an example checklist application, so that they understand how the method works and what is required of them as participants.

Step 5: Product/Device Familiarisation Phase

To ensure that the analysis is as comprehensive as possible, it is recommended that the analysts involved spend some time familiarising themselves with the device in question. This might also involve consultation with the associated documentation (e.g., instruction/user manual), watching a demonstration of the device being operated, or being taken through a walkthrough of device operation. It is imperative that the analysts become proficient with the device or product in question before undertaking the checklist analysis, as analysts with limited experience in the operation of the device may make inappropriate observations.

Step 6: Check Item on Checklist against Product

The analyst should take the first point on the checklist and check it against the product or system under analysis. For example, the first item in Ravden and Johnson's checklist asks, "Is each screen clearly identified with an informative title or description?" The analysts should then proceed to check each screen and its associated title and description. The options given are "Always," "Most of the time," "Some of the time," and "Never." Using subjective judgement, the analyst should rate the device under analysis according to the checklist item. Space is also provided for the analysts to comment on particularly good or bad aspects of the device under analysis. Step 6 should be repeated until each item on the checklist has been dealt with.

Step 7: Analyse Data

If multiple participants are being used, then the data should be analysed accordingly. Normally, mean values for each checklist category (visual clarity, consistency, etc.) are calculated.

ADVANTAGES

1. Checklists are quick and simple to apply, and incur only a minimal cost.
2. Checklists offer an immediately useful output.

Interface Evaluation

3. Checklists are based upon established knowledge about human performance (Stanton and Young, 1999).
4. Checklist methods require very little training and preparation.
5. Resource usage is very low.
6. Checklists are flexible and can easily be modified in order to use them for other devices/systems. Stanton and Young (1999) suggest that the Ravden and Johnson checklist (1989), originally designed for HCI, is easily adapted to cater for the usability of other devices.
7. A number of different checklists are available.
8. Most checklists are generic and can be applied in any domain for the analysis of products, devices, or interfaces.

Disadvantages

1. Checklists are an unsophisticated method of device evaluation.
2. Checklists ignore context.
3. The data gathered are totally subjective and what may represent poor design for one rater may represent good design for another.
4. Checklists have poor reliability and consistency.
5. Considerable work is required to develop a new checklist.
6. There is no guarantee that participants' true responses are given.

Related Methods

Various Human Factors checklists exist, such as Woodson, Tillman, and Tillman's (1992) human engineering checklists, and Ravden and Johnson's (1989) HCI checklist. Task analysis methods (e.g., HTA) are often used to create task lists for participants to follow during the analysis. Other forms of data collection are also often used, including observation, walkthrough, and demonstrations.

Approximate Training and Application Times

Checklists require only minimal training and the application time is typically low, although this is dependent upon the task(s) involved in the analysis. In an analysis of 12 ergonomics methods, Stanton and Young (1999) report that checklists are one of the quickest methods to train, practice, and apply.

Reliability and Validity

While Stanton and Young (1999) report that checklists performed quite poorly on intra-rater reliability, they also report that inter-rater reliability and predictive validity of checklists was good.

Tools Needed

Checklists can be applied using pen and paper; however, participants must have access to some form of the device or interface being assessed. This could either be the device itself, functional drawings of the device, mock-ups, or a prototype version of the device. The checklist being used is also required, normally in paper form.

Example

Extracts from a usability checklist evaluation of the Garmin 305 Forerunner training device are presented in Tables 9.2 and 9.3. Participants completed a checklist evaluation for the Forerunner device using Ravden and Johnson's HCI checklist method. The checklist analysis was completed based on programming the device for a training run and then completing the run while wearing the device.

TABLE 9.2
Visual Clarity Checklist Evaluation Extract

Section 1: Visual Clarity Information displayed on the screen should be clear, well organized, unambiguous, and easy to read.	Always	Most of the Time	Some of the Time	Never	Comments
1. Is each screen clearly identified with an informative title or description?		✓			
2. Is important information highlighted on the screen?			✓		
3. When the user enters information on the screen, is it clear: (a) where the information should be entered? (b) in what format it should be entered?	N/A	N/A	N/A	N/A	
4. Where the user overtypes information on the screen, does the system clear the previous information so that it does not get confused with the updated input?	N/A	N/A	N/A	N/A	
5. Does information appear to be organized logically on the screen (e.g., menus organized by probable sequence of selection or alphabetically)?	✓				
6. Are different types of information clearly separated from each other on the screen (e.g., instructions, control options, data displays)?	✓				
7. Where a large amount of information is displayed on one screen, is it clearly separated into sections on the screen?	✓				
8. Are columns of information clearly aligned on the screen (e.g., columns of alphanumeric left-justified, columns of integers right-justified)?		✓			
9. Are bright or light colours displayed on a dark background, and vice versa?	✓				
10. Does the use of colour help to make the displays clear?			✓		Without backlight, Display is unreadable in the dark or under shadow conditions due to colours used.
11. Where colour is used, will all aspects of the display be easy to see if used on a monochrome or low-resolution screen, or if the user is colour-blind?		✓			
12. Is the information on the screen easy to see and read?			✓		Some of the information presented on the screen is very small and may be hard to read for some users

TABLE 9.2 (CONTINUED)
Visual Clarity Checklist Evaluation Extract

Section 1: Visual Clarity Information displayed on the screen should be clear, well organized, unambiguous, and easy to read.	Always	Most of the Time	Some of the Time	Never	Comments
13. Do screens appear uncluttered?			✓		Some of the screens do appear cluttered at times. This is due to the sheer amount of information that is presented on such a small display
14. Are schematic and pictorial displays (e.g., figures and diagrams) clearly drawn and annotated?		✓			
15. Is it easy to find the required information on a screen?		✓			Menu titles are logical most of times

TABLE 9.3
System Usability Problems Checklist Evaluation Extract

Section 10: System Usability Problems

When using the system, did you experience problems with any of the following?	No Problems	Minor Problems	Major Problem	Comments
1. Working out how to use the system		✓		Initially found it difficult to find some features/functionality
2. Lack of guidance on how to use the system	✓			
3. Poor system documentation	✓			
4. Understanding how to carry out the tasks		✓		
5. Knowing what to do next		✓		
6. Understanding how the information on the screen relates to what you are doing	✓			
7. Finding the information you want	✓			
8. Information which is difficult to read clearly			✓	Without backlight display cannot be read in the dark or in shadowy conditions
9. Too many colours on the screen	✓			
10. Colours which are difficult to look at for any length of time	✓			
11. An inflexible, rigid system structure	✓			
12. An inflexible help (guidance) facility	N/A	N/A	N/A	
13. Losing track of where you are in the system or of what you are doing or have done	✓			
14. Having to remember too much information while carrying out a task	✓			

TABLE 9.3 (CONTINUED)
System Usability Problems Checklist Evaluation Extract

Section 10: System Usability Problems

When using the system, did you experience problems with any of the following?	No Problems	Minor Problems	Major Problem	Comments
15. System response times that are too quick for you to understand what is going on	✓			
16. Information which does not stay on the screen long enough for you to read it	✓			
17. System response times that are too slow			✓	*Often system takes to long to find satellite coverage.... feedback depicting progress is inadequate and inconsistent*
18. Unexpected actions by the system		✓		*Feedback bar when locating satellite coverage often jumps from 90% complete to 25-50% complete*
19. An input device which is difficult or awkward to use		✓		*Buttons are stiff and awkward...often need multiple presses before they work*
20. Knowing where or how to input information	✓			
21. Having to spend too much time inputting information		✓		*Problems with buttons means that multiple presses are often required for one button press tasks*
22. Having to be very careful in order to avoid errors	✓			
23. Working out how to correct errors	✓			
24. Having to spend too much time correcting errors	✓			
25. Having to carry out the same type of activity in different ways	✓			

FLOWCHART

(See Flowchart 9.1.)

RECOMMENDED TEXTS

Ravden, S. J., and Johnson, G. I. (1989). *Evaluating usability of human-computer interfaces: A practical method*. Chichester: Ellis Horwood.

Stanton, N. A. and Young, M. S. (1999). *A guide to methodology in ergonomics: Designing for human use*. London: Taylor & Francis.

HEURISTIC ANALYSIS

BACKGROUND AND APPLICATIONS

Heuristic analysis offers a quick and simple approach to device or interface evaluation. Conducting a heuristic analysis involves analysts performing a range of representative tasks with the product,

Interface Evaluation

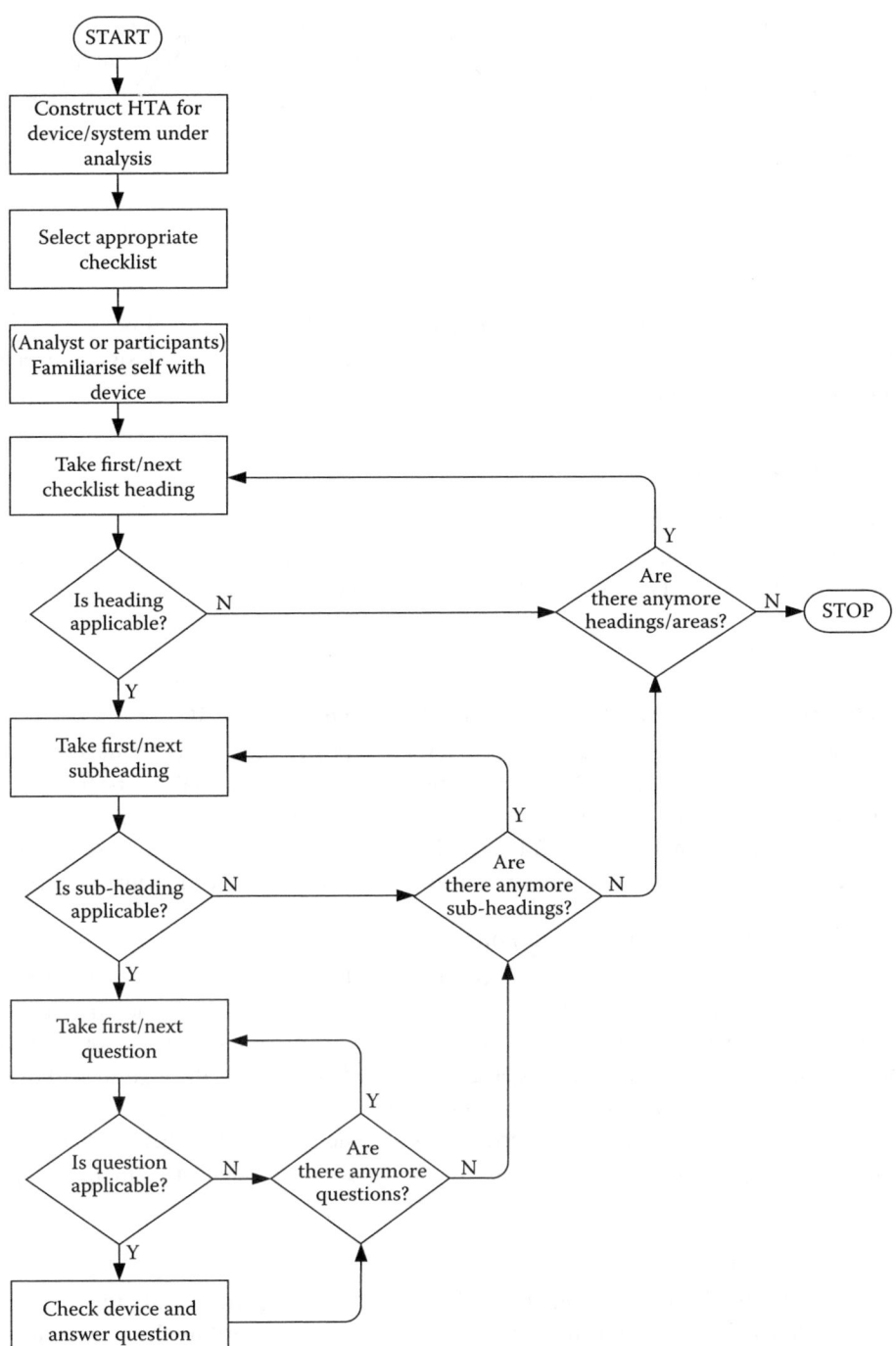

FLOWCHART 9.1 Checklist flowchart.

device, or interface in question and providing positive and negative subjective opinions based upon their interactions with it. Its simplicity and flexibility allow it to be used to assess anything ranging from usability and design-induced error to interface design and layout. Observations and opinions are recorded as the analyst interacts with the device. Heuristic-type analyses are typically conducted throughout the design process in order to evaluate design concepts and propose remedial measures for any problems encountered. The popularity of heuristic analysis lies in its simplicity and the fact that it can be conducted quickly, easily, and with only minimal resource usage, at any stage throughout the design process.

Domain of Application
Generic.

Application in Sport

Heuristic analysis can be used within a sporting context to evaluate sports technology, devices, and performance aids. It can be used throughout the design process to evaluate design concepts or for the evaluation of operational designs.

Procedure and Advice

Step 1: Define Aims of the Analysis
The flexibility of heuristic analysis makes it especially important to clearly define what the aims of the analysis are beforehand. This ensures that time is not wasted looking at unwanted aspects of the device or product under analysis. The aims of the analysis should also make it easier to identify a set of suitable tasks during step 2.

Step 2: Identify Set of Representative Tasks for the Device in Question
Once the aims of the analysis are clearly defined, a set of representative tasks for the device or product under analysis should be identified. It is recommended that heuristic analyses be based upon analysts performing an exhaustive set of tasks with the device in question. The tasks defined should then be placed in a task list. It is normally useful to conduct an HTA for this purpose, based on the operation of the device in question. The HTA output then acts as a task list for the heuristic analysis.

Step 3: Product/Device Familiarisation Phase
To ensure that the analysis is as comprehensive as possible, it is recommended that the analysts/participants involved spend some time to familiarise themselves with the device in question. This might also involve consultation with the associated documentation (e.g., instruction/user manual), watching a demonstration of the device being operated, being taken through a walkthrough of device operation or engaging in user trials with the device. It is imperative that the analysts become proficient with the device or product in question before undertaking the heuristic analysis, as analysts with limited experience in the operation of the device or product may make inappropriate observations.

Step 4: Perform Task(s)
Once familiar with the device under analysis, the analysts/participants should then perform each task from the task list developed during step 2 and offer opinions regarding the design and the heuristic categories required (e.g., usability, error potential). During this stage, any good points or bad points identified during their interactions with the device should be recorded. Each opinion offered should be recorded in full. It is normally standard practise for analysts/participants to record their own observations; however, it is also acceptable for an observer to record them.

Step 5: Propose Design Remedies
Once the analyst has completed all of the tasks from the task list, remedial measures for any of the negative observations should be proposed and recorded.

Interface Evaluation

Advantages

1. Heuristic analysis offers a very simple, quick, and low cost approach to usability assessment.
2. It can be used to look at a range of factors, including usability, errors, interface design, and workload.
3. Very little training (if any) is required.
4. It can be applied throughout the design process on design mock-ups, concepts, and prototypes.
5. The output derived is immediately useful, highlighting problems associated with the device in question.
6. It provides remedial measures for the problems identified.
7. When applied with a large number of participants the outputs are potentially very powerful.

Disadvantages

1. The method suffers from poor reliability, validity, and comprehensiveness.
2. Devices designed on end-user recommendations only are typically flawed.
3. It is entirely subjective. What represents good design to one analyst may represent poor design to another.
4. It has a totally unstructured approach.

Related Methods

The heuristics method is similar to the user trial approach to design evaluation. HTA, observational study, and walkthrough analysis are often used as part of a heuristic analysis.

Approximate Training and Application Times

The heuristics method requires very little training, and the associated application time is typically very low, although this is dependent upon the device and the range of tasks being assessed.

Reliability and Validity

Due to its unstructured nature, heuristic analysis suffers from poor reliability and validity (Stanton and Young, 1999).

Tools Needed

A heuristic analysis is conducted using pen and paper only. The device under analysis is required in some form, e.g., functional diagrams, mock-up, prototype, or the actual device itself.

Example

The example in Table 9.4 is a heuristic analysis of the Garmin Forerunner 305 training watch. The participant performed a series of tasks with the device (including programming the device, using the device for a series of training runs, and downloading training information from the device) and then completed a heuristic evaluation focussing on device usability and interface design and layout.

TABLE 9.4
Garmin Forerunner Device Heuristic Evaluation Example

	Heuristic Analysis: Garmin Forerununner 305 Training Aid	
	Comment	**Remedial Measures**
✓	Buttons are large and distinct	
✓	Labelling is clear and appropriate	
✓	Audible alert pacing system removes the need to look at screen when running and is a particularly effective training aid	
✘	Buttons can be quite tough to depress	Reduce stiffness of buttons
✓	Display is large and clear	
✓	Loud audible feedback on button presses	
✘	Display is unreadable in the dark or under shadow without backlight	Use permanent backlit display
✘	Backlit display is difficult to activate when running at pace	Use permanent backlit display Reduce stiffness of buttons
✓	Contrast is adjustable	
✓	Backlight can be set to permanent	
✓	Menu system works well and structure is logical	
✓	Menus are clear and use large text	
✘	Bar showing depth of menu can be illegible at times	Increase slider bar legibility (size and contrast)
✘	Clock display is very small and difficult to read; it is almost impossible to read the clock display when running	Increase size of clock display or build clock into menu if it doesn't fit on display
✘	Clock display is located on different screen and user has to change mode to view the actual time; this is difficult when running at pace	Build clock display into main running display
✓	Training displays all large and clear	
✓	High level of consistency throughout interface	
✓	Audible training alerts loud and clear	
✓	Three different display layouts offered	
✘	Feedback when logging onto satellites is poor, inconsistent, and often illogical, e.g., almost complete then only 50% complete	Use more informative feedback, e.g., percent logged on display or counter; also make display consistent and logical
✘	Display does not inform user that pace alert function is being used; user may begin run unaware that pace alert function is still active from last run	Use pace alert function icon on standard display

FLOWCHART

(See Flowchart 9.2.)

RECOMMENDED TEXT

Stanton, N. A., and Young, M. S. (1999). *A guide to methodology in ergonomics: Designing for human use.* London: Taylor & Francis.

Interface Evaluation

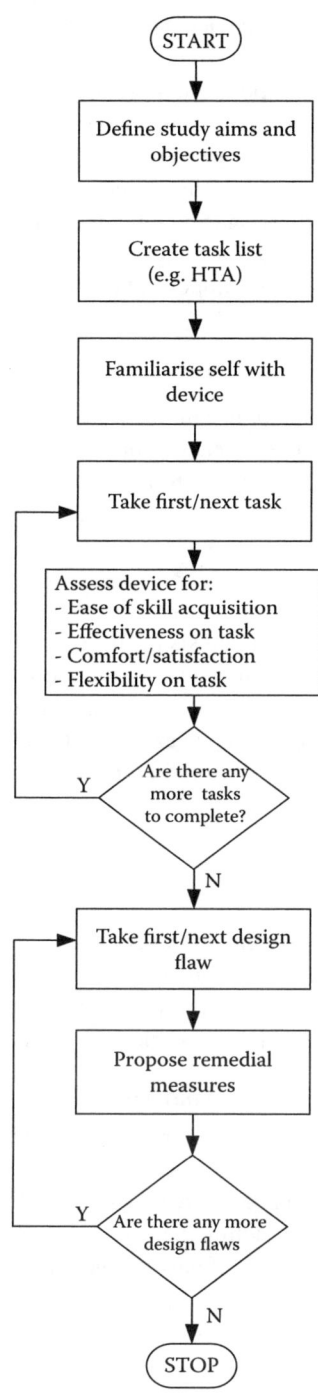

FLOWCHART 9.2 Heuristics flowchart.

LINK ANALYSIS

BACKGROUND AND APPLICATIONS

Link analysis is used to identify and represent associations between elements of a system during task performance; these associations may be links between interface components, between human operators, between human operators and devices, or between human operators and elements of an interface. Links are defined as movements of attentional gaze or position between parts of the system, or communication with other system elements, and are analysed in terms of their nature, frequency, and importance during the task under analysis. Link analysis uses a links matrix and spatial diagrams, similar to social network diagrams, in order to represent the links within the system under analysis. Specifically aimed at aiding the design of interfaces and systems, link analyses' most obvious use is in the area of workspace-layout optimisation (Stanton and Young, 1999). The outputs can be used to suggest revised layouts of the components for the device, based on the premise that links should be minimised in length, particularly if they are important or frequently used.

Domain of Application

Link analysis was originally developed for use in the design and evaluation of process control rooms (Stanton & Young, 1999); however, it is a generic method and can be applied in any domain. For example, Stanton and Young (1999) describe an application of the method to the evaluation and redesign of in-car stereo interfaces.

Application in Sport

There is great potential for applying link analysis in sport, not only as a device and interface design and evaluation approach, but also as a performance evaluation tool to analyse the links present during task performance. Various links that emerge during sporting performance can be analysed using this approach, such as passes made between players, links between game areas (e.g., return of serve placements in tennis), and links between players in terms of contact during the game.

Procedure and Advice

Step 1: Define Analysis Aims

First, the aims of the analysis should be clearly defined. For example, the aims of the analysis may be to evaluate the passing links between players during different game (e.g., defensive vs. attacking) scenarios, or it may be to evaluate the placement of returns of service in a tennis game. The analysis aims should be clearly defined so that appropriate scenarios are used and relevant data are collected.

Step 2: Define Task(s) or Scenario under Analysis

The next step involves clearly defining the task or scenario under analysis. It is recommended that the task be described clearly, including the different actors involved, the task goals, and the environment within which the task is to take place. HTA may be useful for this purpose if there is sufficient time available. It is important to clearly define the task, as it may be useful to produce link analyses for different task phases, such as the first half and second half, or attacking and defensive phases of a particular game.

Step 3: Collect Data

The next step involves collecting the data that are to be used to construct the link analysis. This typically involves observing the task under analysis and recording the links of interest that occur during task performance. Typically, the frequency, direction (i.e., from actor A to actor B), type, and content of the links are recorded.

Step 4: Validate Data Collected

More often than not it is useful to record the task in question and revisit the data to check for errors or missing data. This involves observing the task again and checking each of the links recorded originally.

Interface Evaluation

Step 5: Construct Link Analysis Table

Once the data are checked and validated, the data analysis component can begin. The first step in this process involves the construction of a link analysis table. This involves constructing a simple matrix and entering the direction (i.e., from and to) and frequency of links between each of the actors and device/interface elements involved. The finished matrix depicts the direction and frequency of links during the task under analysis.

Step 6: Construct Link Analysis Diagram

Next, a link analysis diagram should be constructed. The link analysis diagram depicts each actor/device/interface element and the links that occurred between them during the scenario under analysis. Within the diagram, links are represented by directional arrows linking the actor/device/interface element involved. The frequency of links can be represented either numerically in terms of number of arrows, or via the thickness of the arrows. An example link analysis diagram representing the passing links in the build up to a goal scored during a group game in the Euro 2008 soccer championships is presented in Figure 9.2.

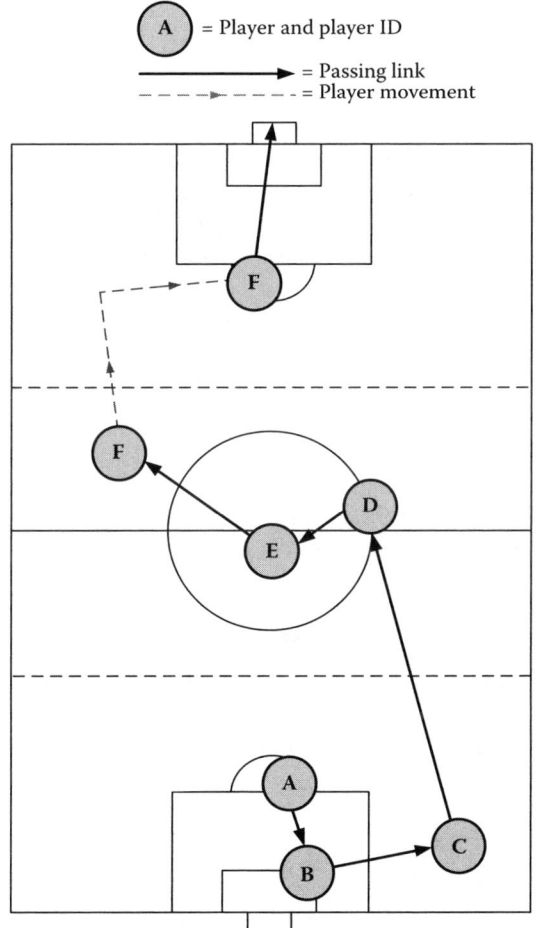

FIGURE 9.2 Link analysis diagram example for goal scored during Euro 2008 group game.

Step 7: Offer Redesign Proposals/Analyse Link Analysis Data

If an interface or device display is being evaluated, it is normally appropriate to suggest redesign proposals based on the links identified (although this is not compulsory as part of a link analysis). The redesign is typically based upon reducing the distance between the linked interface components, particularly the most important and frequently used linked components. Alternatively, if associations within a particular game are being evaluated, then the links should be analysed statistically at this point.

ADVANTAGES

1. The link analysis diagram provides a powerful means of representing the associations present in a particular system.
2. Link analysis is simple and quick to apply and requires only minimal training.
3. The output is immediately useful.
4. It offers redesign proposals.
5. It has a range of potential uses in a sporting context, including as an interface/device evaluation approach, a coaching aid, and/or a performance evaluation tool.
6. Link analysis is a generic approach and can be applied in any domain.
7. Link analysis can be used throughout the design process to evaluate and modify design concepts.

DISADVANTAGES

1. The data collection phase can be time consuming.
2. The use of HTA adds considerable time to the analysis.
3. For large, complex tasks involving many actors and devices, the analysis may be difficult to conduct and the outputs may become large and unwieldy.
4. It only considers the basic physical relationship between the user and the system. Cognitive processes and error mechanisms are not typically accounted for.
5. The output is not easily quantifiable.

RELATED METHODS

A link analysis normally requires an initial task description (e.g., HTA) to be created for the task under analysis. In addition, an observation or walkthrough analysis of the task(s) under analysis is normally undertaken in order to establish the links between components in the system.

APPROXIMATE TRAINING AND APPLICATION TIMES

Link analysis requires only minimal training and the application time is typically low. In conclusion to a comparison of 12 ergonomics methods, Stanton and Young (1999) report that the link analysis method is relatively fast to train and practise, and also that execution time is moderate compared to the other methods (e.g., SHERPA, layout analysis, repertory grids, checklists, and TAFEI).

RELIABILITY AND VALIDITY

In conclusion to the comparison study described above, Stanton and Young (1999) reported that link analysis performed particularly well on measures of intra-rater reliability and predictive validity. They also reported, however, that the method was let down by poor inter-rater reliability.

Interface Evaluation

TOOLS NEEDED

When conducting a link analysis the analyst should have access to the device under analysis, and can use pen and paper only. For the observation part of the analysis, a video recording device is normally required; although the links can be recorded live, it is normally necessary to return to the recording of the scenario to validate the data obtained. An eye tracker device can also be used to record eye fixations during task performance. A drawing software package, such as Microsoft Visio, is often used to create the link analysis diagrams.

EXAMPLE

The link analysis method can be used to analyse any links present within a system. For the purposes of this example, we present two examples of how the approach may be used for the analysis of sports performance (the approach can also be used for the evaluation of sports product/device interfaces). The first example presented involved using the method for analysing the passing links leading up to goals scored in soccer games. Analysing the passing sequences leading up to goals being scored has been commonly undertaken as a means of assessing soccer team performance. Reep and Benjamin (1968; cited in Hughes and Franks, 2005), for example, analysed, among other things, the passing sequences in 3,213 UK Football Association (FA) games taking place between 1953 and 1968. According to Hughes and Franks (2005), Reep and Benjamin's main findings demonstrated that approximately 80% of the goals scored in the games analysed arose from passing sequences of three passes or less. Hughes and Franks (2005) analysed passing sequences, goals, and shots during the 1990 and 1994 FIFA World Cup finals and found that 84% of goals during the 1990 World Cup, were scored and 80% in the 1994 World Cup Games, as a result of team possessions consisting of four passes or less. Grant, Williams, and Reilly (1999) conducted an analysis on the 171 goals scored during the 1998 FIFA World Cup. As part of their study they analysed, amongst other things, the area of the pitch in which possession was regained prior to goals being scored, and also the number of passes made before goals were scored. Grant, Williams, and Reilly (1999) found that passing was involved in 47.2% of the goals scored from open play. The mode value for the number of passes before goals being scored was 3.

The authors recently used a link analysis–style approach to analyse the passing links or sequences involved in the build up to the goals scored in the Euro 2008 soccer championships. The aim of the analysis was to analyse the number of continuous (i.e., uninterrupted) passing links that occurred in the build up to the 77 goals scored throughout the tournament (penalty shootout goals were not counted). Each successful pass between players on the same team was classified as a link. The number of continuous links in the build up to goals scored was analysed (i.e., if the opposition intercepted the ball or the pass did not reach its intended recipient the chain of links was broken). Based on observational study of television coverage of each of the games played, link tables and diagrams were constructed for each of the goals scored. Unfortunately, the use of network television coverage for the analysis may have led to a small number of links being missed (i.e., when the camera momentarily was not focussing on the game) for some of the goals analysed. An example of a link diagram for a goal scored in one of the group games is presented in Figure 9.3.

The findings indicated that, on average, there were 3 (2.75) continuous links or passes involved in the build up to the goals scored during the European Championships 2008 soccer tournament. The analysis also found that 45 of the 77 goals scored involved multiple links (i.e., two or more passes). The mean number of passes involved in the build up to goals scored by each team, along with the total goals scored by each team and the greatest number of passing links leading up to goals scored for each team during the tournament, is presented in Figure 9.4.

The second example focuses on the passing links made by individual players during soccer games. The analysis presented involved a comparison of two players playing on the right hand side of midfield for an International soccer team in a recent International friendly game. The first player, hereafter referred to as player A, played on the right hand side of midfield during the first half, and the second

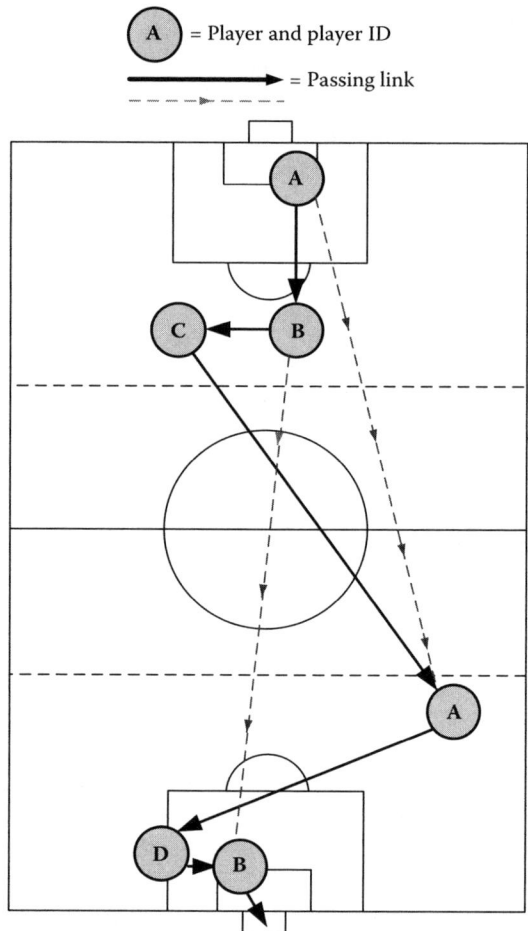

FIGURE 9.3 Link diagram for goal scored in Euro 2008 group game.

player, hereafter referred to as player B, played in the same position in the second half (player A having been moved to the left hand side of midfield). A link analysis–style approach was used to compare both players' passing performance. Both unsuccessfully and successfully completed passes were recorded based on live observation. Each pass's origin location and destination location was recorded. The analysis produced is presented in Figure 9.5. Within the figure, successfully completed passes (i.e., successfully received by intended recipient) are shaded black. Unsuccessful passes (i.e., intercepted by opposition player or misplaced) are shaded in grey. The arrows show the origin and destination of the passes made, with the beginning of the arrows representing the passing point of origin, and the arrow point representing the end location of the pass. As demonstrated by the analysis, player B made more successful passes than player A (33 vs. 6). The diagram also shows how player B made more passes into the attacking third of the pitch than player A did. Player B also had a lower percentage of unsuccessful passes (19.51%) than player A (45%). The link analysis method is therefore useful for representing individual player passing performance during team games. Aside from the link table, the link diagram provides a powerful, immediately useful spatial representation of passing performance.

FLOWCHART

(See Flowchart 9.3.)

Interface Evaluation

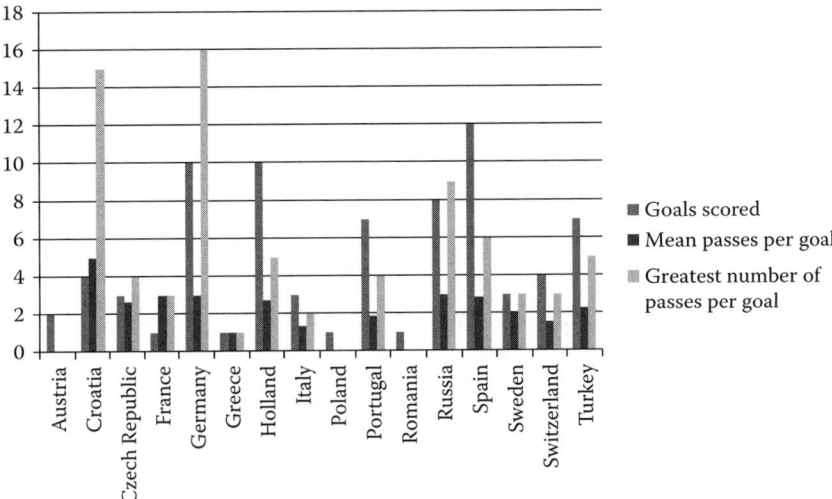

FIGURE 9.4 Goals scored and mean number of passes leading to goals per team for the Euro 2008 soccer championships.

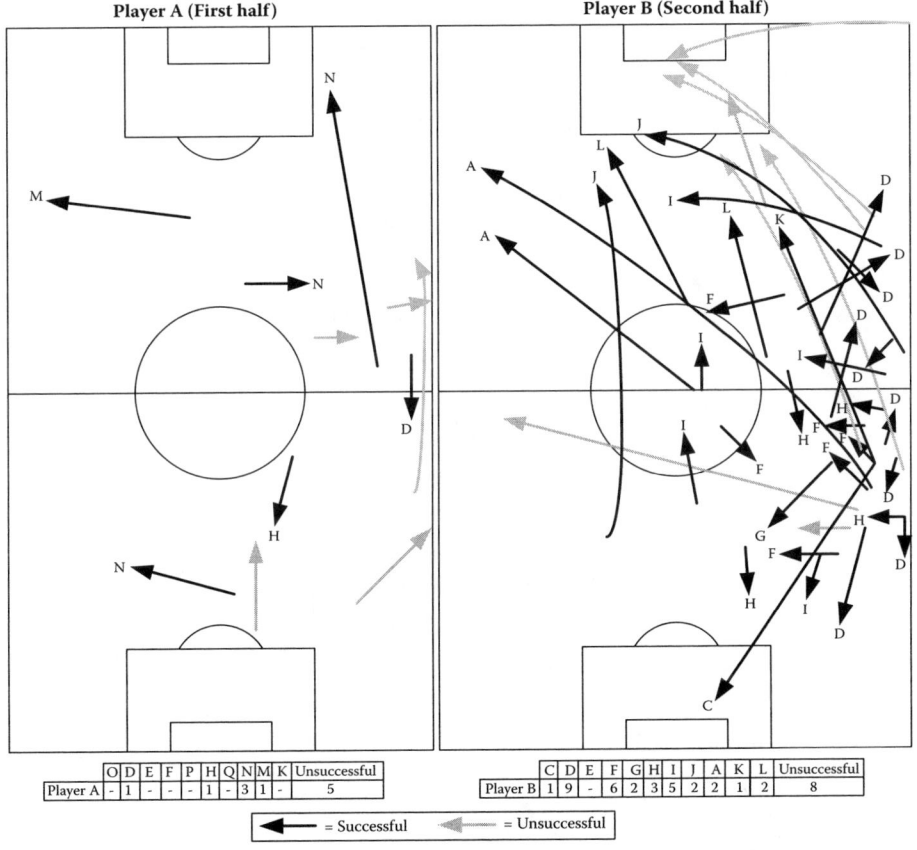

FIGURE 9.5 Link analysis diagram and tables showing comparison of player passing; the darker arrows depict the origin, destination, and receiving player for each successful pass made, and the lighter arrows depict the origin, destination, and intended receiving player for each unsuccessful pass.

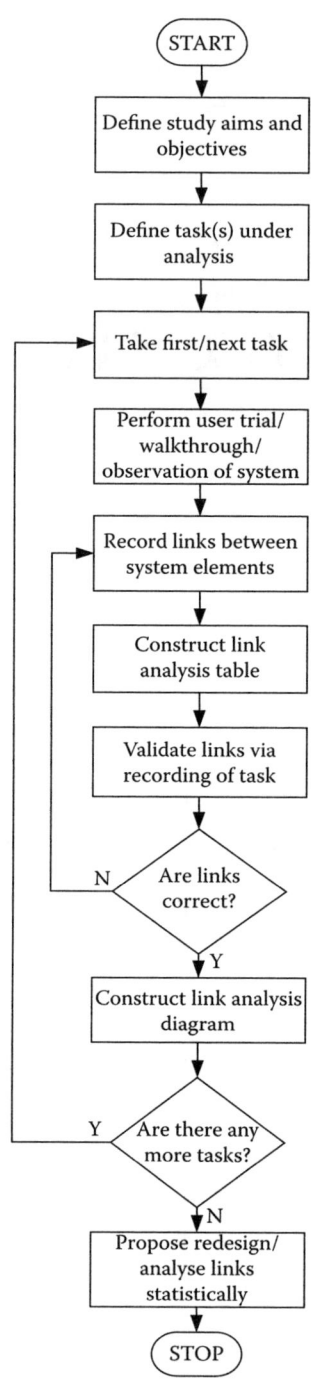

FLOWCHART 9.3 Link analysis flowchart.

Recommended Texts

Drury, C. G. (1990). Methods for direct observation of performance. In *Evaluation of human work: A practical ergonomics methodology*, 2nd ed., eds. J. Wilson and E. N. Corlett, 45–68. London: Taylor & Francis.

Stanton, N. A., and Young, M. S. (1999). *A guide to methodology in ergonomics: Designing for human use*. London: Taylor & Francis.

LAYOUT ANALYSIS

BACKGROUND AND APPLICATIONS

Layout analysis is used to evaluate and redesign interface layouts on functional groupings and their importance, frequency, and sequence of use. The theory behind layout analysis is that device interfaces should mirror the users' structure of the task and that the conception of the interface as a task map greatly facilitates design (Easterby, 1984). A layout analysis involves arranging all of the components of the interface into functional groupings and then organising them based on their importance of use, sequence of use, and frequency of use. Following this, components within each functional group are then reorganised, again according to importance, sequence, and frequency of use. At the end of the process, an interface layout redesign is produced based upon the user's model of the task and the importance, sequence, and frequency of use of the different functional components of the interface.

DOMAIN OF APPLICATION

Layout analysis is a generic approach that can be applied in any domain.

APPLICATION IN SPORT

Layout analysis can be used in a sporting context to evaluate and redesign sports technology, device, and performance aid interfaces. It can be used throughout the design process to evaluate design concepts or for the evaluation of existing devices.

PROCEDURE AND ADVICE

Step 1: Define Aims of the Analysis

First, the aims of the analysis should be clearly defined. For example, the aims of the analysis may be to evaluate a particular device's interface or to redesign an inadequate design concept. The analysis aims should be clearly defined so that an appropriate set of tasks is selected and a relevant analysis is produced.

Step 2: Identify Set of Representative Tasks for the Device in Question

Once the aims of the analysis are clearly defined, a set of representative tasks for the device or product under analysis should be identified. It is recommended that layout analyses be based upon analysts performing an exhaustive set of tasks with the device in question. The tasks defined should then be placed in a task list. It is normally useful to conduct an HTA for this purpose, based on the operation of the device in question. The HTA output then acts as a task list for the sequence, importance, and frequency of use analyses.

Step 3: Create Schematic Diagram

Next, the analysis can begin. First, the analyst should create a schematic diagram for the device under analysis. This diagram should contain each (clearly labelled) interface element. An example of a schematic interface diagram for the Garmin 305 Forerunner training device is presented in Figure 9.6.

Step 4: Familiarisation with Device

It is recommended that the analysts involved spend some time in familiarising themselves with the device in question. This might involve consultation with associated documentation (e.g., instruction manual), performing a series of tasks with the device, watching a demonstration of the device

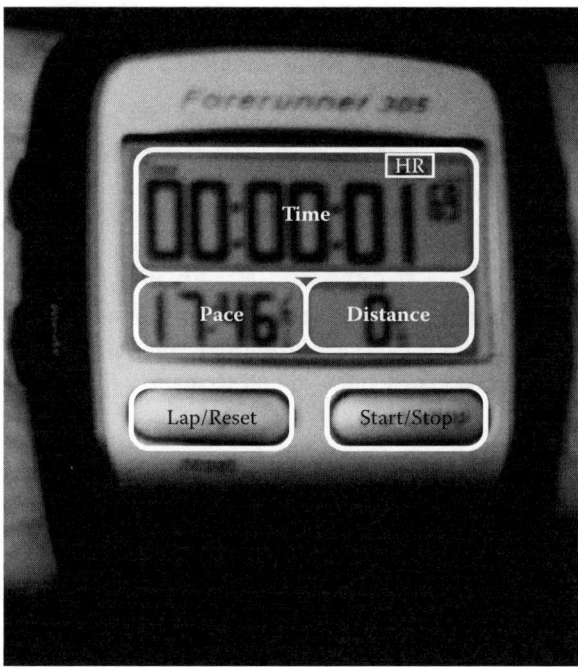

FIGURE 9.6 Garmin Forerunner 305 schematic diagram.

being operated, or being taken through a walkthrough of device operation. It is imperative that the analysts become proficient with the device or product in question before undertaking the layout analysis, as analysts with limited experience in the operation of the device or product may make inappropriate observations.

Step 5: Arrange Interface Components into Functional Groupings

Next, the analyst arranges the interface components into functional groupings. Each interface element should be grouped according to its function in relation to the device under analysis. For example, the interface components of the main display on the Garmin 305 Forerunner training device were arranged into time, pace, distance, lap, and start/stop functional groups.

Step 6: Arrange Functional Groupings Based on Importance of Use

Next, the analyst should arrange the functional groupings based on importance of use. This rearrangement is based on the analyst's subjective judgement; however, the most important interface elements are normally made the most prominent on the redesigned interface.

Step 7: Arrange Functional Groupings Based on Sequence of Use

Next, the analyst should arrange the functional groupings based on sequence of use during typical device operation. This rearrangement is based on the analyst's subjective judgement; however, those interface elements that are used together are normally placed in close proximity to one another on the redesigned interface.

Step 8: Arrange Functional Groupings Based on Frequency of Use

Next, the analyst should arrange the functional groupings based on frequency of use during typical device operation. This rearrangement is based on the analyst's subjective judgement; however, those interface elements that are used the most frequently are normally made more prominent on

Interface Evaluation

the redesigned interface. At the end of step 8, the analyst has redesigned the device according to the end user's model of the task (Stanton and Young, 1999).

ADVANTAGES

1. Layout analysis offers a quick, easy to use, and low cost approach to interface evaluation and design.
2. The output is immediately useful, offering a redesign of the interface under analysis based upon importance, sequence, and frequency of use of the interface elements.
3. It has low resource usage.
4. It requires very little training.
5. It can be applied in any domain.
6. The outputs are easy to communicate to designers.
7. It works just as well with paper diagrams, design concepts, mock-ups, and prototypes.

DISADVANTAGES

1. It has poor reliability and validity (Stanton and Young, 1999).
2. The redesign of the interface components is based to a high degree on the analyst's subjective judgement.
3. Literature regarding layout analysis is sparse.
4. If an initial HTA is required, application time can rise significantly.
5. Conducting a layout analysis for complex interfaces may be very difficult and time consuming, and the outputs may be complex and unwieldy.

RELATED METHODS

Layout analysis is similar to link analysis in its approach to interface design. HTA is also typically used to inform sorting of interface elements based on importance, sequence, and frequency of use.

APPROXIMATE TRAINING AND APPLICATION TIMES

The training and application times associated with layout analysis are low. In conclusion to a comparison study of 12 ergonomics methods, Stanton and Young (1999) report that little training is required for layout analysis and that it is amongst the quickest of 12 methods to apply. For large, complex interfaces, however, the application time may be considerable, particularly if HTA is used initially for the task description.

RELIABILITY AND VALIDITY

The layout analysis method suffers from poor levels of reliability and validity. In conclusion to a comparison study of 12 ergonomics methods, Stanton and Young (1999) report poor statistics for intra-rater reliability and predictive validity for layout analysis.

TOOLS NEEDED

Layout analysis can be conducted using pen and paper, providing the device or pictures of the device under analysis are available. It is useful to use a drawing software package, such as Microsoft Visio, to produce the layout analysis outputs.

EXAMPLE

A layout analysis was conducted for the Garmin 305 Forerunner training device. This involved two analysts completing the layout analysis based on their experiences with the device during training and race events. The layout analysis outputs are presented in Figure 9.6.

Based on importance of use during running events, the interface was reorganised as presented in Figure 9.7. The analysts both felt that "current pace" was more important than "time" during both racing and training events, and that "time" was only slightly more important than "distance ran."

Next, the interface was reorganised based on sequence of use, as presented in Figure 9.8. In this case, the analysts determined that "pace" is used most often in conjunction with "distance ran," and that "distance ran" is used most often in conjunction with "time" expired.

Finally, the interface was reorganised based on frequency of use, as presented in Figure 9.9. In this case the analysts suggested that the "pace" reading is used the most frequently, followed by "distance ran" and then "time."

Based on the three interface reorganisations, the redesign for the Forerunner training device interface was devised. One important function is that the analysts strongly recommended that the display component of the interface (i.e., pace, time, heart rate, and distance readings) be customisable based on personal preference, as they felt that different runners of different ability levels may require different interface set-ups. This is represented in Figure 9.10.

FLOWCHART

(See Flowchart 9.4.)

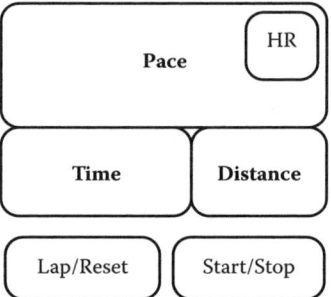

FIGURE 9.7 Interface reorganised based on importance of use during running events.

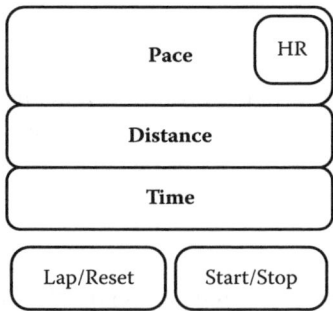

FIGURE 9.8 Interface reorganised based on sequence of use.

Interface Evaluation

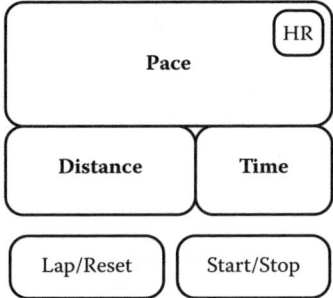

FIGURE 9.9 Interface reorganised based on frequency of use.

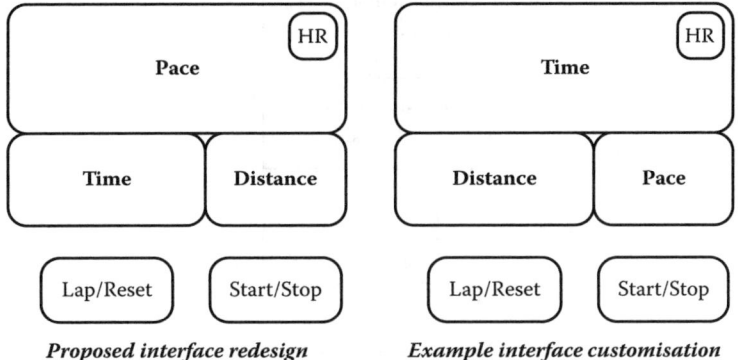

Proposed interface redesign   *Example interface customisation*

FIGURE 9.10 Interface redesign based on layout analysis.

RECOMMENDED TEXT

Stanton, N. A., and Young, M. S. (1999). *A guide to methodology in ergonomics*. London: Taylor & Francis.

INTERFACE SURVEYS

BACKGROUND AND APPLICATION

Kirwan and Ainsworth (1992) describe the interface surveys method, which is used to assess the physical aspects of man-machine interfaces. The method involves the use of survey-based analysis to consider the following interface elements:

- Controls and displays
- Labelling
- Coding consistency
- Operator modification
- Sightline
- Environmental aspects

The interface surveys are used to pinpoint design inadequacies for a particular interface or design concept. A brief summary of the control and displays, labelling, coding consistency, and operator modification survey methods is given below.

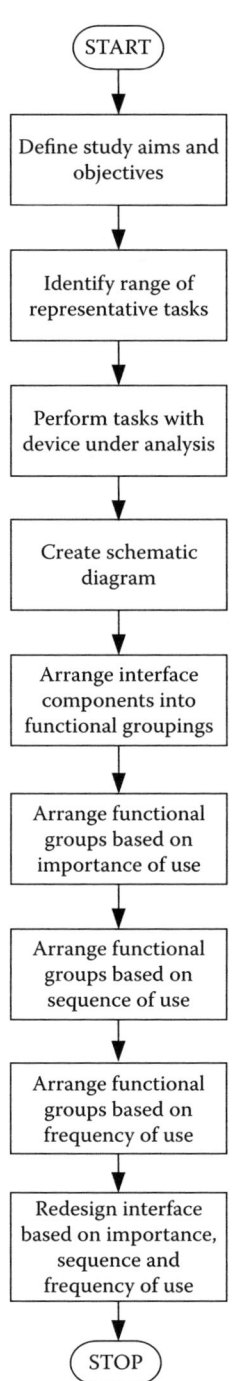

FLOWCHART 9.4 Layout analysis flowchart.

Control and Display Survey

The control and display survey is used to evaluate the controls and displays provided by a particular device or interface. The process first involves recording all parameters that can be controlled and displayed, and then creating a list of all controls and displays. Developing this list involves examining each control, and recording exactly what the control is controlling, its location, type of control, and any other relevant details, such as movement (e.g., up/down, rotary, left to right, etc.). Likewise, each display should be investigated in the same manner, e.g., display type, what is being displayed, location, etc. According to Kirwan and Ainsworth (1992) the list should then be sorted into a hierarchical list containing the system, sub-system, and parameter. The control and display list can then be used as a checklist to ensure that the user is presented with adequate information and provided with the appropriate controls in order to perform the task. If required (depending upon the scope of the analysis), the appropriate guidelines or standards can also be applied, in order to check that the system controls and displays adhere to the relevant guidelines/standards.

Labelling Surveys

Labelling surveys are used to examine the labelling used on the interface under analysis in terms of reference, wording, size, position, and colour (Kirwan and Ainsworth, 1992). It may also be useful to make a subjective judgement on the clarity and ease of identification of each label identified. Any missing or confusing labels should also be recorded. Again, depending upon available resources, the labels identified can also be compared to the associated labelling standards and guidelines for the system under analysis.

Coding Consistency Survey

Coding surveys are used to analyse any coding used on the interface under analysis. Typical types of coding used are colour coding (e.g., green for go, red for stop), positional coding, size coding, and shape coding (Kirwan and Ainsworth, 1992). The coding analysis is used to highlight ambiguous coding and aspects where additional coding may be required (Kirwan and Ainsworth, 1992). The analyst should systematically work through the interface, recording each use of coding, its location, the feature that is coded, description, relevance, instances where coding could be used but is not, instances of counteracting coding, and any suggested revisions in terms of coding to the interface.

Operator Modification Survey

End users often add temporary modifications to the interface or use work-arounds in order to bypass design inadequacies. For example, operators use labels or markings (e.g., Post-It notes or stickies) to highlight where specific controls should be positioned or place objects such as paper cups over redundant controls. The modifications made by the end users offer an intriguing insight into the usability of the interface, often highlighting bad design, poor labelling, and simpler procedures (i.e., missing out one or two actions). Kirwan and Ainsworth (1992) suggest that such information can be gathered quickly through a survey of the operational system. The information gathered can be used to inform the design of similar systems or interfaces. Conducting an operator modification survey simply involves observing a representative set of tasks being performed using the system under analysis, and recording any instances of operator modification. The use of interviews is also useful, to help understand why the modification occurred in the first place.

DOMAIN OF APPLICATION

The interface survey method is generic can be applied in any domain.

APPLICATION IN SPORT

Interface surveys can be applied in a sporting context to evaluate sports technology, devices, and performance aids. It can be used throughout the design process to evaluate design concepts or for the evaluation of operational designs.

Procedure and Advice

Step 1: Define Analysis Aims

First, the aims of the analysis should be clearly defined. For example, the aims of the analysis may be to evaluate the labelling on a particular device or to assess the controls and displays used. The analysis aims should be clearly defined so that appropriate interface surveys are selected and relevant data are collected.

Step 2: Define Task(s) or Scenario under Analysis

The next step involves clearly defining the task or scenario under analysis. It is recommended that the task be described clearly, including the different actors involved, the task goals, and the environment within which the task is to take place. HTA may be useful for this purpose if there is sufficient time available.

Step 3: Data Collection

The data collection phase involves completing each survey for the system under analysis. There are a number of ways to accomplish this. Access to the system under analysis is normally required, although Kirwan and Ainsworth (1992) suggest that the relevant data can sometimes be collected from drawings of the system under analysis. It is recommended that a walkthrough of the system under analysis be conducted, involving as representative a set of tasks of the full functionality of the system as possible. Observational study of task performance with the system under analysis is also very useful. For the operator modification surveys, interviews with system operators are required, and for the environmental survey, on-line access to the operating system is required.

Step 4: Complete Appropriate Surveys

Once the data collection phase is complete, the appropriate surveys should be completed and analysed accordingly. The results are normally presented in tabular form.

Step 5: Propose Remedial Measures

Once the surveys are completed, it is often useful to propose any remedial measures designed to remove any problems highlighted by the surveys. Such recommendations might offer countermeasures for the system under analysis in terms of design inadequacies, error potential, poor coding, operator modifications, etc.

Advantages

1. Each of the surveys described are easy to apply, requiring very little training.
2. The surveys are generic and can be applied in any domain.
3. The output of the surveys offers a useful analysis of the interface under analysis, highlighting instances of bad design and problems arising from the man-machine interaction.
4. Standards and guidelines can be used in conjunction with the methods in order to ensure comprehensiveness.
5. If all of the surveys are applied, the interface in question is subjected to a very exhaustive analysis.

Disadvantages

1. The application of the surveys can be highly time consuming.
2. An operational system is required for most of the methods. The use of such methods early on in the design process would be limited.
3. Reliability is questionable.

Interface Evaluation

4. Interface surveys do not account for context or performance shaping factors.
5. While the surveys address the design inadequacies of the interface, no assessment of performance is given.

Related Methods

Interface surveys use various data collection methods, including observational study, questionnaires, and interviews with SMEs. Task analysis is also often used as part of the interface survey approach.

Approximate Training and Application Times

Only minimal training is required for each of the survey methods. The application time, although dependent upon the device/system/interface under analysis, is typically high.

Reliability and Validity

No data regarding the reliability and validity of the method are presented in the literature. There may be problems associated with the intra- and inter-rater reliability of the method. For example, different analysts may derive different results for the same interface, and also the same analyst may derive different results when using the method for the same device or system on different occasions.

Tools Needed

Most of the surveys described can be applied using pen and paper. The environmental survey requires the provision of equipment capable of measuring the relevant environmental conditions, including noise, temperature, lighting, and humidity levels. The sightline survey requires the appropriate measurement equipment, such as a tape measures and rulers.

Example

Control, display, and labelling surveys were used to analyse the Garmin 305 Forerunner training device. Extracts of each survey are presented in to Tables 9.5 to 9.7.

Flowchart

(See Flowchart 9.5.)

Recommended Text

Kirwan, B., and Ainsworth, L. K. (1992). *A guide to task analysis*. London: Taylor & Francis.

TABLE 9.5
Control Survey Analysis for Garmin Forerunner Training Device

		Control			
Name	**Type**	**Function**	**Feedback?**		**Comments**
Power button	Push button	Turn device on/off Operates display backlight	Audible alert	✘	Button, has dual functionality (on/off and display backlight); however, this is not labelled clearly
				✘	Operation is inconsistent as it requires one push to switch device on, but a hold and depress to switch device off
				✓	Audible alert is loud and clear.
Mode button	Push button	Switches between modes on device back button	Audible alert	✘	Button can often be stiff and awkward to operate
				✘	Dual functionality not labelled
				✓	Audible alert is loud and clear
Start/stop button	Push button	Starts/stops lap timer	Audible alert	✓	Button is large and clearly labelled
				✓	Audible alert is loud and clear
				✓	Button is large and clearly labelled
Lap/reset button	Push button	Sets lap function Resets timer	Audible alert	✓	Dual functionality of button is clearly labelled
				✓	Audible alert is loud and clear
Enter button	Push button	Selection of items from menu, etc.	Audible alert	✘	Button is stiff and awkward, sometimes requiring multiple presses before it works
				✓	Dual functionality of button is clearly labelled
				✓	Audible alert is loud and clear
Up/down buttons	Push button	Scroll up and down through menus Toggle between display types	Audible alert	✓	Audible alert is loud and clear
				✓	Buttons clearly labelled

Interface Evaluation

TABLE 9.6
Display Survey Analysis for Garmin Forerunner Training Device

Name	Type	Function		Display Comments
Main display	Numerical text	Set up display pace, time, distance traveled, etc.	✓	Display is large and clear for watch device
			✓	Numerical items (e.g., pace, time, distance) are large and take up a large proportion of the display
			✗	Some elements of the display are too small, such as the standard clock display, menu bars, etc
			✗	Without backlight, display is unreadable in the dark or under shadow
			✓	Three display configurations available
			✗	Battery strength is not displayed at all times (i.e., when running)
			✓	Contrast is adjustable
			✓	Backlight length is adjustable and can be set to permanent
			✓	Icons used are logical and appropriate

TABLE 9.7
Labelling Survey Analysis for Garmin Forerunner Training Device

Label	Description	Clarity	Error Potential
On/off label	Standard power button icon	Label is clear; however, button, has dual functionality and other function. is not labelled	User may not recognise dual functionality of button
Mode button, label	Text label "mode"	Text label is clear. Button also acts as "back" button but this is not labelled	User may not understand what modes are available; user may not realise that button also acts as a back button
Lap/reset button	Text label "lap" and "reset"	Text label is clear. Dual functionality is also clearly labelled	N/A
Up/down buttons	Standard up/down icons	Buttons are clearly labelled with standard up/down icons	N/A
Start/stop button	Text label "start/stop"	Text label is clear	N/A
Enter button	Text label "enter"	Text label is clear	N/A

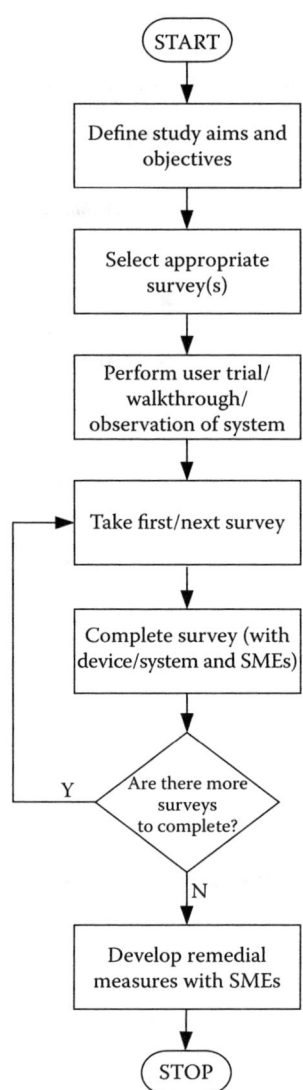

FLOWCHART 9.5 Interface surveys flowchart.

10 Human Factors Methods Integration
Case Study

INTRODUCTION

The aim of this concluding chapter is threefold: first, to present an example of a comprehensive Human Factors methodology for evaluating sports performance; second, to demonstrate how Human Factors methods can be applied in a sporting context; and third, and more importantly, to demonstrate how the methods described in this book can be combined and integrated in order to provide a more exhaustive analysis of human performance. Applying the methods described in this book in isolation is perfectly acceptable, and while their utility when applied alone is assured, Human Factors methods are increasingly being applied together as part of frameworks or integrated suites of methods for the evaluation of complex sociotechnical systems (e.g., Stanton et al., 2005; Walker et al., 2006). Often scenarios are so complex and multi-faceted, and analysis requirements so diverse, that various methods need to be applied since one method in isolation cannot cater for the scenario and analysis requirements. The EAST framework (Stanton et al., 2005) described in Chapter 8 is one such example that was developed to examine the work of distributed teams in complex sociotechnical systems. Frameworks or toolkits of methods have also been recommended and applied in the analysis of other Human Factors concepts; for example, for the analysis of human error (Kirwan, 1998a, 1998b), situation awareness (Salmon et al., 2006), and mental workload (Young and Stanton, 2004).

Through the case study presented in this chapter we aim to demonstrate how some of the methods described in this book can be used together to provide a multi-perspective analysis of sports performance. Of course, it is impractical to attempt to show all of the possible combinations of methods available, as the framework used is dependent upon the specific problem faced; however, this chapter will succeed in its purpose if it demonstrates that with intelligent application, a toolkit approach can be developed from the methods covered in this book. For this purpose, we illustrate the principle of a toolkit approach with reference to a case study drawn from the analysis of a fell running race event.

INTEGRATING HUMAN FACTORS METHODS

There are well over 100 Human Factors methods available covering all manner of concepts. In this chapter, an approach based on method integration is proposed. The aim is to show how existing methods can be combined in useful ways to analyse complex, multi-faceted scenarios involving humans and technology. Methods integration has a number of compelling advantages, because not only does the integration of existing methods bring reassurance in terms of a validation history, but it also enables the same data to be analysed from multiple perspectives. These multiple perspectives, as well as being inherent in the scenario that is being described and measured, also provide a form of internal validity. Assuming that the separate methods integrate on a theoretical level, then their application to the same data set offers a form of "analysis triangulation."

At the most basic level, when analysing human performance within complex sociotechnical systems, the descriptive constructs of interest can be distilled down to simply

- *Why* (the goals of the system, sub-system[s], and actor[s])
- *Who* (the actors performing the activity are)
- *When* (activities take place and which actors are associated with them)
- *Where* (activities and actors are physically located)
- *What* (activities are undertaken, what knowledge/decisions/processes/devices are used, and what levels of workload are imposed)
- *How* (activities are performed and how actors communicate and collaborate to achieve goals)

More than likely, none of the Human Factors methods described can, in isolation, cater for these constructs in their entirety. Using an integrated suite of methods, however, allows scenarios to be analysed exhaustively from the perspectives described. In the past, we have taken a "network of networks" approach to analysing activities taking place in complex sociotechnical systems. This allows activity to be analysed from three different but interlinked perspectives: the task, social, and knowledge networks that underlie teamwork activity. Task networks represent a summary of the goals and subsequent tasks being performed within a system. Social networks analyse the organisation of the team and the communications taking place between the actors working in the team. Finally, knowledge networks describe the information and knowledge (distributed situation awareness) that the actors use and share in order to perform the teamwork activities in question. This "network of networks" approach to understanding collaborative endeavour is represented in Figure 10.1.

SUMMARY OF COMPONENT METHODS

For the purposes of this study, we were interested in the tasks undertaken; the situation awareness, knowledge, and decision-making processes used by fell runners; the level of mental workload imposed on them during task performance; and their interactions with other runners and technological devices during the race. Accordingly we used an adapted version of the EAST approach described in Chapter 8, which included HTA (Annett et al., 1971), OSDs (Stanton et al., 2005), SNA (Driskell and Mullen, 2004), the NASA-TLX (Hart and Staveland, 1988), the propositional

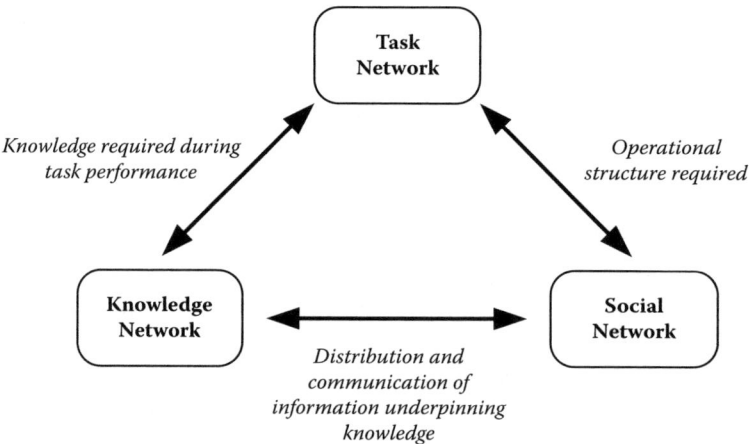

FIGURE 10.1 Network of networks approach. *Source*: Adapted from Houghton, R. J., Baber, C., Cowton, M., Stanton, N. A., and Walker, G. H. (2008). WESTT (Workload, Error, Situational Awareness, Time and Teamwork): An analytical prototyping system for command and control. *Cognition Technology and Work* 10(3):199–207.

Human Factors Methods Integration

network approach (Salmon et al., 2009), and the CDM (Klein, Calderwood, and McGregor, 1989). The framework used is represented in Figure 10.2.

This integrated approach allowed us to consider the following facets of fell runner performance (depicted in Figure 10.3):

- The physical and cognitive tasks involved during the race (HTA, OSD, CDM)
- Fell runner decision making during the race (CDM, propositional networks)
- Fell runner communications during the race (with other runners, race officials, and technology) (SNA, OSD)
- Fell runner workload during the race (NASA-TLX)
- Fell runner situation awareness during the race (propositional networks)

The process begins with the conduct of an observational study (Chapter 2) of the scenario under analysis. HTA (Chapter 3) is then used to describe the goals, sub-goals, and operations involved during the scenario. The resultant task network identifies the actors involved, what tasks are being performed, and the temporal structure of tasks. The HTA output is then used to construct an OSD (Chapter 3) of the task, detailing all activities and interactions between actors involved. The social network is embodied by SNA (Chapter 8), which considers the associations between agents during the scenario. It is important to note here that agents may be human or technological, and so SNA caters too for human-machine interactions. The CDM (Chapter 4) focuses on the decision-making processes used by runners during task performance and the information and knowledge underpinning their decision making. Based on these data, the propositional network approach (Chapter 6) represents situation awareness during the scenario, from both the point of view of the overall system, and the individual actors performing activities within the system. Thus, the type, structure, and distribution of knowledge throughout the scenario is represented. Finally, the NASA-TLX (Chapter 7) provides an indication of the level of workload experienced by actors during the scenario. The overlap between methods and the constructs they access is explained by the multiple perspectives

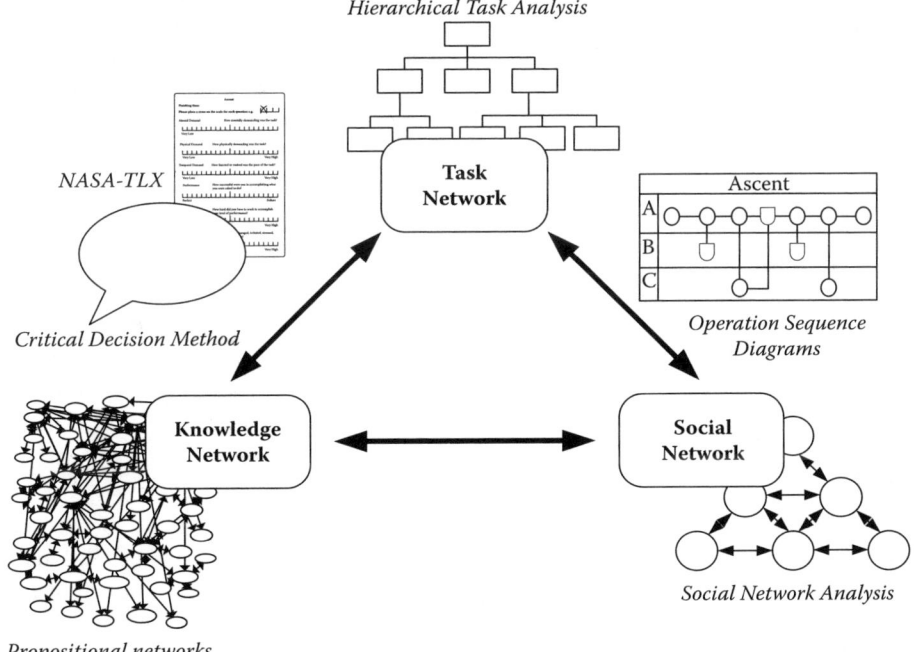

FIGURE 10.2 Network of network approach overlaid with methods applied during case study.

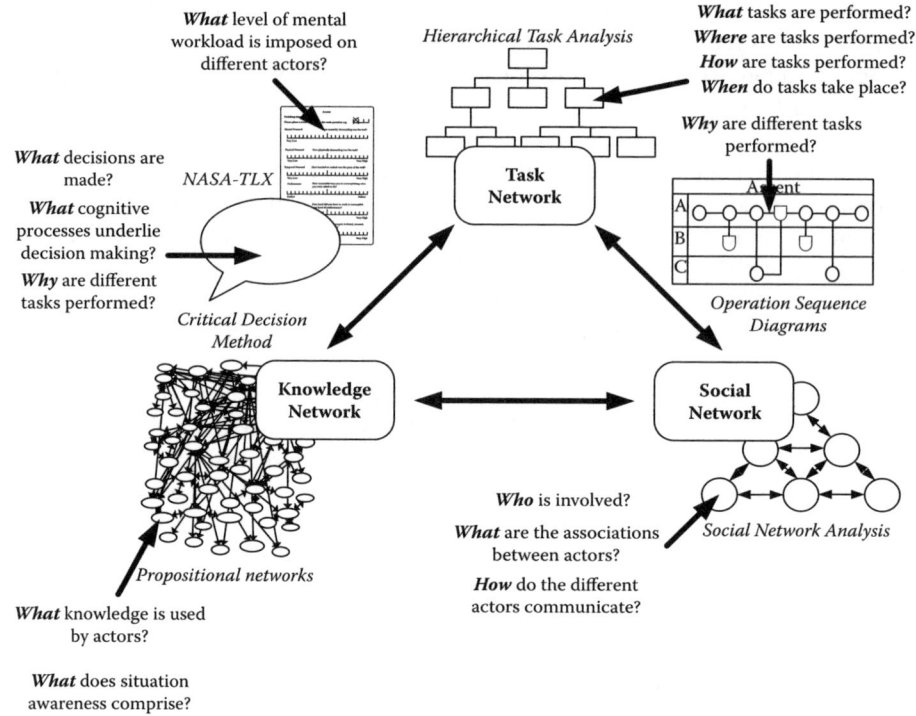

FIGURE 10.3 Network of network approach overlaid with questions potentially addressed.

provided on issues such as the "who" and the "what." For example, the HTA deals with "what" tasks and goals; the CDM deals with "what" decisions are required to achieve goals; and the propositional networks deal with "what" knowledge or situation awareness underpins the tasks being performed and decisions being made. Each of these methods is a different but complementary perspective on the same descriptive construct, and a different but complementary perspective on the same data derived from observation and interview, which is an example of analysis triangulation. This is represented in Figure 10.3. The internal structure of the integrated framework applied is represented in Figure 10.4.

METHODOLOGY

The methodology described above was used to analyse a local fell race. The race was approximately 5.3 miles long and involved an 885-foot ascent from the starting point up to a local monument situated on the Cleveland way, and then a descent back down to the finish.

Participants

A total of 256 senior runners took part in the race. Two analysts completed the race as participant observers, and 10 participants took part in the study. Based upon finishing times, the participants were placed into the four following groups: elite (sub 37 minutes finishing time), amateur (37–44 minutes finishing time), recreational (44–48 minutes finishing time), and other (over 48 minutes finishing time). Upon completion of the race, the runners completed CDM and NASA-TLX questionnaires. Due to the nature of the event and the study procedure, it was not possible to collect demographic data for the participants involved.

Human Factors Methods Integration

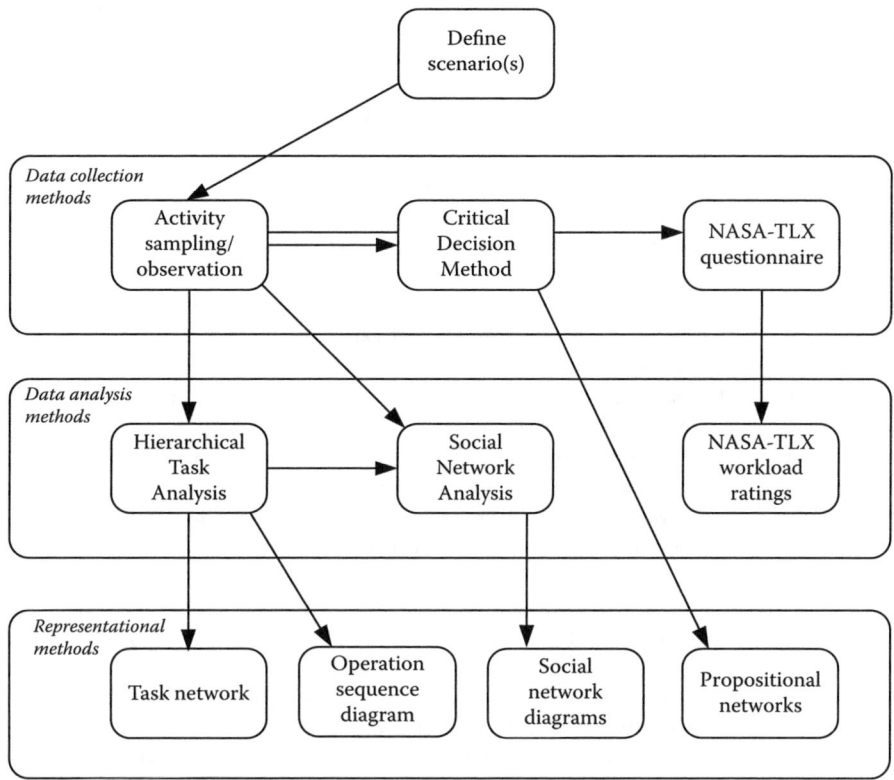

FIGURE 10.4 Internal structure of integrated Human Factors methods framework.

Materials

NASA-TLX and CDM pro-formas were used for data collection purposes. The HTA software tool (Salmon et al., 2009) was used to create the HTA. Microsoft Visio was used to construct the task network, OSD, propositional networks, and social network diagrams presented in this chapter. Some of the runners also used GPS-based timing and heart rate monitoring devices, such as the Garmin 305 Forerunner described in Chapter 9.

Procedure

Initially, an advertisement providing details of the study was placed on the running club website. Upon registration for the race, runners who expressed an interest in taking part in the study were given a study pack, containing an overview of the study, consent forms, and NASA-TLX and CDM pro-formas. Upon completion of the race, participants were asked to complete three NASA-TLX pro-formas: one for a typical training run of a similar distance, one for the ascent part of the race, and one for the descent part of the race. Participants were also asked to complete two CDM pro-formas: one for the ascent part of the race, and one for the descent part of the race. The CDM probes used are presented in Table 10.1. An HTA for the race was constructed based on the two participating analysts' observation of the event. An OSD was developed based on the HTA description of the event. A task network, summarising the goals and tasks contained in the HTA, was then constructed based on the initial HTA description. Propositional networks were constructed for each participant based on a content analysis of their CDM responses. SNAs were conducted based on the two participating analysts, who completed a short SNA questionnaire (i.e., who did you communicate with, how often, and who were the most important communications with?) upon completion of the race.

TABLE 10.1
CDM Probes

	Ascent/Descent
Goal specification	What were your specific goals during this part of the race?
Decisions	What decisions did you make during this part of the race?
Cue identification	What features were you looking for when you formulated your decisions?
	How did you know that you needed to make the decisions? How did you know when to make the decisions?
Expectancy	Were you expecting to make these sorts of decisions during the course of the event? Describe how this affected your decision-making process.
Conceptual	Are there any situations in which your decisions would have turned out differently?
Influence of uncertainty	At any stage, were you uncertain about either the reliability or the relevance of the information that you had available?
Information integration	What was the most important piece of information that you used to make your decisions?
Situation awareness	What information did you have available to you at the time of the decisions?
Situation assessment	Did you use all of the information available to you when making decisions?
	Was there any additional information that you might have used to assist you in making decisions?
Options	Were there any other alternatives available to you other than the decisions you made?
Decision blocking—stress	Was their any stage during the decision-making process in which you found it difficult to process and integrate the information available?
Basis of choice	Do you think that you could develop a rule, based on your experience, that could assist another person to make the same decisions successfully?
Analogy/generalisation	Were you at any time reminded of previous experiences in which similar/different decisions were made?

Source: Adapted from O'Hare, E., Wiggins, M., Williams, A., and Wong, W. (2000). Cognitive task analysis for decision centred design and training. In *Task analysis*, eds. J. Annett and N. A. Stanton, 170–90. London: Taylor & Francis.

Results

The HTA for the fell race is presented in Figure 10.5.

The task network, which provides a summary of the main goals and tasks involved, is presented in Figure 10.6.

The CDM outputs represent a summary of the key decisions made by the runners during the race and the information underlying them. An example CDM output table, elicited from a runner from the "elite" group, is presented in Table 10.2.

Propositional networks were constructed for the ascent and descent portions of the race based on content analyses of participant CDM responses. Example propositional networks for one of the participants are presented in Figure 10.7.

Information element usage by the different runners was analysed using the propositional network data. The information element usage during the race by the four different runner groups is presented in Table 10.3. Table 10.3 shows the total pool of information elements reported by participants and individual participant usage of them based on the propositional network analysis.

Social network diagrams were constructed for the ascent and descent portions of the race based on an SNA questionnaire completed by two runners post trial. Example social network diagrams for one of the runners are presented in Figure 10.8.

Finally, NASA-TLX workload ratings were collected from each participant for a typical training run of a similar distance, and for the ascent and descent portions of the race. The mean NASA-TLX workload ratings are presented in Figure 10.9.

Human Factors Methods Integration

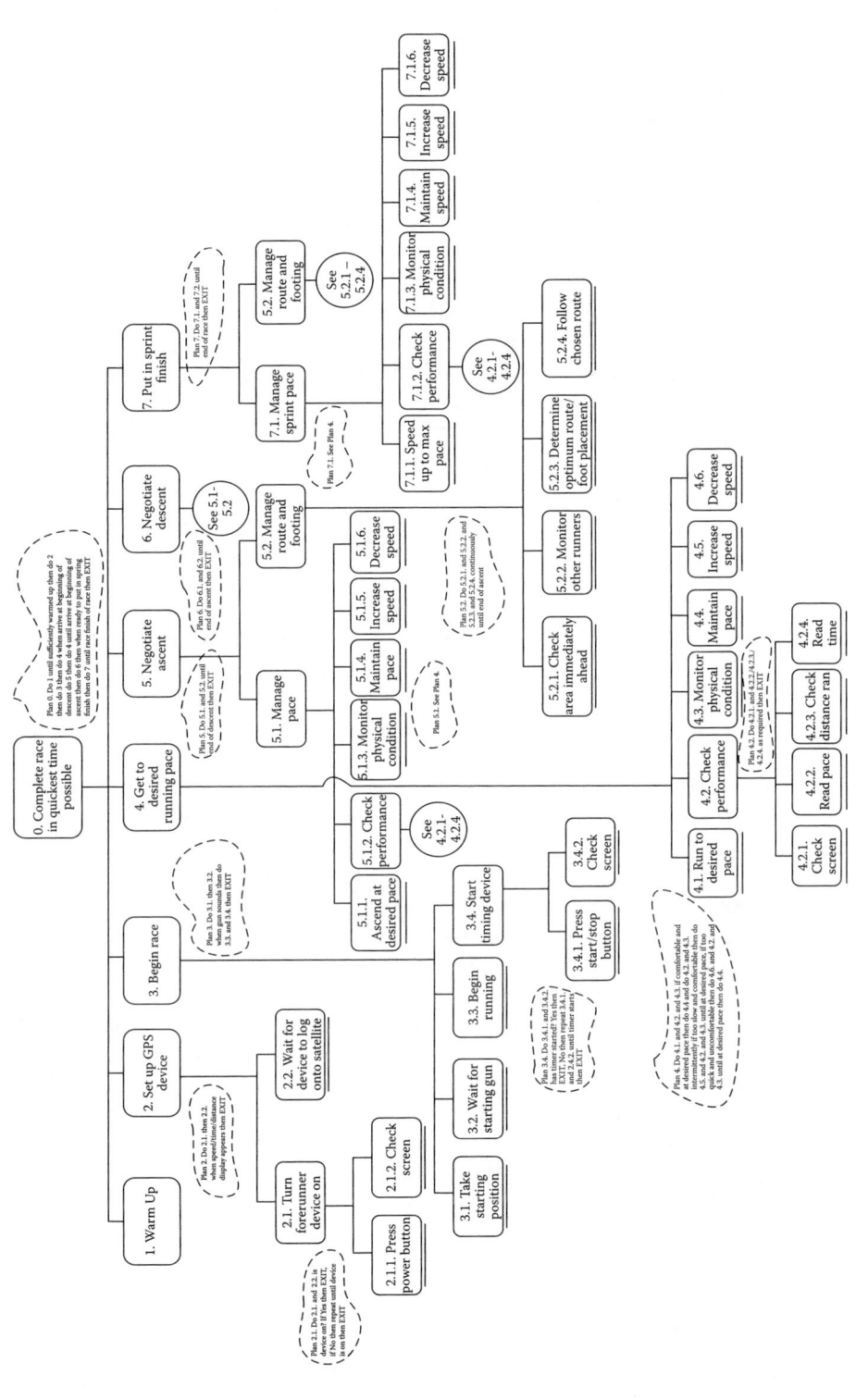

FIGURE 10.5 Fell race HTA.

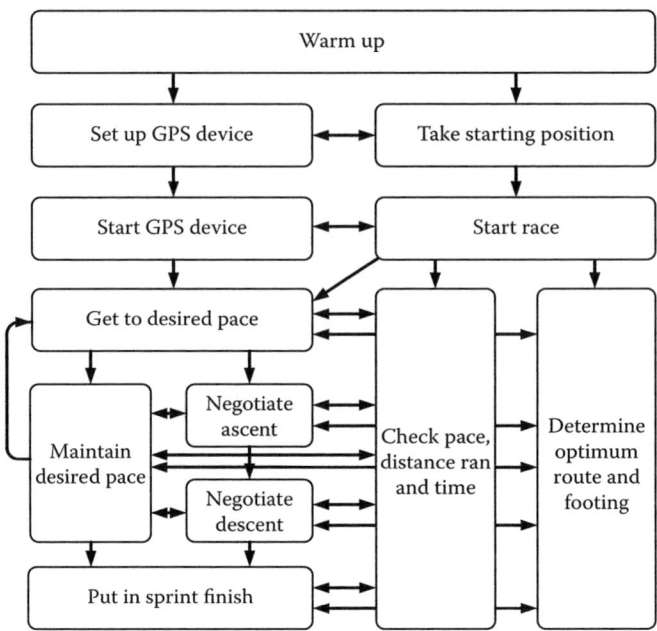

FIGURE 10.6 Task model of fell race.

The mean ratings for each NASA-TLX dimension for the ascent and descent portions of the race are presented for each runner group in Figures 10.10 and 10.11.

DISCUSSION

The analysis presented provides an interesting insight into the goals, situation awareness, and decision-making processes adopted by fell runners of different abilities. The HTA and ensuing task network depict the task structure of the race in terms of how goals and tasks relate to each other functionally and temporally. In this case, the key tasks involved the physical tasks of running the race at an optimum pace, and the cognitive tasks of checking pace, distance ran, and time; calculating the distance remaining; identifying an optimum route; and making key decisions (e.g., when to increase/decrease pace). The technological aspect of the task is also demonstrated, with the runners involved having to set up and then monitor their GPS device throughout the race. The linkage between the main tasks is demonstrated in Figure 10.6. For example, the maintenance of desired pace is informed by the runner's physical condition and the GPS running device, which gives pace, time, and distance ran readings. Maintaining one's desired pace is also contingent upon the selection and negotiation of an optimum route.

The CDM responses indicate that the decision-making process involved during fell running can be broadly decomposed into two forms of decision making: diagnostic decision making (Klien, 1992) and course of action selection (Klein, 2008). Essentially runners used a continuous process of diagnostic decision making to evaluate the current situation and then modified their chosen course of action, or adopted a new course of action, based on their situational diagnosis. Diagnostic decision making in this case refers to the decisions made when diagnosing "what is going on," and involves making decisions during situation assessment, such as what is my current pace, how hard is the runner in front working, how hard am I working? According to Klein (1992), diagnosis involves the decision maker generating one or more potential hypotheses and evaluating them. Klein's previous work in the military domain indicates that diagnostic decisions are more important than decisions regarding course of action selection, since appropriate courses of action are selected based on

TABLE 10.2
Example Elite Runner "Ascent" CDM Output

	Ascent
Goal specification	*What were your specific goals during this part of the race?* • To run as hard as possible to the monument (top of ascent) and get there in as short a time as possible as I knew I could recover on the downhill which followed. In fact I treated it like a 2.5-mile race.
Decisions	*What decisions did you make during this part of the race?* • To maintain a high relative perceived effort (RPE). Also ensure that I reached bottlenecks (e.g., gates) before the runners alongside me. I also tried to stay ahead/with competitors who I knew I was fitter than.
Cue identification	*What features were you looking for when you formulated your decisions?* • Bottlenecks, and fellow competitors I was familiar with. I also glanced at my heart rate from time to time and if it was below 160 I made an extra effort. *How did you know that you needed to make the decisions? How did you know when to make the decisions?* • If I was overtaken by someone who I knew I was fitter than or if my heart rate dropped below 160 BPM.
Expectancy	*Were you expecting to make these sorts of decisions during the course of the event?* • Yes. Just as expected. *Describe how this affected your decision-making process.*
Conceptual	*Are there any situations in which your decisions would have turned out differently?* • No.
Influence of uncertainty	*At any stage, were you uncertain about either the reliability or the relevance of the information that you had available?* • No.
Information integration	*What was the most important piece of information that you used to make your decisions?* • RPE and heart rate to ensure I was running as hard as possible to the monument.
Situation awareness	*What information did you have available to you at the time of the decisions?* • RPE, heart rate, time taken to get to the monument.
Situation assessment	*Did you use all of the information available to you when making decisions?* • No, I didn't use split times at various reference points, which I had recorded in previous years. *Was there any additional information that you might have used to assist you in making decisions?* • Not that I'm aware of.
Options	*Were there any other alternatives available to you other than the decisions you made?* • No.
Decision blocking—stress	*Was their any stage during the decision-making process in which you found it difficult to process and integrate the information available?* • No.
Basis of choice	*Do you think that you could develop a rule, based on your experience, that could assist another person to make the same decisions successfully?* • Yes, if RPE is below 19 or heart rate is below 160 (this is dependent on heart rate maximum) then run harder. Also stay with competitors you are expected to beat.
Analogy/generalisation	*Were you at any time reminded of previous experiences in which similar/different decisions were made?* • All the time.

318 Human Factors Methods and Sports Science: A Practical Guide

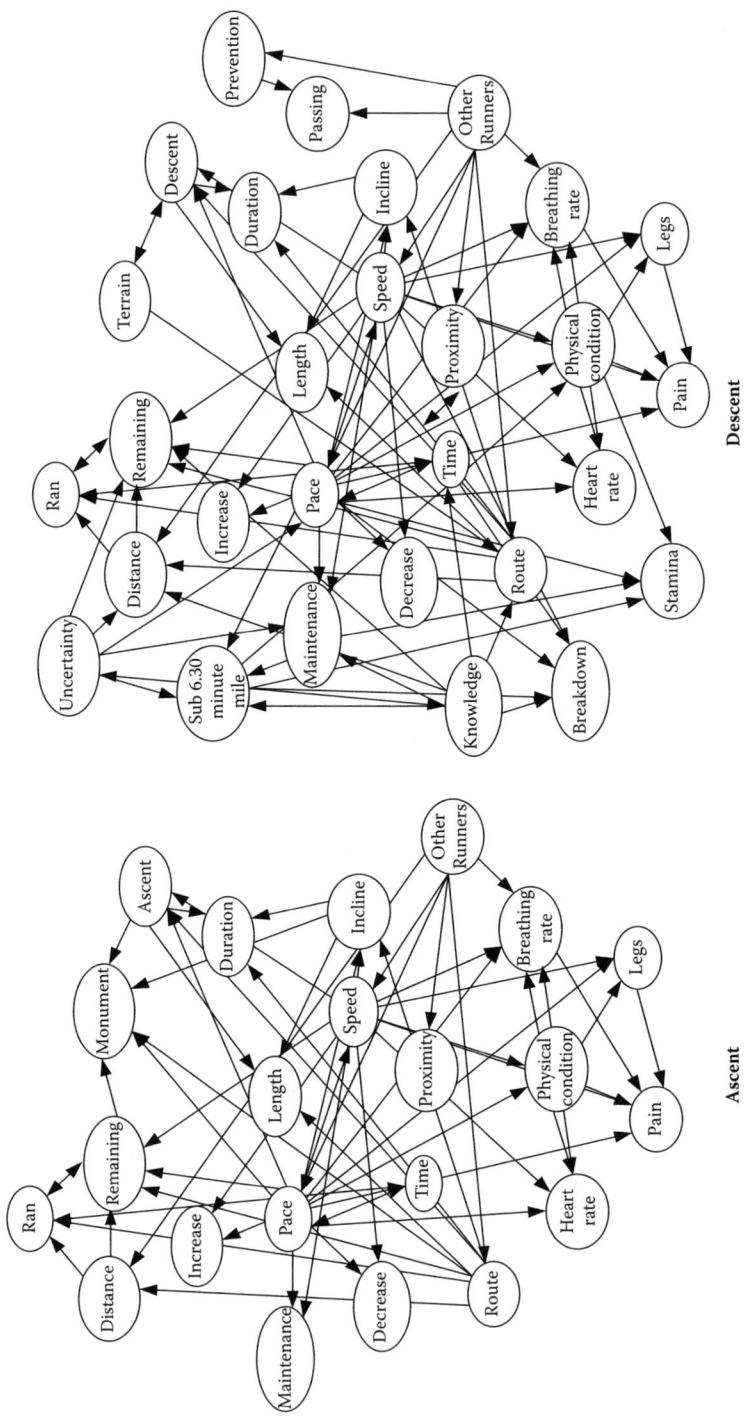

FIGURE 10.7 Ascent and descent propositional networks for amateur runner.

TABLE 10.3
Information Element Usage by Runner Group

	Elite	Amateur	Recreational	Other
Monument	X	X	X	
Pace	X	X	X	X
Speed	X	X	X	X
Time	X	X	X	X
Other runners	X	X	X	
Proximity	X	X	X	X
Route	X	X	X	X
Heart rate	X	X		
Breathing rate	X			
Pain	X	X	X	
Legs	X	X	X	
Physical condition	X		X	
Incline	X			
Ascent	X	X	X	
Maintenance	X		X	
Increase	X	X	X	
Decrease	X	X	X	
Distance	X	X	X	X
Ran	X		X	
Remaining	X	X	X	
Knowledge	X	X	X	X
Position	X	X	X	
Duration	X		X	X
Recovery	X	X	X	X
Race	X	X	X	X
Relative perceived effort	X		X	
Bottlenecks	X	X	X	
Fitness levels	X	X		X
Familiarity	X	X	X	X
Overtake	X	X	X	
Split times	X	X	X	
Reference points	X	X	X	
Previous races	X	X	X	X
Max. heart rate	X	X		
Expectations	X	X		
Descent	X	X	X	
Leaders	X	X	X	
Stress	X	X	X	
Relaxation	X	X	X	
Hindered	X	X	X	
Performance	X	X	X	X
Ahead	X	X	X	
Behind	X	X		X
Tape	X	X	X	
Marshalls	X	X	X	
Burn out	X	X	X	
Walk			X	
Experience	X	X	X	X
Concentration (loss)	X	X	X	X
Terrain	X	X	X	

320   Human Factors Methods and Sports Science: A Practical Guide

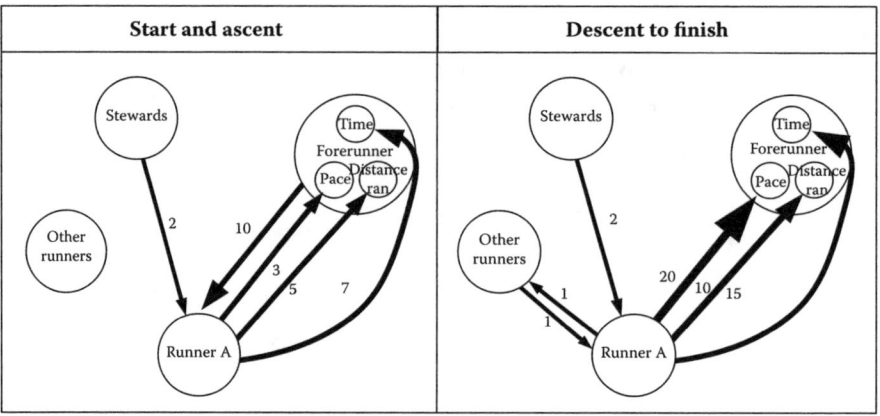

FIGURE 10.8 Example social network diagram showing associations between runner, GPS device, other runners, and stewards.

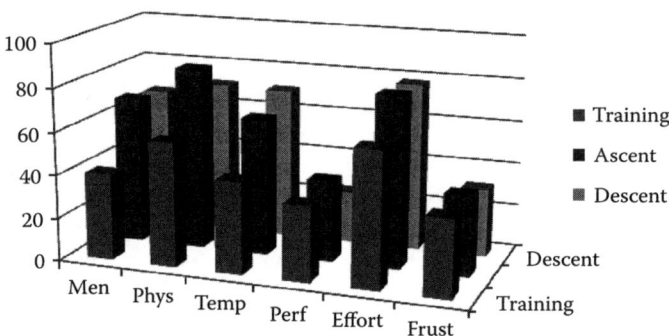

FIGURE 10.9 Mean workload ratings for typical training run and race ascent and descent.

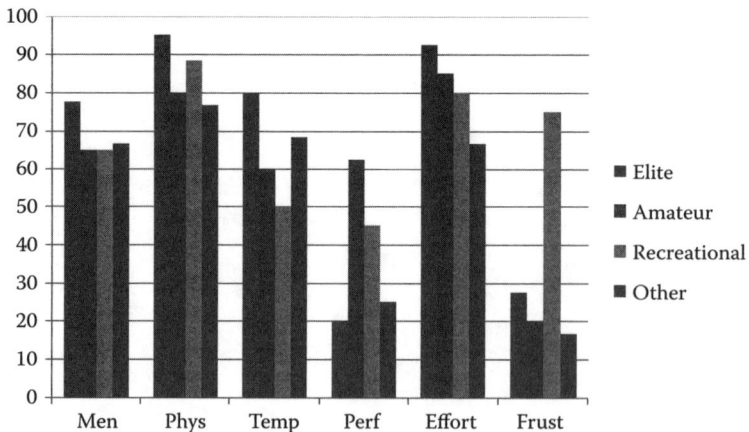

FIGURE 10.10 Ascent mean workload ratings per runner group.

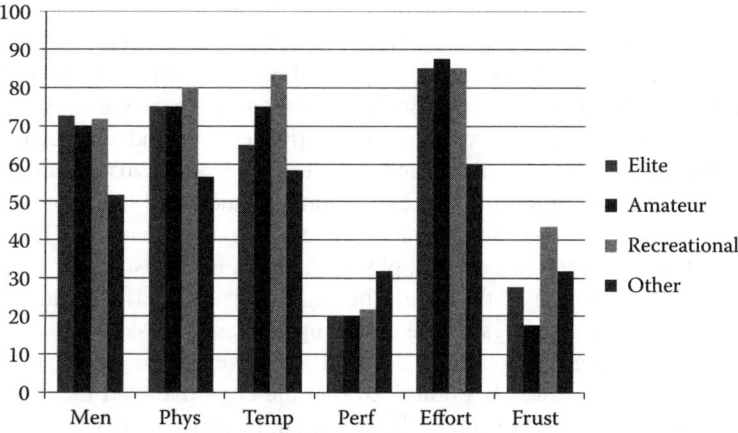

FIGURE 10.11 Descent mean workload ratings per runner group.

an accurate situation assessment (involving diagnostic decision making). Klein suggested that, by achieving a clear situation assessment, the appropriate course of action to use becomes manifest. The main diagnostic decisions reported included determining current pace, rate of physical exertion, heart rate, condition, capability, time, position, and other runners' current workload, condition, and capability. The main course of action decisions that runners reported included route selection, foot placement, race strategy selection (i.e., when to attack or attempt to pass other runners), and pace selection (i.e., when to increase/decrease pace). Following diagnostic decision making, courses of action were chosen and enacted. This cyclical nature of decision making was apparent in all runner CDM responses.

The CDM outputs also demonstrated the differences in race strategies used by the different runner groups. The elite runners decomposed the race into a series of mini races. For example, the first runner from the elite group reported how he treated the ascent portion of the race as a 2.5-mile race, as he knew that he could recover and maintain a quick pace throughout the descent portion of the race. The second runner from the elite group, on the other hand, decomposed the race into three stages: setting up a good opening position, maintaining this position while ensuring he did not burn out, and then improving his position towards the end. No strategies of this sort were reported by the other runners. The use of stored rules for course of action selection was also reported by the elite runners. In this case, the elite runners used situational diagnosis and a comparison with stored rules for course of action selection purposes. For example, one elite runner used rate of perceived exertion and heart rate levels to determine how hard he should run. If both values were above or below a certain threshold amount, then pace was increased/decreased accordingly. Amateur and recreational runners were more likely to base their course of action selection/modification on how they felt physically (i.e., discomfort levels) and other runner behaviour rather than rate of perceived exertion and heart rate levels.

While the types of decision were similar across all runner groups, the information and situation awareness (identified through the CDM and propositional network analyses) underpinning diagnostic decision making and course of action selection was very different. The propositional network analysis indicated that elite runner situation awareness was richer in terms of the number of information elements used and the relationships between the information elements. Elite runners were using more information elements and expressed more relationships between the information being used than the amateur, recreational, and other group runners did.

The SNAs highlight the importance placed on the GPS running watch device by the runners involved. In the example presented, the runner mainly interacts with the GPS in both the ascent and descent phases of the race. It is interesting to note that during the ascent phase the runner is less concerned with checking his pace (due to the incline and difficulty in achieving a good pace) and is receiving more instruction to increase his pace from the GPS device (as depicted by the 10 communications from the GPS device). During the descent phase, however, the runner makes 20 checks of his pace, as he is racing downhill and needs to maintain as quick a pace as possible. As the runner is coming closer to the finish, he also makes more checks of the time expired and distance ran, presumably in order to determine when to embark on his sprint finish. The SNA output also highlights the important role of technology in fell running performance. The analysis indicated that they key agent within a network of runners, stewards, and GPS device was the technological agent: the GPS watch.

The workload ratings indicate that mental, physical, effort, and frustration ratings were greater, on average, for the ascent portion of the race. The higher physical, effort, and frustration ratings were expected, due to the greater physical demand imposed and the slower progress (which leads to frustration) when ascending as opposed to descending; however, the greater mental demand ratings were not expected, since it was hypothesised that the cognitive load incurred by route selection during the fast-paced descent portion of the race would lead to greater mental demand ratings. Temporal demand was, on average, rated as higher during the descent portion of the race. This reflects the fact that the descent portion is a much quicker part of the race and also the fact that runners are getting closer to the finish of the race. For the comparison across groups, the elite runner group provided greater ratings for all dimensions aside from frustration for the ascent phase; however, similar ratings were given by all groups for the descent phase.

In isolation the results generated are of utility; however, further insight can be gained by viewing the outputs in unison. For example, Figure 10.12 presents an example of the multi-perspective view

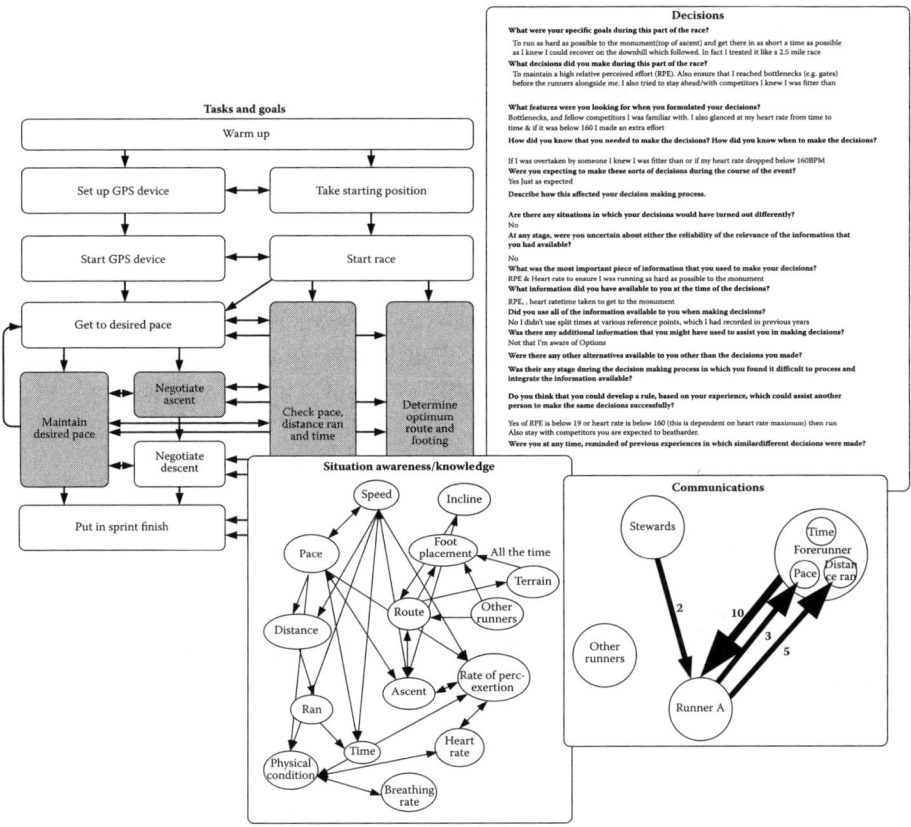

FIGURE 10.12 Multi-perspective output example.

on activity provided by the different outputs. In this case, the tasks, decisions, situation awareness, and communications involved for the ascent portion of the race are represented.

CONCLUSIONS

The analysis presented indicates that the main decisions involved in fell running are diagnostic decisions regarding current performance (e.g., pace, position, rate of exertion) and course of action selection decisions (e.g., route selection, foot placement, race strategy, changes of pace) regarding responses to situational diagnosis. The analysis suggests that the decision-making processes adopted by elite runners are different than those of runners of less ability, and also that elite runner situation awareness is richer, in terms of the information comprising it, and more "connected" in terms of their understanding of the relationships between different concepts. The importance of different components of the running system was emphasised; for example, the important role of technology was highlighted, in particular the provision of pace, time, and distance ran information by GPS-based devices. Further, the use of other runners in determining race strategy was also highlighted. Finally, fell runner workload was found to be greater during the ascent phase of the race.

This study is presented as a proof of concept study for Human Factors methods integration. The individual outputs provided by the integrated Human Factors methods approach used for this case study provide a number of distinct but overlapping perspectives on the fell race scenario analysed. The important point to make is that further combinations and recombinations of separate Human Factors methods enables insights previously generated to be explored further. The application described demonstrates the utility of applying integrated suites of Human Factors methods to tackle the evaluation of complex sociotechnical system performance, such as that seen in the sporting domains. The approach presented summarises the task, social, and knowledge structures involved into a form that enables comparisons on key metrics across actors and scenarios. Beyond this, the method demonstrates how the individual Human Factors methods described in this book can be combined to fully encompass the boundaries of complex sociotechnical systems.

References

Adamic, L. A., Buyukkokten, O., and Adar, E. (2003). A social network caught in the web. *First Monday* 8(6).
Adams, M. J., Tenney, Y. J., and Pew, R. W. (1995). Situation awareness and the cognitive management of complex systems. *Human Factors* 37(1):85–104.
Adams, P., and David, G. C. (2007). Light vehicle fuelling errors in the UK: The nature of the problem, its consequences and prevention. *Applied Ergonomics* 38(5):499–511.
Ahlstrom, U. (2005). Work domain analysis for air traffic controller weather displays. *Journal of Safety Research* 36(2):159–69.
Andersson, H., Ekblom, B., and Kustrup, P. (2008). Elite football on artificial turf versus natural grass: Movement patterns, technical standards, and player impressions. *Journal of Sports Sciences* 26(2):113–22.
Anderson, J.R. (1983). The Architecture of Cognition. Cambridge, MA: Harvard University Press.
Annett, J. (2003). Hierarchical task analysis. In *The handbook of task analysis for human-computer interaction*, eds. D. Diaper and N. A. Stanton, 67–82. Mahwah, NJ: Lawrence Erlbaum Associates.
———. (2004). Hierarchical task analysis. In *Handbook of Human Factors and ergonomics methods*, eds. N. A. Stanton, A. Hedge, K. Brookhuis, E. Salas, and H. Hendrick, 33-1–33-7. Boca Raton, FL: CRC Press.
Annett, J., Duncan, K. D., Stammers, R. B., and Gray, M. J. (1971). Task analysis. Department of Employment Training Information Paper 6. HMSO, London.
Annett, J., and Stanton, N. A. (1998). Task analysis. *Ergonomics* 41(11):1529–36.
———. (2000). *Task analysis*. London, Taylor & Francis.
Artman, H., and Garbis, C. (1998). Situation awareness as distributed cognition. In *Cognition and cooperation: Proceedings of Ninth Conference of Cognitive Ergonomics*, eds. T. Green, L. Bannon, C. Warren, Buckley, 151–6). Limerick, Ireland.
Baber, C. (1996). Repertory Grid Theory and its application to product evaluation. In *Usability in industry*, eds. P. Jordan, et al., 157–65. London: Taylor & Francis.
Baber, C., and Stanton, N. A. (1994). Task analysis for error identification: A methodology for designing error-tolerant consumer products. *Ergonomics* 37(11):1923–41.
———. (1996a). Human error identification techniques applied to public technology: Predictions compared with observed use. *Applied Ergonomics* 27(2):119–31.
———. (1996b). Observation as a technique for usability evaluation. In *Usability evaluation in industry*, eds. P. W. Jordan, B. Thomas, B. A. Weerdmeester, and I. McClelland, 85–94. London: Taylor & Francis.
———. (1999). Analytical prototyping. In *Interface technology: The leading edge,* eds. J. M. Noyes and M. Cook. Baldock: Research Studies Press.
———. (2001). Analytical prototyping of personal technologies: Using predictions of time and error to evaluate user interfaces. In *Proceedings of IFIP INTERACT01: Human-Computer Interaction 2001*, 585–92. Tokyo, Japan.
———. (2002). Task analysis for error identification: Theory, method and validation. *Theoretical Issues in Ergonomics Science* 3(2):212–27.
Bainbridge, L. (1972). An analysis of a verbal protocol from a process control task. Unpublished Ph.D. thesis, University of Bristol, 1972.
———. (1987). Ironies of automation. *Automatica* 19(6):775–9.
———. (1995). Verbal protocol analysis. In *Evaluation of human work: A practical ergonomics methodology*, eds. J. R. Wilson and E. N. Corlett, 161–79. London: Taylor & Francis.
Baker, D. (2004). Behavioural observation scales (BOS). In *Handbook of Human Factors and ergonomics methods*, eds. N. A. Stanton, A. Hedge, K. Brookhuis, E. Salas, and H. Hendrick. Boca Raton, FL: CRC Press.
Baker, D. P., Salas, E., and Cannon-Bowers, J. A. (1998). Team task analysis: Lost but hopefully not forgotten. *Industrial and Organizational Psychologist* 35(3):79–83.
Bartlett, F. C. (1932). *Remembering: A study in experimental and social psychology*. Cambridge: Cambridge University Press.
Bass, A., Aspinall, J., Walters, G., and Stanton, N. A. (1995). A software toolkit for hierarchical task analysis. *Applied Ergonomics* 26(2):147–51.
BBC Sport (2008). The laws of scrummaging. http://news.bbc.co.uk/sport1/hi/rugby_union/rules_and_equipment/4205180.stm. Accessed January 17, 2009.

Bedny, G., and Meister, D. (1999). Theory of activity and situation awareness. *International Journal of Cognitive Ergonomics* 3(1):63–72.

Bell, B. S., and Kozlowski, S. W. J. (2002). A typology of virtual teams: Implications for effective leadership. *Group Organisation Management* 27(1):14–49.

Billings, C. E. (1995). Situation awareness measurement and analysis: A commentary. *Proceedings of the International Conference on Experimental Analysis and Measurement of Situation Awareness*, Embry-Riddle Aeronautical University Press, FL.

Bisantz, A. M., Roth, E., Brickman, B., Gosbee, L. L., Hettinger, L., and McKinney, J. (2003). Integrating cognitive analyses in a large-scale system design process. *International Journal of Human-Computer Studies* 58(2):177–206.

Blandford, A., and Wong, B. L. W. (2004). Situation awareness in emergency medical dispatch. *International Journal of Human-Computer Studies* 61(4):421–52.

Bolstad, C. A., Riley, J. M., Jones, D. G., and Endsley, M. R. (2002). Using goal directed task analysis with Army brigade officer teams. In *Proceedings of the 46th Annual Meeting of the Human Factors and Ergonomics Society*, 472–6. Baltimore, MD.

Bowers, C., & Jentsch, F. (2004). Team workload analysis. . In N. A. Stanton, A. Hedge, K. Brookhuis, E. Salas, & H. Hendrick (Eds.), *Handbook of Human Factors and Ergonomics Methods*. Boca Raton, USA, CRC Press.

Brooke, J. (1996). SUS: A "quick and dirty" usability scale. In *Usability evaluation in industry*, eds. P. W. Jordan, B. Thomas, B. A. Weerdmeester, and A. L. McClelland. London: Taylor & Francis.

Bruseberg, A., and Shepherd, A. (1997). Job design in integrated mail processing. In *Engineering psychology and cognitive ergonomics. Volume two: Job design and product design*, ed. D. Harris, 25–32. Aldershot, UK: Ashgate Publishing.

Burford, B. (1993). Designing adaptive automatic telling machines. Unpublished MSc thesis, University of Birmingham.

Burke, S. C. (2004). Team task analysis. In *Handbook of Human Factors and ergonomics methods*, eds. N. A. Stanton, A. Hedge, K. Brookhuis, E. Salas, and H. Hendrick, 56.1–56.8). Boca Raton, FL: CRC Press.

Burns, C. M., and Hajdukiewicz, H. R. (2004). *Ecological interface design*. Boca Raton, FL: CRC Press.

Carron, A. V., Bray, S. R., and Eys, M. A. (2002). Team cohesion and team success in sport. *Journal of Sports Sciences* 20(2):119–26.

Cha, D. W. (2001). Comparative study of subjective workload assessment techniques for the evaluation of ITS-orientated human-machine interface systems. *Journal of Korean Society of Transportation* 19(3):450–8.

Chin, J. P., Diehl, V. A., and Norman, K. L. (1988). Development of an instrument measuring user satisfaction of the human-computer interface. CHI'88.

Chow, G. M., and Feltz, D. L. (2008). Exploring the relationships between collective efficacy, perceptions of success, and team attributions. *Journal of Sports Sciences* 26(11):1179–89.

Collier, S. G., and Folleso, K. (1995). SACRI: a measure of situation awareness for nuclear power control rooms. In: D.J. Garland and M.R. Endsley, Editors, Experimental Analysis and Measurement of Situation Awareness, Embry-Riddle University Press, Daytona Beach, FL (1995), pp. 115–122.

Crandall, B., Klein, G., and Hoffman, R. (2006). Working minds: A practitioner's guide to cognitive task analysis. Cambridge, MA: MIT Press.

Crawford, J. O., Taylor, C., and Po, N. L. W. (2001). A case study of on-screen prototypes and usability evaluation of electronic timers and food menu systems. *International Journal of Human Computer Interaction* 13(2):187–201.

Davidson, A., and Trewartha, G. (2008). Understanding the physiological demands of netball: A time-motion investigation. *International Journal of Performance Analysis in Sport* 8(3):1–17.

Dekker, A. H. (2002). Applying social network analysis concepts to military C4ISR architectures. *Connections* 24(3):93–103.

Dekker, S. W. A. (2002). Reconstructing human contributions to accidents: The new view on human error and performance. *Journal of Safety Research* 33(3):371–85.

Diaper, D., and Stanton, N. S. (2004). *The handbook of task analysis for human-computer interaction*. Mahwah, NJ: Lawrence Erlbaum Associates.

Dixon, S. J. (2008). Use of pressure insoles to compare in-shoe loading for modern running shoes. *Ergonomics* 51:1503–14.

Doggart, L., Keane, S., Reilly, T., and Stanhope, J. (1993). A task analysis of Gaelic football. In *Science and football II*, eds. T. Reilly, J. Clarys, and A. Stibbe, 186–9. London: E and FN Spon.

References

Dominguez, C. (1994). Can SA be defined? In *Situation Awareness: Papers and annotated bibliography*. Report AL/CF-TR-1994-0085, eds. M. Vidulich, C. Dominguez, E. Vogel, and G. McMillan. Wright-Patterson Airforce Base, OH: Air Force Systems Command.

Driskell, J. E., and Mullen, B. (2004). Social network analysis. In *Handbook of Human Factors and ergonomics methods*, eds. N. A. Stanton, A. Hedge, K. Brookhuis, E. Salas, and H. Hendrick, 58.1–58.6. Boca Raton, FL: CRC Press.

Drury, C. (1990). Methods for direct observation of performance. In *Evaluation of human work*, eds. J. R. Wilson and E. N. Corlett, 45–68. London: Taylor & Francis.

Durso, F. T., Hackworth, C. A., Truitt, T., Crutchfield, J., and Manning, C. A. (1998). Situation awareness as a predictor of performance in en route air traffic controllers. *Air Traffic Quarterly* 6(1):1–20.

Easterby, R. (1984). Tasks, processes and display design. In *Information design*, eds. R. Easterby and H. Zwaga, 19–36. Chichester: John Wiley and Sons.

Embrey, D. E. (1986). SHERPA: A systematic human error reduction and prediction approach. Paper presented at the International Meeting on Advances in Nuclear Power Systems, Knoxville, TN, USA.

Endsley, M. R. (1989). *Final report: Situation awareness in an advanced strategic mission (NOR DOC 89-32)*. Hawthorne, CA: Northrop Corporation.

———. (1990). Predictive utility of an objective measure of situation awareness. In *Proceedings of the Human Factors Society 34th Annual Meeting*, 41–5. Santa Monica, CA: Human Factors Society.

———. (1993). A survey of situation awareness requirements in air-to-air combat fighters. *The International Journal of Aviation Psychology* 3(2):157–68.

———. (1995a). Measurement of situation awareness in dynamic systems. *Human Factors* 37(1):65–84.

———. (1995b). Towards a theory of situation awareness in dynamic systems. *Human Factors* 37(1):32–64.

———. (2000). Theoretical underpinnings of situation awareness: A critical review. In *Situation awareness analysis and measurement*, eds. M. R. Endsley and D. J. Garland. Mahwah, NJ: Laurence Erlbaum Associates.

———. (2001). Designing for situation awareness in complex systems. In *Proceedings of the Second International Workshop on Symbiosis of Humans, Artifacts and Environment*, Kyoto, Japan.

Endsley, M. R., Bolte, B., and Jones, D. G. (2003). *Designing for situation awareness: An approach to user-centred design*. London: Taylor & Francis.

Endsley, M. R., Holder, C. D., Leibricht, B. C., Garland, D. C., Wampler, R. L., and Matthews, M. D. (2000). *Modelling and measuring situation awareness in the infantry operational environment. (1753)*. Alexandria, VA: Army Research Institute.

Endsley, M. R., and Jones, W. M. (1997). *Situation awareness, information dominance, and information warfare. Technical report 97-01*. Belmont, MA: Endsley Consulting.

Endsley, M. R., and Kiris, E. O. (1995). *Situation awareness global assessment technique (SAGAT) TRACON air traffic control version user guide*. Lubbock TX: Texas Tech University.

Endsley, M. R., and Robertson, M. M. (2000). Situation awareness in aircraft maintenance teams. *International Journal of Industrial Ergonomics* 26(2):301–25.

Endsley, M. R., & Rodgers, M. D. (1994). Situation Awareness Global Assessment Technique (SAGAT): en route air traffic control version user's guide. Lubbock, TX, Texas Tech University.

Endsley, M. R., Selcon, S. J., Hardiman, T. D., and Croft, D. G. (1998). A comparative evaluation of SAGAT and SART for evaluations of situation awareness. In *Proceedings of the Human Factors and Ergonomics Society Annual Meeting*, 82–6. Santa Monica, CA: Human Factors and Ergonomics Society.

Endsley, M. R., Sollenberger, R., and Stein, E. (2000). Situation awareness: A comparison of measures. In *Proceedings of the Human Performance, Situation Awareness and Automation: User-Centered Design for the New Millennium*. Savannah, GA: SA Technologies, Inc.

Eysenck, M. W., and M.T. Keane. (1990). *Cognitive psychology: A student's handbook*. Hove, UK: Lawrence Erlbaum.

Federal Aviation Administration. (1996). Report on the interfaces between flightcrews and modern flight deck systems, Federal Aviation Administration, Washington DC, USA.

Ferarro, V. F., Sforza, C., Dugnani, S., Michielon, G., and Mauro, F. (1999). Morphological variation analysis of the repeatability of soccer offensive schemes. *Journal of Sport Sciences* 17:89–95.

Fidel, R., and Pejtersen, A. M. (2005). Cognitive work analysis. In *Theories of information behavior: A researcher's guide*, eds. K. E. Fisher, S. Erdelez, and E. F. McKechnie. Medford, NJ: Information Today.

Fiore, S. M., and Salas, E. (2006). Team cognition and expert teams: Developing insights from cross-disciplinary analysis of exceptional teams. *International Journal of Sports and Exercise Psychology* 4:369–75.

———. (2008). Cognition, competition, and coordination: The "why" and the "how" of the relevance of the Sports Sciences to learning and performance in the military. *Military Psychology* 20(1):S1–S9.

Fiore, S. M., Salas, E., Cuevas, H. M., and Bowers, C. A. (2003). Distributed coordination space: Toward a theory of distributed team process and performance. *Theoretical Issues in Ergonomics Science* 4(3-4):340–64.

Flanagan, J. C. (1954). The Critical Incident Technique. *Psychological Bulletin* 51(4):327–58.

Fleishman, E. A., and Zaccaro, S. J. (1992). Toward a taxonomy of team performance functions. In *Teams: Their training and performance*, eds. R. W. Swezey and E. Salas, 31–56. Norwood, NJ: Ablex.

Fox, J., Code, S. L., and Langfield-Smith, K. (2000). Team mental models: Techniques, methods and analytic approaches. *Human Factors* 42(2):242–71.

Fracker, M. (1991). *Measures of situation awareness: Review and future directions (Rep. No.AL-TR-1991-0128)*. Wright Patterson Air Force Base, OH: Armstrong Laboratories, Crew Systems Directorate.

Garland, D. J., & Endsley, M. R. (1995). *Experimental analysis and measurement of situation awareness*. Daytona Beach, FL: Embry-Riddle, Aeronautical University Press.

Glendon, A. I., and McKenna, E. F. (1995). *Human safety and risk management*. London: Chapman and Hall.

Goodwin, G. F. (2008). Psychology in sports and the military: Building understanding and collaboration across disciplines. *Military Psychology* 20(1):S147–S153.

Gorman, J. C., Cooke, N., and Winner, J. L. (2006). Measuring team situation awareness in decentralised command and control environments. *Ergonomics* 49(12-13):1312–26.

Gould, P. R., and Gatrell, A. (1980). A structural analysis of a game: The Liverpool v. Manchester United Cup final of 1977. *Social Networks* 2:247–67.

Grant, A. G., Williams, A. M., and Reilly, T. (1999). Communications to the fourth World Congress of Science and Football. *Journal of Sports Sciences* 17(10):807–40.

Gregoriades, A., and Sutcliffe, A. (2007). Workload prediction for improved design and reliability of complex systems. *Reliability Engineering and System Safety* 93(4):530–49.

Hackman, J. R. (1987). The design of work teams. In *Handbook of organizational behavior*, ed. J. Lorsch, 315–42. Englewood Cliffs, NJ: Prentice-Hall.

Hajdukiewicz, J. R. (1998). Development of a structured approach for patient monitoring in the operating room. Masters thesis. University of Toronto.

Hajdukiewicz, J. R., and Vicente, K. J. (2004). A theoretical note on the relationship between work domain analysis and task analysis. *Theoretical Issues in Ergonomics Science* 5(6):527–38.

Hancock, P. A., and Verwey, W. B. (1997). Fatigue, workload and adaptive driver systems. *Accident Analysis and Prevention* 29(4):495–506.

Hanton, S., Fletcher, D., and Coughlan, G. (2005). Stress in elite sport performers: A comparative study of competitive and organizational stressors. *Journal of Sports Sciences* 23(10):1129–41.

Harris, D., Stanton, N. A., Marshall, A., Young, M. S., Demagalski, J., and Salmon, P. M. (2005). Using SHERPA to predict design induced error on the flight deck. *Aerospace Science and Technology* 9(6):525–32.

Hart, S. G., and Staveland, L. E. (1988). Development of a multi-dimensional workload rating scale: Results of empirical and theoretical research. In *Human mental workload*, eds. P. A. Hancock and N. Meshkati. Amsterdam, The Netherlands: Elsevier.

Hauss, Y., and Eyferth, K. (2003). Securing future ATM-concepts' safety by measuring situation awareness in air traffic control. *Aerospace Science and Technology* 7(6):417–27.

Hazlehurst, B., McMullen, C. K., and Gorman, P. N. (2007). Distributed cognition in the heart room: How situation awareness arises from coordinated communications during cardiac surgery. *Journal of Biomedical Informatics* 40(5):539–51.

Helmreich, R. L. (2000). On error management: Lessons from aviation. *British Medical Journal* 320:781–5.

Helmreich, R. L., and Foushee, H. C. (1993). Why crew resource management? Empirical and theoretical bases of human factors training in aviation. In *Cockpit resource management*, eds. E. Wiener, B. Kanki, and R. Helmreich, 3–45. San Diego, CA: Academic Press.

Helsen, W., and Bultynck, J. B. (2004). Physical and perceptual-cognitive demands of top-class refereeing in association football. *Journal of Sports Sciences* 22:179–89.

Helsen, W., Gilis, B., and Weston, M. (2006). Errors in judging "offside" in association football: Test of the optical error versus the perceptual flash-lag hypothesis. *Journal of Sports Sciences* 24(5):521–8.

Hess, B. (1999). Graduate student cognition during information retrieval using the World Wide Web: A pilot study. *Computers and Education* 33(1):1–13.

Higgins, P. G. (1998). Extending cognitive work analysis to manufacturing scheduling. OzCHI '98, 236–43. Adelaide, Australia, Nov. 30–Dec. 4.

Hilburn, B. (1997). Dynamic decision aiding: The impact of adaptive automation on mental workload. In *Engineering psychology and cognitive ergonomics*, ed. D. Harris, 193–200. Aldershot: Ashgate.

Hilliard, A., and Jamieson, G. A. (2008). Winning solar races with interface design. *Ergonomics in Design: The Quarterly of Human Factors Applications* 16(2):6–11.

References

Hodginkson, G. P., and Crawshaw, C. M. (1985). Hierarchical task analysis for ergonomics research. An application of the method to the design and evaluation of sound mixing consoles. *Applied Ergonomics* 16(4):289–99.

Hogg, D. N., Folleso, K., Strand-Volden, F., and Torralba, B. (1995). Development of a situation awareness measure to evaluate advanced alarm systems in nuclear power plant control rooms. *Ergonomics* 38(11):2394–2413.

Hollnagel, E. (1998). *Cognitive reliability and error analysis method—CREAM*, 1st ed. Oxford, England: Elsevier Science.

———. (2003). *Handbook of cognitive task design*. Mahwah, NJ: Erlbaum.

Hopkins, A. (2000). *Lessons from Longford: The Esso gas plant explosion*. Sydney: CCH.

———. (2005). *Safety, culture and risk: The organisational causes of disasters*. Sydney: CCH.

Houghton, R. J., Baber, C., Cowton, M., Stanton, N. A. and Walker, G. H. (2008). WESTT (Workload, Error, Situational Awareness, Time and Teamwork): An analytical prototyping system for command and control. *Cognition Technology and Work* 10(3):199–207.

Houghton, R. J., Baber, C., McMaster, R., Stanton, N. A., Salmon, P., Stewart, R., and Walker, G. (2006). Command and control in emergency services operations: A social network analysis. *Ergonomics* 49(12-13):1204–25.

Hughes, M. D., and Franks, I. M. (2004). *Notational analysis of sport: Systems for better coaching and performance in sport*, 2nd ed. London: Routledge.

———. (2005). Possession length and goal scoring in soccer. *Journal of Sport Sciences* 23:509–14.

Hutchins, E. (1995). *Cognition in the wild*. Cambridge, MA: MIT Press.

Isaac, A., Shorrock, S. T., Kennedy, R., Kirwan, B., Anderson, H., and Bove, T. (2002). Technical review of human performance models and taxonomies of error in air traffic management (HERA). Eurocontrol Project report. HRS/HSP-002-REP-01.

James, N. (2006). Notational analysis in soccer: Past, present and future. *International Journal of Performance Analysis in Sport* 6(2):67–81.

James, N., and Patrick, J. (2004). The role of situation awareness in sport. In *A cognitive approach to situation awareness: theory and application*, eds. S. Banbury and S. Tremblay, 297–316. Aldershot, UK: Ashgate.

Jamieson, G. A., and Vicente, K. J. (2001). Ecological interface design for petrochemical applications: Supporting operator adaptation, continuous learning, and distributed, collaborative work. *Computers and Chemical Engineering* 25(7):1055–74.

Jansson, A., Olsson, E., and Erlandsson, M. (2006). Bridging the gap between analysis and design: Improving existing driver interfaces with tools from the framework of cognitive work analysis. *Cognition, Technology and Work* 8(1):41–49.

Jeannott, E., Kelly, C., Thompson, D. (2003). The development of Situation Awareness measures in ATM systems. EATMP report. HRS/HSP-005-REP-01.

Jenkins, D. P., Salmon, P. M., Stanton, N. A., and Walker, G. H. (in press). A systemic approach to accident analysis: A case study of the Stockwell shooting. Submitted to *Ergonomics*.

Jenkins, D. P., Stanton, N. A., Walker, G. H., and Salmon, P. M. (2008). *Cognitive work analysis: Coping with complexity*. Aldershot, UK: Ashgate.

Johnson, C. (1999). Why human error modelling has failed to help systems development. *Interacting with Computers* 11(5):517–24.

Johnson C. W., and de Almeida, I. M. (2008). Extending the borders of accident investigation: Applying novel analysis techniques to the loss of the Brazilian space launch vehicle VLS-1 V03. *Safety Science* 46(1):38–53.

Jones, D. G., and Endsley, M. R. (2000). Can real-time probes provide a valid measure of situation awareness? *Proceedings of the Human Performance, Situation Awareness and Automation: User Centred Design for the New Millennium Conference*, October.

Jones, D. G., and Kaber, D. B. (2004). Situation awareness measurement and the situation awareness global assessment technique. In *Handbook of Human Factors and ergonomics methods*, eds. N. Stanton, A. Hedge, H. Hendrick, K. Brookhuis, and E. Salas, 42.1–42.7). Boca Raton, FL: CRC Press.

Jones, N. M. P., James, N., and Mellalieu, S. D. (2008). An objective method for depicting team performance in elite professional rugby union. *Journal of Sports Sciences* 26(7):691–700.

Jordet, G. (2009). Why do English players fail in soccer penalty shootouts? A study of team status, self-regulation, and choking under pressure. *Journal of Sports Sciences* 27(2):97–106.

Kaber, D. B., Perry, C. M., Segall, N., McClernon, C. K., and Prinzel, L. P. (2006). Situation awareness implications of adaptive automation for information processing in an air traffic control-related task. *International Journal of Industrial Ergonomics* 36:447–62.

Kahneman, D. (1973). *Attention and effort*. Englewood Cliffs, NJ: Prentice-Hall.

Karwowski, W. (2001). *International encyclopaedia of ergonomics and human factors*. London: Taylor & Francis.

Kidman, L. and Hanrahan, S. J. (2004). *The coaching process: A practical guide to improve your effectiveness*. Sydney: Thompson Learning.

Kieras, D. (2003). GOMS models for task analysis. In *The handbook of task analysis for human-computer interaction*, eds. D. Diaper and N. A. Stanton, 83–117. Mahwah, NJ: Lawrence Erlbaum Associates.

Kilgore, R., and St-Cyr, O. (2006). The SRK Inventory: A tool for structuring and capturing a worker competencies analysis. *Human Factors and Ergonomics Society Annual Meeting Proceedings, Cognitive Engineering and Decision Making*, 506–9.

Kirakowski, J. (1996). The software usability measurement inventory: Background and usage. In *Usability evaluation in industry*, eds. P. Jordan, B. Thomas, and B. Weerdmeester. London: Taylor & Francis.

Kirwan, B. (1992a). Human error identification in human reliability assessment. Part 1: Overview of approaches. *Applied Ergonomics* 23(5):299–318.

———. (1992b). Human error identification in human reliability assessment. Part 2: Detailed comparison of techniques. *Applied Ergonomics* 23(6):371–81.

———. (1994). *A guide to practical human reliability assessment*. London: Taylor & Francis.

———. (1996). The validation of three Human Reliability Quantification techniques—THERP, HEART and JHEDI: Part 1—Technique descriptions and validation issues, *Applied Ergonomics* 27(6):359–73.

———. (1998a). Human error identification techniques for risk assessment of high-risk systems—Part 1: Review and evaluation of techniques. *Applied Ergonomics* 29(3):157–77.

———. (1998b). Human error identification techniques for risk assessment of high-risk systems—Part 2: Towards a framework approach. *Applied Ergonomics* 29(5):299–319.

Kirwan, B., and Ainsworth, L. K. (1992). *A guide to task analysis*. London: Taylor & Francis.

Kirwan, B., Evans, A., Donohoe, L., Kilner, A., Lamoureaux, T., Atkinson, T., and MacKendrick, H. (1997) Human Factors in the ATM system design life cycle. FAA/Eurocontrol ATM RandD Seminar, Paris, France, June 16–20.

Klein, G. (1992). Decision making in complex military environments. Fairborn, OH: Klein Associates Inc. Prepared under contract, N66001-90-C-6023 for the Naval Command, Control and Ocean Surveillance Center, San Diego, CA.

Klein, G. (2000). Cognitive task analysis of teams. In *Cognitive task analysis*, eds. J. M. Schraagen, S. F. Chipman, and V. L. Shalin, 417–31. Mahwah, NJ: Lawrence Erlbaum Associates.

Klein, G. (2008). Naturalistic decision making. *Human Factors*, 50:3, pp. 456 – 460.

Klein, G., and Armstrong, A. A. (2004). Critical decision method. In *Handbook of Human Factors and ergonomics methods*, eds. N. A. Stanton, A. Hedge, E. Salas, H. Hendrick, and K. Brookhaus, 35.1–35.8. Boca Raton, FL: CRC Press.

Klein, G., Calderwood, R, and McGregor, D. (1989). Critical decision method for eliciting knowledge. *IEEE Transactions on Systems, Man and Cybernetics* 19(3):462–72.

Klinger, D. W., and Hahn, B. B. (2004). Team decision requirement exercise: Making team decision requirements explicit. In *Handbook of Human Factors methods*, eds. N. A. Stanton, A. Hedge, K. Brookhuis, E. Salas, and H. Hendrick. Boca Raton, FL: CRC Press.

Kuperman, G. G. (1985). Pro-SWAT applied to advanced helicopter crew station concepts. *Human Factors and Ergonomics Society Annual Meeting Proceedings* 29(4):398–402.

Lake, M. J. (2000). Determining the protective function of sports footwear. *Ergonomics* 43(10):1610–21.

Lane, R., Stanton, N. A., and Harrison, D. (2007). Applying hierarchical task analysis to medication administration errors. *Applied Ergonomics* 37(5):669–79.

Lawton, R., and Ward, N. J. (2005). A systems analysis of the Ladbroke grove rail crash. *Accident Analysis and Prevention* 37:235–44.

Lee, J. D. (2006). Human factors and ergonomics in automation design. In *Handbook of human factors and ergonomics*, ed. G. Salvendy, 1570–97. New York: John Wiley and Sons.

Lees, A., Rojas, J., Cepero, M., Soto, V., and Gutierrez, M. (2000). How the free limbs are used by elite high jumpers in generating vertical velocity. *Ergonomics* 43(10):1622–36.

Lehto, J. R., and Buck, M. (2007). *Introduction to Human Factors and Ergonomics for engineers*. Boca Raton, FL: CRC Press.

References

Lemyre, P. N., Roberts, G. C., and Stray-Gundersen, J. (2007). Motivation, overtraining, and burnout: Can self-determined motivation predict overtraining and burnout in elite athletes? *European Journal of Sport Science* 7(2):115–26.

Luximon, A., and Goonetilleke, R. S. (2001). A simplified subjective workload assessment technique. *Ergonomics* 44(3):229–43.

Ma, R., and Kaber, D. B. (2007). Situation awareness and driving performance in a simulated navigation task. *Ergonomics* 50:1351–64.

Macquet, A. C., and Fleurance, P. (2007). Naturalistic decision-making in expert badminton players. *Ergonomics* 50(9):433–50.

Marsden, P., and Kirby, M. (2004). Allocation of functions. In *Handbook of Human Factors and ergonomics methods*, eds. N. A. Stanton, A. Hedge, K. Brookhuis, E. Salas, and H. Hendrick. Boca Raton, FL: CRC Press.

Martin, J. A., Smith N. C., Tolfrey K., and Jones A. M. (2001). Activity analysis of English premiership rugby football union refereeing. *Ergonomics* 44(12):1069–75.

Matthews, M. D., Strater, L. D., and Endsley, M. R. (2004). Situation awareness requirements for infantry platoon leaders. *Military Psychology* 16(3):149–61.

Megaw, T. (2005). The definition and measurement of mental workload. In *Evaluation of human work*, eds. J. R. Wilson and N. Corlett, 525–53. Boca Raton, FL: CRC Press.

McClelland, I., and Fulton Suri, J. F. (2005). Involving people in design. In *Evaluation of human work: A practical ergonomics methodology*, eds. J. R Wilson and E. Corlett, 281–334. Boca Raton, FL: CRC Press.

McFadden, K. L., and Towell, E. R. (1999). Aviation human factors: A framework for the new millennium. *Journal of Air Transport Management* 5:177–84.

McGarry, T., and Franks, I. M. (2000). On winning the penalty shoot-out in soccer. *Journal of Sports Sciences* 18(6):401–9.

McGuinness, B., and Foy, L. (2000). A subjective measure of SA: The Crew Awareness Rating Scale (CARS). Paper presented at the Human Performance, Situational Awareness and Automation Conference, Savannah, Georgia, Oct. 16–19.

McIntyre, R. M., and Dickinson, T. L. (1992). Systemic assessment of teamwork processes in tactical environments. Naval Training Systems Centre, Contract No. N16339-91-C-0145, Norfolk, VA, Old Dominion University.

McLean, D. A. (1992). Analysis of the physical demands of international rugby union. *Journal of Sports Sciences* 10(3):285–96.

McMorris, T., Delves, S., Sproule, J., Lauder, M., and Hale, B. (2005). Incremental exercise, perceived task demands and performance of a whole body motor task. *British Journal of Sports Medicine* 39:537–41.

McPherson, S. L., and Kernodle, M. (2007). Mapping two new points on the tennis expertise continuum: Tactical skills of adult advanced beginners and entry-level professionals during competition. *Journal of Sports Sciences* 25(8):945–59.

Meister, D. (1989) Conceptual aspects of Human Factors. Baltimore: John Hopkins University Press.

Militello, L. G., and Hutton, J. B. (2000). Applied cognitive task analysis (ACTA): A practitioner's toolkit for understanding cognitive task demands. In *Task analysis*, eds. J. Annett and N. S. Stanton, 90–113. London: Taylor & Francis.

Miller, C. A., and Vicente K. J. (2001). Comparison of display requirements generated via hierarchical task and abstraction-decomposition space analysis techniques. *International Journal of Cognitive Ergonomics* 5(3):335–55.

Miller, G. A., Galanter, E., and Pribram, K. H. (1960). *Plans and the structure of behaviour*. New York: Holt.

Montgomery, A. (2008). The goal that never was—Is this the most ridiculous refereeing decision ever? Daily Mail, September 21. http://www.dailymail.co.uk/sport/football/article-1058801/The-goal--ridiculous-refereeing-decision-ever.html. Accessed February 22, 2009.

Murrell, K. F. H. (1965). *Human performance in industry*. New York: Reinhold Publishing.

Nachreiner, F. (1995). Standards for ergonomics principles relating to the design of work systems and to mental workload. *Applied Ergonomics* 26(4):259–63.

Naikar, N. (2006). Beyond interface design: Further applications of cognitive work analysis. *International Journal of Industrial Ergonomics* 36(5):423–38.

Naikar, N., and Sanderson, P. M. (1999). Work domain analysis for training-system definition. *International Journal of Aviation Psychology* 9(3):271–90.

———. (2001). Evaluating design proposals for complex systems with work domain analysis. *Human Factors* 43(4):529–42.

Naikar, N., Hopcroft, R., and Moylan, A. (2005). Work domain analysis: Theoretical concepts and methodology. Defence Science and Technology Organisation Report, DSTO-TR-1665, Melbourne, Australia.

Naikar, N., Moylan, A., and Pearce, B. (2006). Analysing activity in complex systems with cognitive work analysis: Concepts, guidelines, and case study for control task analysis. *Theoretical Issues in Ergonomics Science* 7(4):371–94.

Naikar, N., Pearce, B., Drumm, D., and Sanderson, P. M. (2003). Technique for designing teams for first-of-a-kind complex systems with cognitive work analysis: Case study. *Human Factors* 45(2):202–17.

Naikar, N., and Saunders, A. (2003). Crossing the boundaries of safe operation: A technical training approach to error management. *Cognition Technology and Work* 5:171–80.

Neerincx, M. A. (2003). Cognitive Task Load Analysis: Allocating tasks and designing support. In *Handbook of cognitive task design*, ed. E. Hollnagel, 281–305. Mahwah, NJ: Lawrence Erlbaum Associates, Inc.

Neisser, U. (1976). *Cognition and reality: Principles and implications of cognitive psychology*. San Francisco: Freeman.

Nelson, W. R, Haney, L. N, Ostrom, L. T, and Richards, R. E. (1998). Structured methods for identifying and correcting potential human errors in space operations. *Acta Astronautica* 43:211–22.

Nicholls, A., and Polman, R. (2008). Think aloud: Acute stress and coping strategies during golf performances. *Anxiety, Stress and Coping* 21(3):283–94.

Noakes, T. D. (2000). Exercise and the cold. *Ergonomics* 43(10):1461–79.

Norman, D. A. (1988). *The psychology of everyday things*. New York: Basic Books.

Noyes, J. M. (2006). Verbal protocol analysis. In *International encyclopaedia of Ergonomics and Human Factors,* 2nd ed., ed. W. Karwowski, 3390–92. London: Taylor & Francis.

Nygren, T. E. (1991). Psychometric properties of subjective workload measurement techniques: Implications for their use in the assessment of perceived mental workload. *Human Factors* 33(1):17–33.

O'Donoghue, P., and Ingram, B. A. (2001). Notational analysis of elite tennis strategy. *Journal of Sports Sciences* 19:107–15.

O'Hare, D., Wiggins, M., Williams, A., and Wong, W. (2000). Cognitive task analysis for decision centred design and training. In *Task analysis*, eds. J. Annett and N. A. Stanton, 170–90), London: Taylor & Francis.

Openheim, A. N. (2000). *Questionnaire design, interviewing and attitude measurement*. London: Continuum.

Orasanu, J., and Fischer, U. (1997). Finding decisions in natural environments: The view from the cockpit. In *Naturalistic decision making*, eds. C. E. Zsambok and G. Klein, 343–57. Mahwah, NJ: Lawrence Erlbaum.

Oudejans, R. R. D., Bakker, F. C., and Beek, P. J. (2007). Helsen, Gilis and Weston (2006) err in testing the optical error hypothesis. *Journal of Sports Sciences* 25(9):987–90.

Paris, C. R., Salas, E., and Cannon-Bowers, J. A. (2000). Teamwork in multi-person systems: A review and analysis. *Ergonomics* 43(8):1052–75.

Pasmore, W., Francis, F., Haldeman, J., and Shani, A. (1982). Sociotechnical systems: A North American reflection on empirical studies of the seventies. *Human Relations* 35(12):1179–1204.

Patrick, J., Gregov, A., and Halliday, P. (2000). Analysing and training task analysis. *Instructional Science* 28(4)57–79.

Patrick, J., James, N., Ahmed, A., and Halliday, P. (2006). Observational assessment of situation awareness, team differences and training implications. *Ergonomics* 49(12-13):393–417.

Patrick, J., Spurgeon, P., and Shepherd, A. (1986.) *A guide to task analysis: Applications of hierarchical methods*. Birmingham: An Occupational Services Publication.

Pedersen, H. K, and Cooke, N. J. (2006). From battle plans to football plays: Extending military team cognition to football. *International Journal of Sport and Exercise Psychology* 4:422–46.

Piso, E. (1981) Task analysis for process-control tasks: The method of Annett et al. applied. *Occupational Psychology* 54:347–54.

Pocock, S., Fields, R. E., Harrison, M. D., and Wright, P. C. (2001). THEA—A reference guide. University of York Technical Report.

Pocock, S., Harrison, M., Wright, P., and Johnson, P. (2001). THEA—A technique for human error assessment early in design. In *Human-computer interaction: INTERACT'01*. ed. M. Hirose, 247–54. IOS Press.

Potter, S. S., Roth, E. M., Woods, D. D., and Elm, W. C. (2000). Bootstrapping multiple converging cognitive task analysis techniques for system design. In *Cognitive task analysis*, eds. J. M. Schraagen, S. F. Chipman, and V. L. Shaling, 317–40. Mahwah, NJ: Lawrence Erlbaum Associates.

Purvis, A., and Tunstall, H. (2004). Effects of sock type on foot skin temperature and thermal demand during exercise. *Ergonomics* 47(15);1657–68.

Rahnama, N., Lees, A., and Bambaecichi, E. (2005). A comparison of muscle strength and flexibility between the preferred and non-preferred leg in English soccer players. *Ergonomics* 48(11-14):1568–75.

Rasmussen, J. (1982). Human error: A taxonomy for describing human malfunction in industrial installations. *Journal of Occupational Accidents* 4:311–33.

———. (1986). *Information processing and human-machine interaction: An approach to cognitive engineering*. New York: North-Holland. http://www.ischool.washington.edu/chii/portal/literature.html. Accessed August 14, 2008.

———. (1997). Risk management in a dynamic society: A modelling problem. *Safety Science* 27(2/3):183–213.

Ravden, S. J., and Johnson, G. I. (1989). *Evaluating usability of human-computer interfaces: A practical method*. Chirchester: Ellis Horwood.

Reason, J. (1990). *Human error*. New York: Cambridge University Press.

———. (1997). *Managing the risks of organisational accidents*. Burlington, VT: Ashgate Publishing Ltd.

———. (2000). Human error: Models and management. *British Medical Journal* 320:768–70.

Reid, G. B., and Nygren, T. E. (1988). The subjective workload assessment technique: A scaling procedure for measuring mental workload. In *Human mental workload*, eds. P. S. Hancock and N. Meshkati. Amsterdam, The Netherlands: Elsevier.

Reifman, A. (2006). Network analysis of basketball passing patterns II. In *Proceedings of the International Workshop and Conference on Network Science 2006*, Bloomington, USA.

Reilly, T. (1986). Fundamental studies on soccer. In *Sportwissenschaft und sportpraxis*, 114–21). Hamburg: Verlag Ingrid Czwalina.

———. (1997). Energetics of high intensity exercise (soccer) with particular reference to fatigue. *Journal of Sports Sciences* 15:257–63.

Reimer, T., Park, E. S., and Hinsz, V. B. (2006). Shared and coordinated cognition in competitive and dynamic task environments: An information-processing perspective for team sports. *International Journal of Exercise and Sport Psychology* 4:376–400.

Riley, J., and Meadows, J. (1995). The role of information in disaster planning: A case study approach. *Library Management* 16(4):18–24.

Riley, J. M., Endsley, M. R., Bolstad, C. A., and Cuevas, H. M. (2006). Collaborative planning and situation awareness in army command and control. *Ergonomics* 49:1139–53.

Roberts, S. P., Trewartha, G., Higgitt, R. J., El-Abd, J., and Stokes, K. A. (2008). The physical demands of elite English rugby union. *Journal of Sports Sciences* 26(8):825–33.

Rojas, F. J., Cepero, M., Ona, A., Gutierrez, M. (2000). Kinematic adjustments in the basketball jump shot against an opponent. *Ergonomics* 43(10):1651–60.

Roth, E. M. (2008). Uncovering the requirements of cognitive work. *Human Factors* 50(3):475–80.

Roth, E. M., Patterson, E. S., and Mumaw, R. J. (2002). Cognitive engineering: Issues in user-centered system design. In *Encyclopaedia of software engineering*, 2nd Ed., ed. J. J. Marciniak, 163–79). New York: Wiley-Interscience, John Wiley and Sons.

Royal Australian Air Force. (2001). *The report of the F-111 Deseal/Reseal Board of Inquiry*. Canberra, ACT: Air Force Head Quarters.

Rubio, S., Diaz, E., Martin, J., & Puente, J. M. (2003). Evaluation of subjective mental workload: a comparison of SWAT, NASA TLX and workload profile methods. *Applied Psychology*, 53:1, pp. 61–86.

Rumar, K. (1990). The basic driver error: Late detection. *Ergonomics* 33(10/11):1281–90.

Salas, E. (2004). Team methods. In *Handbook of Human Factors methods*, eds. N. A. Stanton, A. Hedge, K. Brookhuis, E. Salas, and H. Hendrick, Boca Raton, FL: CRC Press.

Salas, E., Burke, C. S., and Samman, S. N. (2001). Understanding command and control teams operating in complex environments. *Information Knowledge Systems Management* 2(4):311–23.

Salas, E., and Cannon-Bowers, J. A. (2000). The anatomy of team training. In *Training and retraining: A handbook for business, industry, government and the military*, eds. S. Tobias and J. D. Fletcher, 312–35. New York: Macmillan Reference.

Salas, E., Cooke, N. J., and Rosen, M. A. (2008). On teams, teamwork, and team performance: Discoveries and developments. *Human Factors* 50(3):540–47.

Salas, E., Prince, C., Baker, P. D., and Shrestha, L. (1995). Situation awareness in team performance. *Human Factors* 37:123–36.

Salas, E., Sims, D. E., and Burke, C. S. (2005). Is there a big five in teamwork? *Small Group Research* 36(5):555–99.

Salmon, P. M., Gibbon, A. C., Stanton, N. A., Jenkins, D. P., and Walker, G. H. (2009). Network analysis and sport: A social network analysis of the England soccer teams passing performance. Unpublished manuscript.

Salmon, P. M., Jenkins, D. P., Stanton, N. A., and Walker, G. H. (in press). Goals versus constraints: A theoretical and methodological comparison of Hierarchical Task Analysis and Cognitive Work Analysis. *Theoretical Issues in Ergonomics Science*.

Salmon, P. M., Regan, M., and Johnston, I. (2007). Managing road user error in Australia: Where are we now, where are we going and how are we going to get there? In *Multimodal safety management and human factors: Crossing the borders of medical, aviation, road and rail industries*, ed. J. Anca, 143–56. Aldershot, UK: Ashgate.

Salmon, P. M., Stanton, N. A., Jenkins, D. P., Walker, G. H. (in press). Hierarchical task analysis versus cognitive work analysis: comparison of theory, methodology, and contribution to system design. *Theoretical Issues in Ergonomics Science*. Accepted for publication 29th May 2009.

Salmon, P. M., Stanton, N. A., Walker, G. H., Jenkins, D. P., Ladva, D., Rafferty, L., Young, M. S. (2009). Measuring situation awareness in complex systems: Comparison of measures study. *International Journal of Industrial Ergonomics*, 39, pp. 490-500.

Salmon, P. M., Stanton, N. A., Jenkins, D. P., Walker, G. H., Rafferty, L. and Revell, K. (in press). Decisions, decisions … and even more decisions: The impact of digitisation on decision making in the land warfare domain. *International Journal of Human Computer Interaction*.

Salmon, P. M., Stanton, N. A., Regan, M., Lenne, M., and Young, K. (2007). Work domain analysis and road transport: Implications for vehicle design. *International Journal of Vehicle Design* 45(3):426–48.

Salmon, P. M, Stanton, N. A., Walker, G. H., Baber, C., Jenkins, D. P. and McMaster, R. (2008). What really is going on? Review of situation awareness models for individuals and teams. *Theoretical Issues in Ergonomics Science* 9(4):297–323.

Salmon, P., Stanton, N., Walker, G., and Green, D. (2006). Situation awareness measurement: A review of applicability for C4i environments. *Applied Ergonomics* 37(2):225–38.

Salmon, P. M., Stanton, N. A., Walker, G. H., and Jenkins, D. P. (2009). *Distributed situation awareness: Advances in theory, measurement and application to teamwork*. Aldershot, UK: Ashgate.

———. (2009). Analysing the analysis in task analysis: The hierarchical task analysis software tool. Unpublished manuscript.

Salmon, P. M., Stanton, N. A., Walker, G. H., Jenkins, D. P., Baber, C., and McMaster, R. (2008). Representing situation awareness in collaborative systems: A case study in the energy distribution domain. *Ergonomics* 51(3):367–84.

Salmon, P. M., Stanton, N. A., Young, M. S., Harris, D., Demagalski, J., Marshall, A., Waldon, T., and Dekker, S. (2002). Using existing HEI techniques to predict pilot error: A comparison of SHERPA, HAZOP and HEIST. In *Proceedings of International Conference on Human-Computer Interaction in Aeronautics, HCI-Aero 2002*, eds. S. Chatty, J. Hansman, and G. Boy, 126–130. Menlo Park, CA: AAAI Press.

Salvendy, G. (1997). *Handbook of human factors and ergonomics*. New York: John Wiley and Sons.

Sanders, M. S., and McCormick, E. J. (1993). *Human factors in engineering and design*. New York: McGraw-Hill.

Sanderson, P. M. (2003). Cognitive work analysis. In *HCI models, theories, and frameworks: Toward an interdisciplinary science*, ed. J. Carroll. New York: Morgan-Kaufmann.

Sarter, N. B., and Woods, D. D. (1991). Situation awareness—A critical but ill-defined phenomenon. *International Journal of Aviation Psychology* 1(1):45–57.

Schaafstal, A., and Schraagen, J. M. (2000). Training of troubleshooting: A structured, analytical approach. In *Cognitive task analysis*, eds. J. M. Schraagen, S. F. Chipman, and V. L. Shaling, 57–71. Mahwah, NJ: Lawrence Erlbaum Associates.

Schraagen, J. M., Chipman, S. F., and Shalin, V. L. (2000). *Cognitive task analysis*. Mahwah, NJ: Lawrence Erlbaum Associates.

Scoulding, A., James, N., and Taylor, J. (2004). Passing in the soccer world cup 2002. *International Journal of Performance Analysis in Sport* 4(2):36–41.

Seamster, T. L., Redding, R. E., and Kaempf, G. L. (2000). A skill-based cognitive task analysis framework, In *Cognitive task analysis*, eds. J. M. Schraagen, S. F. Chipman, and V. L. Shaling, 135–46. Mahwah, NJ: Lawrence Erlbaum Associates.

Sebok, A. (2000). Team performance in process control: Influence of interface design and staffing levels. *Ergonomics* 43(8):1210–36.

Senders, J., and Moray, N. (1991). *Human error: Cause, prediction and reduction*. Hillsdale, NJ: Lawrence Erlbaum Associates.

Shadbolt, N. R., and Burton, M. (1995). Knowledge elicitation: A systemic approach. In *Evaluation of human work: A practical ergonomics methodology*, eds. J. R. Wilson and E. N. Corlett, 406–40.

Sharit, J. (2006). Human error. In *Handbook of human factors and ergonomics*, ed. G. Salvendy, 708–60. New York: John Wiley and Sons.

Shepherd, A. (2001). *Hierarchical task analysis*. London: Taylor & Francis.

Shorrock, S. T., Kirwan, B. (2002). Development and application of a human error identification tool for air traffic control. *Applied Ergonomics* 33:319–36.

Shu, Y., and Furuta, K. (2005). An inference method of team situation awareness based on mutual awareness. *Cognition Technology and Work* 7(4):272–87.

Shute, V. J., Torreano, L. A., and Willis, R. E. (2000). Tools to aid cognitive task analysis. In *Cognitive task analysis*, eds. J. M. Schraagen, S. F. Chipman, and V. L. Shaling. Hillsdale, NJ: Lawrence Erlbaum Associates.

Siemieniuch, C. E., and Sinclair, M. A. (2006). Systems integration. *Applied Ergonomics* 37(1):91–110.

Skillicorn, D. B. (2004). Social network analysis via matrix decompositions: al Qaeda. Unpublished manuscript.

Smith, K., and Hancock, P. A. (1995). Situation awareness is adaptive, externally directed consciousness. *Human Factors* 37(1):137–48.

Smith. M., and Cushion, C. J. (2006). An investigation of the in-game behaviours of professional, top-level youth soccer coaches. *Journal of Sport Sciences* 24(4):355–66.

Smolensky, M. W. (1993). Toward the physiological measurement of situation awareness: The case for eye movement measurements. In *Proceedings of the Human Factors and Ergonomics Society 37th Annual Meeting*. Santa Monica: Human Factors and Ergonomics Society.

Sonnenwald, D. H., Maglaughlin, K. L., and Whitton, M. C. (2004). Designing to support situation awareness across distances: An example from a scientific collaboratory. *Information Processing and Management* 40(6):989–1011.

Sparks, S., Cable, N., Doran, D., and Maclaren, D. (2005). The influence of environmental temperature on duathlon performance. *Ergonomics* 48(11-14):1558–67.

Spath, D., Braun, M., and Hagenmeyer, L. (2006). Human factors and ergonomics in manufacturing and process control. In *Handbook of human factors and ergonomics*, ed. G. Salvendy, 1597–1626. New York: John Wiley and Sons.

Spath, D., Braun, M., & Hagenmeyer, L. (2007). Human factors and ergonomics in manufacturing and process control. In G. Salvendy (Ed.) *Handbook of human factors and ergonomics* (pp. 1597 – 1626), John Wiley & Sons, New Jersey.

Stammers, R. B., and Astley, J. A. (1987). Hierarchical task analysis: Twenty years on. In *Contemporary ergonomics*, ed. E. D. Megaw, 135–9. London: Taylor & Francis.

Stanton, N., Harris, D., Salmon, P. M., Demagalski, J. M., Marshall, A., Young, M. S., Dekker, S. W. A. & Waldmann, T. (2006). Predicting design induced pilot error using HET (Human Error Template) – A new formal human error identification method for flight decks. *Journal of Aeronautical Sciences*, February, pp. 107-115.

Stanton, N. A. (2006). Hierarchical task analysis: Developments, applications, and extensions. *Applied Ergonomics* 37(1):55–79.

Stanton, N. A., and Baber, C. (1996). A systems approach to human error identification. *Safety Science* 22(1-3):215–28.

———. (1998). A systems analysis of consumer products. In *Human Factors in consumer products*, ed. N. A. Stanton. London: Taylor & Francis.

———. (2002). Error by design: Methods for predicting device usability. *Design Studies* 23(4):363–84.

Stanton, N. A., Chambers, P. R. G., Piggott, J. (2001). Situational awareness and safety. *Safety Science* 39(3):189–204.

Stanton, N., Harris, D., Salmon, P. M., Demagalski, J. M., Marshall, A., Young, M. S., Dekker, S. W. A., and Waldmann, T. (2006). Predicting design induced pilot error using HET (Human Error Template)—A new formal human error identification method for flight decks. *Journal of Aeronautical Sciences* February:107–15.

Stanton, N. A., Hedge, A., Brookhuis, K., Salas, E., and Hendrick, H. (2004). *Handbook of Human Factors methods*. Boca Raton, FL: CRC Press.

Stanton, N. A., Jenkins, D. P., Salmon, P. M., Walker, G. H., Rafferty, L., and Revell, K. (in press). *Digitising command and control: A human factors and ergonomics analysis of mission planning and battlespace management*. Aldershot, UK: Ashgate.

Stanton, N. A., Salmon, P. M., Walker, G., Baber, C., and Jenkins, D. P. (2005). *Human Factors methods: A practical guide for engineering and design*. Aldershot, UK: Ashgate.

Stanton, N. A., Salmon, P. M., Walker, G. H., and Jenkins, D. P. (2009). Genotype and phenotype schemata and their role in distributed situation awareness in collaborative systems. *Theoretical Issues in Ergonomics Science* 10(1):43–68.

Stanton, N. A., and Stevenage, S. V. (1998). Learning to predict human error: Issues of acceptability, reliability and validity. *Ergonomics* 41(11):1737–47.

Stanton, N. A., Stewart, R., Harris, D., Houghton, R. J., Baber, C., McMaster, R., Salmon, P., Hoyle, G., Walker, G., Young, M. S., Linsell, M., Dymott, R., and Green D. (2006). Distributed situation awareness in dynamic systems: Theoretical development and application of an ergonomics methodology. *Ergonomics* 49:1288–1311.

Stanton, N. A., Walker, G. H., Young, M. S., Kazi, T. A., and Salmon, P. (2007). Changing drivers' minds: The evaluation of an advanced driver coaching system. *Ergonomics* 50(8):1209–34.

Stanton, N. A., and Young, M. (1999) A guide to methodology in ergonomics: Designing for human use. London: Taylor & Francis.

———. (2000). A proposed psychological model of driving automation. *Theoretical Issues in Ergonomics Science* 1(4):315–31.

Stanton, N. A., Young, M. S., and McCaulder, B. (1997). Drive-by-wire: The case of driver workload and reclaiming control with adaptive cruise control. *Safety Science* 27(2/3):149–59.

Stewart, R., Stanton, N. A., Harris, D., Baber, C., Salmon, P., Mock, M., Tatlock, K., Wells, L., and Kay, A. (2008). Distributed situation awareness in an airborne warning and control system: Application of novel ergonomics methodology. *Cognition Technology and Work* 10(3):221–29.

Strauch, B. (2005). *Investigating human error: Incidents, accidents and complex systems*. Aldershot, UK: Ashgate.

Svedung, I., and Rasmussen, J. (2002). Graphic representation of accident scenarios: Mapping system structure and the causation of accidents. *Safety Science* 40(5):397–417.

Svensson, J., and Andersson, J. (2006). Speech acts, communication problems, and fighter pilot team performance. *Ergonomics* 49(12-13):1226–37.

Swezey, R. W., Owens, J. M., Bergondy, M. L., and Salas, E. (2000). Task and training requirements analysis methodology (TTRAM): An analytic methodology for identifying potential training uses of simulator networks in teamwork-intensive task environments. In *Task analysis*, eds. J. Annett and N. Stanton, 150–69). London: Taylor & Francis.

Taylor, Lord Justice. (1990). Final report into the Hillsborough Stadium disaster, HMSO.

Taylor, R. M. (1990). Situational awareness rating technique (SART): The development of a tool for aircrew systems design. In *Situational Awareness in Aerospace Operations (AGARD-CP-478)*, 3/1–3/17. Neuilly Sur Seine, France: NATO-AGARD.

Tessitore, A., Meeusen, R., Tiberi, M., Cortis, C., Pagano, R., and Capranica, L. (2005). Aerobic and anaerobic profiles, heart rate and match analysis in older soccer players. *Ergonomics* 48(11-14):1365–77.

Tsang, P. S., and Vidulich, M. A. (2006). Mental workload and situation awareness. In *Handbook of Human Factors and ergonomics*, ed. G. Salvendy, 243–68). New York: John Wiley and Sons.

Tsigilis, N., and Hatzimanouil, D. (2005). Injuries in handball: Examination of the risk factors. *European Journal of Sport Science* 5(3):137–42.

Uhlarik, J., and Comerford, D. A. (2002). A review of situation awareness literature relevant to pilot surveillance functions. (DOT/FAA/AM-02/3). Washington, DC: Federal Aviation Administration, U.S. Department of Transportation.

Van Duijn, M. A. J., and Vermunt, J. K. (2006). What is special about social network analysis? *Methodology* 2(1):2–6.

Vicente, K. J. (1999). *Cognitive work analysis: Toward safe, productive, and healthy computer-based work*. Mahwah, NJ: Lawrence Erlbaum Associates.

Vicente, K. J., and Christoffersen, K. (2006) The Walkerton *E. coli* outbreak: A test of Rasmussen's framework for risk management in a dynamic society. *Theoretical Issues in Ergonomics Science* 7(2):93–112.

Vidulich, M. A. (1989). The use of judgement matrices in subjective workload assessment: The subjective WORkload Dominance (SWORD) technique. In *Proceedings of the Human Factors Society 33rd Annual Meeting*, 1406–10. Santa Monica, CA: Human Factors Society.

Vidulich, M. A., and Hughes, E. R. (1991). Testing a subjective metric of situation awareness. *Proceedings of the Human Factors Society 35th Annual Meeting*, 1307–11. Human Factors Society.

Vidulich, M. A., and Tsang, P. S. (1985). *Collecting NASA workload ratings*. Moffett Field, CA: NASA Ames Research Center.

———. (1986). Technique of subjective workload assessment: A comparison of SWAT and the NASA bipolar method. *Ergonomics* 29(11):1385–98.

Vidulich, M. A., Ward, G. F., and Schueren, J. (1991). Using Subjective Workload Dominance (SWORD) technique for Projective Workload Assessment. *Human Factors* 33(6):677–91.

Waag, W. L., and Houck, M. R. (1994). Tools for assessing situational awareness in an operational fighter environment. *Aviation, Space and Environmental Medicine* 65(5):A13–A19.

References

Walker, G. H. (2004). Verbal Protocol Analysis. In *The handbook of Human Factors and ergonomics methods*, eds. N. A. Stanton, A. Hedge, K. Brookhuis, E. Salas, and H. Hendrick. Boca Raton, FL: CRC Press.

Walker. G. H., Gibson, H., Stanton, N. A., Baber, C., Salmon, P. M., and Green, D. (2006). Event analysis of systemic teamwork (EAST): A novel integration of ergonomics methods to analyse C4i activity. *Ergonomics* 49:1345–69.

Walker, G. H., Stanton, N. A., Baber, C., Wells, L., Gibson, H., Young, M. S., Salmon, P. M., and Jenkins, D. P. (in press). Is a picture (or network) worth a thousand words: Analysing and representing distributed cognition in air traffic control Systems. *Ergonomics*.

Walker, G. H., Stanton, N. A., Kazi, T. A., Salmon, P., and Jenkins, D. P. (in press). Does advanced driver training improve situation awareness? *Applied Ergonomics*.

Walker, G. H., Stanton, N. A., and Young, M. S. (2001). Hierarchical task analysis of driving: A new research tool. In *Contemporary ergonomics*, ed. M. A. Hanson, 435–40. London: Taylor & Francis.

Wasserman, S., and Faust, K. (1994). Social network analysis: Methods and applications. Cambridge: Cambridge University Press.

Watson, M. O., and Sanderson, P. M. (2007). Designing for attention with sound: Challenges and extensions to ecological interface design. *Human Factors* 49(2):331–46.

Watts, L. A., and Monk, A. F. (2000). Reasoning about tasks, activities and technology to support collaboration. In *Task analysis*, eds. J. Annett and N. Stanton, 55–78. London: Taylor & Francis.

Webster, J., Holland, E. J., Sleivert, G., Laing, R. M., and Niven, B. E. (2005). A light-weight cooling vest enhances performance of athletes in the heat. *Ergonomics* 48(7):821–37.

Welford, A. T. (1978). Mental work-load as a function of demand, capacity, strategy and skill. *Ergonomics* 21(3):151–67.

Wellens, A. R. (1993). Group situation awareness and distributed decision-making: From military to civilian applications. In *Individual and group decision making: Current issues*, ed. N. J. Castellan, 267–87. Erlbaum Associates.

Wickens, C. D. (1992). *Engineering psychology and human performance*, 2nd ed. New York: Harper Collins.

Wiegmann, D. A., and Shappell, S. A. (2003). *A human error approach to aviation accident analysis: The human factors analysis and classification system*. Burlington, VT: Ashgate Publishing Ltd.

Wierwille, W. W., and Eggemeier, F. T. (1993). Recommendations for mental workload measurement in a test and evaluation environment. *Human Factors* 35(2):263–82.

Wilson, J. R., and Corlett, N. (1995). *Evaluation of human work—A practical ergonomics methodology*, 2nd ed. London: Taylor & Francis.

———. (2004). *Evaluation of human work*, 3rd ed. London: Taylor & Francis.

Wilson, J. R., and Rajan, J. A. (1995). Human-machine interfaces for systems control. In *Evaluation of human work: A practical ergonomics methodology*, eds. J. R. Wilson and E. N. Corlett, 357–405. London: Taylor & Francis.

Wilson, K. A., Salas, E., Priest, H. A., and Andrews, D. (2007). Errors in the heat of battle: Taking a closer look at shared cognition breakdowns through teamwork. *Human Factors* 49(2):243–56.

Woo, D. M., and Vicente, K. J. (2003). Sociotechnical systems, risk management, and public health: Comparing the North Battleford and Walterton Outbreaks. *Reliability Engineering and System Safety* 80:253–69.

Woodson, W., Tillman, B., and Tillman, P. (1992). *Human factors design handbook*. New York: McGraw Hill, Inc.

Wright, R. L., and Peters, D. M. (2008). A heart rate analysis of the cardiovascular demands of elite level competitive polo. *International Journal of Performance Analysis in Sport* 8(2):76–81.

Yamaoka, T., and Baber, C. (2000). Three-point task analysis and human error estimation. In *Proceedings of the Human Interface Symposium*, 395–98.

Young, M. S., and Stanton, N. A. (2001). Mental workload: Theory, measurement, and application. In *International encyclopaedia of ergonomics and human factors: Volume 1*, ed. W. Karwowski, 507–9. London: Taylor & Francis.

———. (2002). Attention and automation: New perspectives on mental underload and performance. *Theoretical Issues in Ergonomics Science* 3(2):178–94.

———. (2004). Mental workload. In *Handbook of Human Factors and ergonomics methods*, eds. N. A. Stanton, A. Hedge, K. Brookhuis, E. Salas, and H. Hendrick. Boca Raton, FL: CRC Press.

Yu, X., Lau, E., Vicente, K. J., and Carter, M. W. (2002). Toward theory-driven, quantitative performance measurement in ergonomics science: The abstraction hierarchy as a framework for data analysis. *Theoretic Issues in Ergonomic Science* 3(2):124–42.

Zsambok, C. E., Klein, G., Kyne, M., and Klinger, D. W. (1993). How teams excel: A model of advanced team decision making. In *Performance Technology 1992, Selected Proceedings of the 31st NSPI Conference*, 19–27. Chicago, IL: NSPI.

Index

A

Abstraction decomposition space *see* ADS
Abstraction hierarchy *see* AH
Accimaps, 115, **117–124**
  Organisational levels, 117
  Hillsborough disaster, 121
  Hillsborough disaster accimap, 122
ACTA
  Cognitive demands table, 101, 103, 105
  Knowledge audit interview, 101, 102
  Knowledge audit probes, 102–103
  Simulation interview, 101, 103, 104
  Task diagram interview, 101, 102
ADS, 71, **75–77**, 79, 82
AH, 71, 75, 76–77, 82, **83–84**
Applied cognitive task analysis *see* ACTA
Automation, 209

B

British Open, 1, 109, **126–128**

C

Carnoustie, 1, 126
CARS, 190
CDA ,62, 64, 68, 245–246, 248–249, 257–259, **260–270**, 271, 272, 274
CDM, 9, 15, 17, 18, 19, 20, 21, 71, **85–91**, 92–96, 104, 105, 198, 201, 246, 269, 271, 272, 273, 311, 312, 314, 316, 317, 321
  Cognitive probes, 71, 87, 88, 90, 314
  Ascent CDM output, 317
Checklists, 113, 275, 276, **278–285**, 292
CIT, 9, 69
Cognitive task analysis, 4, 5, 6, 9, 42, 57, **69–107**, 245, 247
Cognitive work analysis *see* CWA
Communications, 4, 36, 41, 49, 51, 61, 62, 64, 68, 121, 166, 175, 176, 177, 245, 246–257, 259, 264, 265, 266, 268, 269, 271, 272, 274, 310, 311, 314, 322, 323
Concept maps, **90–100**, 237
  Approach shot concept map, 99
  Concept map about concept map, 97
  Focus question, 93
  Overarching questions, 95
Coordination, 4, 40, 49, 62, 63, 167, 243, 244, 245, 246, 248–249, 257, 258, 260, 262, 265–268, 271
Coordination demands analysis *see* CDA
Crew awareness rating scale *see* CARS
Critical decision method *see* CDM
Crowd, 1, 2, 52, 53, 54, 83, 84, 202, 203, 204
CWA, 70, **71–85**, 104
  Control task analysis, 74, 77, 86
  Decision ladder, 79, 85
  Social organisation and cooperation analysis, 74, 79–80
  Strategies analysis, 74, 77, 86
  Worker competencies analysis, 74, 80–81
  Work domain analysis, 71, 75–77, 8–84

D

Data collection methods, 4, 5, **9–34**, 42, 50, 105, 121, 126, 130, 133, 142, 158, 257, 259, 274, 305
Decision making, 2, 3, 4, 5, 9, 10, 13, 15, 16, 17, 18, 19, 20, 21, 25, 31, 40, 57, 68, 70, 71, 85, 87, 88, 88, 89, 90, 92, 93, 94, 95, 97, 101, 102, 117, 163, 166, 169, 209, 210, 227, 243, 244, 257, 260, 263–265, 269, 272, 310, 311, 312, 314, 316, 321, 323

E

EAST, 246, 249, 267, **268–274**, 309–323
  Network of networks, 268–270
Event analysis of systemic teamwork *see* EAST
England, 1–2, 32–33, 109
European Championship, 1–2, 32, 43, 109, 210, 293

F

Failure, 2, 52, 110, 111–117, 118–129, 133, 143, 146, 200, 227
Fault tree analysis
  AND gates, 123, 125
  OR gates, 123, 125
  Van de Velde triple bogey fault tree diagram, 126–128

G

Garmin forerunner
  Wrist unit, 278
Germany, 1–2, 252, 295
Golf, 1, 6, 57, 83–87, 98, 99, 109, 111, 126, 128, 130, 134, 135–140, 147, 153
Golfer, 13, 56, 83, 93, 98, 99, 109, 111, 130, 132, 134, 135–140

H

HEI, 4, 6, 7, 9, 10, 30, 31, 35, 36, 40, 42, **109–159**
HET, 116, 133, 134, **141–145**
  Analysis extract, 144
  Consequence analysis, 142
  Criticality analysis, 142
  EEM taxonomy, 134
  Interface analysis, 142
  Ordinal probability analysis, 142
Heuristics, 76, 133, 275, 276, **284–289**

HTA 29, 30, 35, 36, **37–47**, 48, 49, 50, 51, 55, 56, 57, 58, 61, 62, 64, 68, 69, 71, 82, 117, 128, 130, 131, 132, 133, 134, 141, 142, 143, 144, 145, 147, 149, 150, 154, 155, 156, 158, 159, 172, 173, 174, 191, 198, 199, 201, 219, 224, 226, 230, 232, 234, 236, 239, 245, 246, 247, 259, 266, 267, 268, 269, 270, 271, 272, 273, 277, 280, 282, 283, 286, 287, 289, 290, 292, 297, 299, 304, 310, 311, 312, 313, 314, 316
   Analysis boundaries, 40
   Defeat opposition, 44
   Fell race HTA, 315
   Golf HTA extract, 135
   Plans, 37, 41
   Score goal from penalty kick, 46
   Scrum HTA description, 65
   Tackle opposition player, 45
   TOTE unit, 37
Human error, 2, 3, 4, 5, 7, 9, 10, 20, 26, 30, 31, 35, 37, 40, 42, 57, 70, 89, 97, **109–158**, 309
   Classifications, 110–112
   Definition, 110
   Error identifier methods, 116
   Person approach, 113
   Systems perspective approach, 113–115
   Taxonomy–based methods, 116
   Theoretical perspectives, 112–114
Human error identification *see* HEI
Human error template *see* HET
Human error template *see* HET
Human factors, **1–10**, 12, 15, 16, 18, 23, 25, 26, 29, 30, 31, 34–37, 42, 43, 51, 57, 58, 71, 82, 89, 97, 109, 110, 112, 115, 121, 132, 133, 134, 144, 154, 156, 161, 207, 210, 243, 245, 246, 259, 268, 272, 275, 282, 309–310, 313, 323

I

Instantaneous self assessment *see* ISA
Interface evaluation, 275–278
Interface evaluation methods, 275–308
Interface surveys, 275, 277, **301–308**
   Coding consistency survey, 303
   Control and display survey, 302–303, 306–307
   Labelling surveys, 303, 307
   Operator modification survey, 303
Interviews, 9, **10–16**, 17, 18, 19, 20, 21, 22, 27, 37, 40, 42, 43, 50, 63, 69, 76, 77, 79, 81, 82, 85, 88, 89, 90, 98, 101, 104, 105, 120, 125, 126, 133, 142, 143, 149, 158, 168, 169, 172, 173, 174, 178, 190, 201, 257, 259, 266, 267, 271, 272, 303, 304, 305
   Closed questions, 12, 13
   Focus groups, 12
   Open-ended questions, 12, 13
   Piloting, 14
   Probing questions, 12–13
   Semi-structured interviews, 12
   Structured interviews, 12
   Unstructured, 12
ISA, 211, 212, 215, **237–242**
   6 mile run ISA example, 242
   Scale, 239

L

Layout analysis, 275, 277, 292, **297–302**
   Functional groupings, 298
   Schematic diagram example, 298
Link analysis, 251, 275, 276, **289–296**, 299
   Link analysis diagram, 291
   Link analysis diagram example, 291
   Link analysis table, 291
   Euro 2008 example, 293–295

M

Mental workload, 2, 3, 4, 5, 6, 10, 24, 57, 70, 182, **207–210**, 309, 310, 312
   Assessment methods, **207–241**
   Definition, 207
   Framework of interacting stressors, 208
   Overload, 207–208
   Underload, 208
Methods criteria, 7–8
Methods integration, 309–323
Multi-perspective output example, 322

N

NASA TLX, 211, 213, 214, 216, 220, **222–229**, 232, 236, 237, 240, 310, 311, 312, 313, 314, 316
   Pair wise comparison, 225
   Pro-forma, 227
   Score calculation, 225
   Sub-scales, 222
   Weighting procedure, 225
Network of networks, 310–312

O

Observational study, 9, 10, **26–34**, 43, 50, 61, 64, 83, 89, 121, 126, 130, 142, 143, 149, 158, 198, 201, 251, 256, 259, 267, 271, 272, 287, 292, 304, 305, 311
   Direct, 27
   Example transcript, 33
   Participant, 27
   Piloting, 29–30
   Plan, 29
   Remote, 27
Operation sequence diagrams *see* OSD
OSD, 34, 36–37, **59–68**, 272, 273, 274
   Additional analyses, 62
   Operational loading figures, 62
   Template, 63
   Scrum OSD diagram, 66

P

Penalty kick, 1–2, 41, 43, 46, **51–54**, 87
Physiological measures, 210, 211–212, 214, 216, 217, **219–223**, 226, 232, 240
   Endogenous eye blink rate, 219
   Heart rate, 219
   Heart rate variability, 219
   Measures of brain activity, 219
Primary task performance measures, 208, 210, 211, **212–218**, 220, 221, 226, 232, 236

Index

Propositional networks, 57, 58, 98, 168, 171, **196–205**, 246, 269, 270, 272, 273, 274, 310–311, 312, 313, 314, 318, 321
  Ascent and descent propositional networks, 318
  Mathematical network analysis, 200
  Information element usage, 199
  Propositional network about propositional networks, 198
  Relationships between concepts, 198–199
  Soccer propositional networks example, 201–204

Q

Questionnaires, 9, 10, **16–28**, 37, 42, 50, 121, 169, 172, 189, 251, 256, 259, 267, 305
  Construction, 23
  Piloting, 23–24
  Questions, 24

R

Rotated figures task, 217
Rugby, 6, 7, 29, 36, 59, 64, 110, 111, 259, 268

S

SA 2, 3, 4, 5, 6, 7, 8, 9, 10, 13, 16, 17, 18, 19, 20, 21, 25, 26, 31, 42, 57, 58, 70, 87, 89, 92, 94, 97, 98, 101, 102, 103, **161–205**, 243, 244, 245, 246, 257, 258, 260, 263–265, 268, 269, 272, 275, 309, 310, 311, 312, 314, 316, 317, 321, 322, 323
  Assessment methods, 4, 6, 7, **161–205**
  Definition, 161–162
  Distributed situation awareness, 165–166
  Individual models, 162–164
  Team models, 164–166
SA requirements, 89, 98, 101, 164, 165, 166, 167, **168–178**, 179, 181, 182, 184, 185, 186, 188, 194, 200
SA requirements analysis
  Goal directed task analysis, 172
  Make pass sub-goal SA requirements, 175–177
  Passing HTA, 174
  SA requirements specification, 172
  SME interviews, 169–172
SAGAT, 168, 170. **177–184**, 185, 186, 187. 190. 196
  Example SAGAT queries, 183
SART, 168, 171, **187–193**, 196
  3D SART, 187
  Dimensions, 187
  Formula, 189
  Rating scale, 192
SA-SWORD, 168, 171, **187–193**, 196
  Judgement matrix, 194
  Matrix consistency evaluation, 194
SHERPA, 35, 116, **128–141**, 143, 144, 292
  Behaviour taxonomy, 130
  Consequence analysis, 130–131
  Criticality analysis, 132
  External error mode taxonomy, 130–131
  Golf SHERPA extract, 136–140
  Ordinal probability analysis, 131–132
  Recovery analysis, 131
  Remedial measures, 132
  Task classification, 130

Secondary task performance measures, 210–211, **212–218**, 220, 221, 226, 232, 236
Situation awareness *see* SA
Situation awareness global assessment technique *see* SAGAT
Situation awareness rating technique *see* SART
Situation awareness requirements *see* SA requirements
Situation awareness requirements analysis *see* SA requirements analysis
Situation present assessment method *see* SPAM
Situation awareness workload dominance *see* SA SWORD
Social network analysis *see* SNA
SNA, 11, 30, 31, 32, 201, 245, **246–256**, 258, 268, 270, 271, 272, 273, 274, 310, 311, 312, 313, 314, 321–322
  Agent association matrix, 247, 250
  Fell race social network diagram, 320
  Mathematical analysis, 250
  Social network diagram, 247, 250, 253–254
  Soccer example, 252–254
  Sociometric status example, 250, 254
Soccer, 1, 2, 7, 13, 14, 22, 29, 32, 41, 43, 51, 59, 87, 109, 110, 111, 114, 121, 163, 165, 166, 167, 174, 182, 198, 201, 202, 219, 246, 247, 250, 252, 291, 293–295
SPAM, 168, 171, **182–188**
Systematic Human Error Reduction and Prediction Approach *see* SHERPA
System Usability Scale, 27–28
Subjective workload assessment technique *see* SWAT
Subjective workload dominance *see* SWORD
SWAT, 211, 212, 214, 215, 216, 222, 225, 226, **229–233**, 235, 236
  Dimensions, 229, 230
  Rating procedure, 231
  Scale development, 230
  Score calculation, 231
SWORD, 211, 212, 215, **232–238**, 240
  Example rating sheet, 236
  Judgement matrix, 235
  Matrix consistency evaluation, 235
  Paired comparison rating sheet, 234

T

TAFEI, 116, 117, **146–154**, 292
  Boil kettle HTA, 147
  Boil kettle TAFEI diagram, 148
  Error descriptions and design solutions, 149
  Garmin 305 forerunner HTA, 152
  Garmin 305 forerunner SSD, 152
  Garmin 305 transition matrix, 153
  State space diagrams, 147
  Transition matrix, 147–148
Task analysis, 1, 4, 5, 6, 9, 10, 29, 30, 31, **35–68**, 82, 92, 125, 142, 146, 168, 169, 172, 173, 174, 178, 232, 236, 255–261, 270, 273, 281, 305, 311, 312
Task analysis for error identification *see* TAFEI,
Task decomposition, 36, **47–55**, 56
  Categories, 48, 49, 51
  Score goal from penalty kick decomposition, 52–54
Task model, 316
Technique for Human Error Assessment *see* THEA
Teams, 243–245

Teamwork, 1, 2, 4, 5, 10, 26, 39, 57, 62, 165, 301, **243–245**, 255–274
    Assessment, 7, 9, 30, 42, 245
    Assessment methods, 10, 31, 42, **243–274**
    Models, 244–245
    Taxonomy, **257–259**, 266, 267
Team task analysis, 36, 245, 248, **255–268**
    CDA results for scrum task, 262
    KSAs, 258, 263–265
    Scrum example, 259–265
    Scrum HTA, 261
    Scrum KSA analysis, 263–265
Team training, 245, 246, 255
Training, 1, 2, 3, 4, 5, 6, 7, 8, 11, 13, 15, 16, 17, 18, 19, 20, 25, 31, 34, 35, 36, 38, 40, 42–43, 49, 50, 51, 52, 53, 54, 58, 62, 64, 69, 70, 71, 72, 74, 75, 77, 79, 81, 82, 85,86, 87, 89, 91, 93, 94, 95, 97, 98, 101, 104, 105, 109, 113, 114, 115, 117, 118, 119, 120, 123, 126, 130, 132, 133, 137–140, 142, 143, 146, 149–50, 152, 153, 158, 162, 163, 166, 169. 170, 171, 173, 181, 185, 186, 187, 189, 190, 194, 195, 198, 200, 201, 202, 203, 204, 217, 22, 226, 232, 237, 240, 241, 243, 248, 249, 251, 255, 258, 259, 262, 267, 272, 273, 275, 276, 277, 281, 287, 288, 292, 297, 298, 299, 300, 304, 305, 306–307, 313, 314, 320

THEA, 116, 117, **153–159**
    Error analysis, 156
    Error analysis questions, 157
    Scenario description template, 155

V

Verbal Protocol Analysis *see* VPA,
VPA, 36, 37, 40, 42, **51–59**, 61, 168, 201
    Encoding, 57
    Runner VPA extract, 60